Mechanism Synthesis and Analysis

Mechanism Synthesis and Analysis

A. H. SONI

Professor
School of Mechanical and Aerospace Engineering
Oklahoma State University
Stillwater, Oklahoma

SCRIPTA BOOK COMPANY
Washington, D.C.

McGRAW-HILL BOOK COMPANY
New York   St. Louis   San Francisco   Düsseldorf   Johannesburg
Kuala Lumpur   London   Mexico   Montreal   New Delhi   Panama
Paris   São Paulo   Singapore   Sydney   Tokyo   Toronto

Library of Congress Cataloging in Publication Data

Soni, Atmaram H
    Mechanism synthesis and analysis.

    1. Machinery, Kinematics of.   2. Mechanic
movements. I. Title.
TJ175.S66   621.8'11   73-22000
ISBN 0-07-059640-9

Mechanism Synthesis and Analysis

1 2 3 4 5 6 7 8 9 0 K P K P 7 9 8 7 6 5 4

This book was set in Press Roman by Scripta Graphica.
The supervising editor was Glenda Hightower; the designer
was Victor Enfield; the compositor was Isabelle Sneeringer;
and the production supervisor was Keith Wilkinson.
The printer and binder was The Kingsport Press.

To my father, Haribhai Ramji Soni,
and my mother, Maniben H. Soni

Contents

Foreword   *xv*

Preface   *xvii*

Acknowledgments   *xxi*

How to Use This Book for Self-study Program   *xxiii*

UNIT I.   **A SYSTEMATIC APPROACH TO NUMBER SYNTHESIS**   **1**

     Objective 1: To name different types of motions displayed by the different types of joints, frequently called *kinematic pairs*   1
     Objective 2: To build kinematic chains and calculate their degrees of freedom   3
     Objective 3: To build kinematic chains and linkages with other types of kinematic pairs   7
         Performance Test #1   11
         Performance Test #2   13
         Performance Test #3   14
     Supplementary References   15

UNIT II.   **VELOCITY ANALYSIS OF LINKAGES** **17**

     Objective 1: To define vectors and vector properties   17
     Objective 2: To define velocity and its components for a particle moving on a path   22
     Objective 3: To describe the velocity of a particle moving on a path using complex number notation   26
     Objective 4: To derive a velocity relationship for a rigid body displacement   31
         Performance Test #1 –Objectives 1, 2, 3, 4   34
         Performance Test #2 –Objectives 1, 2, 3, 4   35
         Performance Test #3 –Objectives 1, 2, 3, 4   35
     Objective 5: To demonstrate the application of vector polygon technique in performing velocity analysis of single-loop mechanisms   36
     Objective 6: To perform velocity analysis of six-link mechanisms using the vector polygon technique   44
     Objective 7: To state the properties of instant center and demonstrate their application in performing velocity analysis of mechanisms   53
         Performance Test #1 –Objectives 5, 6, 7   59
         Performance Test #2 –Objectives 5, 6, 7   60
         Performance Test #3 –Objectives 5, 6, 7   60
     Supplementary References   61

UNIT III.   **ACCELERATION ANALYSIS OF LINKAGES** **63**

Objective 1: To define acceleration and its components for a particle executing planar and space motion in stationary and moving coordinates   63

Objective 2: To calculate, using complex number notation, the acceleration of a particle moving on a path   74

Objective 3: To describe the acceleration-vector relationship of a rigid body moving in a plane   77

    Performance Test #1 – Objectives 1, 2, 3   80
    Performance Test #2 – Objectives 1, 2, 3   81
    Performance Test #3 – Objectives 1, 2, 3   81

Objective 4: To demonstrate the application of the vector polygon technique in order to perform acceleration analysis of mechanisms .   82

Objective 5: To perform acceleration analysis of a six-link mechanism using the vector polygon technique   90

    Performance Test #1 – Objectives 4, 5   94
    Performance Test #2 – Objectives 4, 5   94
    Performance Test #3 – Objectives 4, 5   95

Supplementary References   96

UNIT IV.   **DISPLACEMENT, VELOCITY, AND ACCELERATION ANALYSIS OF MECHANISMS USING ANALYTICAL METHODS** **97**

Objective 1: To perform displacement, velocity, and acceleration analysis of a four-link mechanism   97

Objective 2: To perform displacement, velocity, and acceleration analysis of a slider-crank mechanism   105

Objective 3: To perform displacement, velocity, and aceleration analysis of an inverted slider-crank mechanism **110**

    Performance Test #1   115
    Performance Test #2   115
    Performance Test #3   116

Supplementary References   116

UNIT V.   **FUNDAMENTALS OF UNIFORM ROTARY TRANSMISSION** **117**

Objective 1: To identify the different ways of synthesizing uniform rotary transmission and to develop the criteria of rotary transmission by direct rolling and sliding contact   117

Objective 2: To design uniform rotary transmission using involute and cycloidal curves   123

Objective 3: To calculate speed ratios for external and internal rolling cylinders and cones   127

    Performance Test #1   133
    Performance Test #2   133
    Performance Test #3   134

Supplementary References   134

UNIT VI.   **GEAR TOOTH TECHNOLOGY** **135**

Objective 1: To define the terminology of spur, helical, bevel, and worm gears   135

Objective 2: To derive relationships to calculate tooth thickness, tooth profile, and contact ratio   144

Objective 3: To calculate the effects of diametral pitch, pressure angle, and speed ratio on interference   147

Objective 4: To name AGMA (American Gear Manufacturers' Association) standards for spur, helical, bevel, and worm gears . . .   150

Performance Test #1   151

Performance Test #2   152

Performance Test #3   152

Supplementary References   152

UNIT VII.   DESIGN OF GEAR TRAINS   153

Objective 1: To design a simple gear train   153

Objective 2: To design a compound gear train   156

Objective 3: To design a reverted compound train   159

Objective 4: To design a planetary gear train   161

Performance Test #1   164

Performance Test #2   165

Performance Test #3   166

Supplementary References   166

UNIT VIII.   CAMS AND FOLLOWERS   167

Objective 1: To identify different types of followers   167

Objective 2: To identify different types of cams   169

Objective 3: To identify the notations of a cam follower system . . .   175

Performance Test #1   178

Performance Test #2   179

Performance Test #3   179

Supplementary References   180

UNIT IX.   MOTION PROGRAMS   181

Objective 1: To identify applications of custom-made motion programs in cam design   181

Objective 2: To synthesize motion programs using the analytical technique   197

Performance Test #1   205

Performance Test #2   205

Performance Test #3   206

Supplementary References   206

UNIT X.   SIZING THE CAMS   207

Objective 1: To derive relationships to calculate coordinates to generate cam profiles for different types of followers   207

Objective 2: To calculate the radius of a roller of a translating or oscillating roller follower   218

Objective 3: To calculate the influence of the pressure angle in sizing a cam with a translating roller follower   223

Performance Test #1   229

Performance Test #2   229

Performance Test #3   230

Supplementary References   230

UNIT XI.   **MOTION ANALYSIS AND SYNTHESIS OF FOUR–BAR, SLIDER–CRANK, AND INVERTED SLIDER–CRANK MECHANISMS** 231

Objective 1: To state Grashoff for a four-link mechanism 231
Objective 2: To identify the conditions that determine limit positions and dead-center positions of a four-link mechanism 234
Objective 3: To identify the conditions that determine limit positions of a slider-crank mechanism 240
Objective 4: To identify the conditions that determine limit positions of an inverted slider-crank mechanism 243
Objective 5: To identify the conditions that determine minimum and maximum transmission angles of four-link, slider-crank, and inverted slider-crank mechanisms 245
Objective 6: To design a drag-link mechanism with an optimum transmission angle 250
Objective 7: To design a crank-rocker mechanism using a minimum transmission angle 257
    Performance Test #1 263
    Performance Test #2 264
    Performance Test #3 265
Supplementary References 265

UNIT XII.   **LINKAGE SYNTHESIS COORDINATING INPUT AND OUTPUT MOTIONS** 267

Objective 1: To synthesize a four-link and a slider-crank mechanism to coordinate their input and output displacements 267
Objective 2: To synthesize a four-link and a slider-crank mechanism for two positions of input and output links 272
Objective 3: To synthesize a four-link and a slider-crank mechanism for three positions using pole techniques 275
Objective 4: To synthesize a four-link and a slider-crank mechanism for three positions using the inversion technique 278
Objective 5: To synthesize a four-link mechanism to coordinate for four positions the motions of its input and output links using the point position reduction technique 281
Objective 6: To synthesize a four-link mechanism using the overlay technique 284
    Performance Test #1 296
    Performance Test #2 296
    Performance Test #3 297
Supplementary References 297

UNIT XIII.   **SYNTHESIS OF FOUR-LINK, SLIDER–CRANK, AND INVERTED SLIDER-CRANK MECHANISMS USING THE ANALYTICAL APPROACH** 299

Objective 1: To synthesize a four-link mechanism to coordinate three positions of output and input 299
Objective 2: To synthesize a four-link mechanism to coordinate three positions of input link and three positions of a coupler point . 302
Objective 3: To synthesize a four-link mechanism to guide a rigid body through its three finitely separated positions 305

Objective 4: To synthesize a function generator mechanism using the least-square technique   308
Objective 5: To synthesize a four-link mechanism for prescribed values of displacement, velocity, and acceleration   311
  Performance Test #1   313
  Performance Test #2   313
  Performance Test #3   313

UNIT XIV.   COUPLER CURVES OF FOUR-LINK, SLIDER-CRANK, AND INVERTED SLIDER-CRANK MECHANISMS   315

Objective 1: To identify the coupler curves of a four-link mechanism   315
Objective 2: To derive relationships to calculate coordinates of a coupler point of a four-line mechanism   316
Objective 3: To derive relationships to calculate coordinates of a coupler point of a slider-crank mechanism   336
Objective 4: To derive relationships to calculate coordinates of a coupler point of an inverted slider-crank mechanism   342
  Performance Test #1   350
  Performance Test #2   350
Supplementary References   350

UNIT XV.   APPLICATION OF COUPLER CURVES IN DESIGN OF SIX-LINK MECHANISMS   351

Objective 1: To name different types of coupler driven six-link mechanisms   351
Objective 2: To name different types of motion programs   352
Objective 3: To design six-link mechanisms with a single dwell . . .   359
Objective 4: To design six-link mechanisms with double dwells . . .   361
Objective 5: To design six-link mechanisms with double strokes . .   365
Objective 6: To design a six-link mechanism with an output link moving with a constant velocity for a finite interval of time   367
Objective 7: To design six-link mechanisms with output links having 720° rotation corresponding to 360° rotation of the input link   369
  Performance Test #1   370
  Performance Test #2   372
  Performance Test #3   373
Supplementary References   373

UNIT XVI.   COUPLER COGNATE MECHANISMS   375

Objective 1: To design coupler cognate mechanisms of four-link and slider-crank mechanisms   375
Objective 2: To design six-link mechanisms for generation of parallel motion   381
  Performance Test #1   383
  Performance Test #2   383
  Performance Test #3   383
  Performance Test #4   384
  Performance Test #5   384
Supplementary References   384

ANSWERS **UNIT I** **385**

Performance Test #1 – Objectives 1, 2, 3   385
Performance Test #2   386
Performance Test #3   388

UNIT II **391**

Performance Test #1 – Objectives 1, 2, 3, 4   391
Performance Test #2 – Objectives 1, 2, 3, 4   392
Performance Test #3 – Objectives 1, 2, 3, 4   393
Performance Test #1 – Objectives 5, 6, 7   394
Performance Test #2 – Objectives 5, 6, 7   396
Performance Test #3 – Objectives 5, 6, 7   398

UNIT III **401**

Performance Test #1 – Objectives 1, 2, 3   401
Performance Test #2 – Objectives 1, 2, 3   402
Performance Test #3 – Objectives 1, 2, 3   404
Performance Test #1 – Objectives 4, 5   406
Performance Test #2 – Objectives 4, 5   409
Performance Test #3 – Objectives 4, 5   412

UNIT IV **415**

Performance Test #1   415
Performance Test #2   415
Performance Test #3   415

UNIT V **416**

Performance Test #1   416
Performance Test #2   416
Performance Test #3   417

UNIT VI **419**

Performance Test #1   419
Performance Test #2   420
Performance Test #3   422

UNIT VII **425**

Performance Test #1   425
Performance Test #2   426
Performance Test #3   427

UNIT VIII **430**

Performance Test #1   430
Performance Test #2   432
Performance Test #3   432

UNIT IX **434**

Performance Test #1   434
Performance Test #2   438
Performance Test #3   441

UNIT X 445

Performance Test #1 445
Performance Test #2 445
Performance Test #3 446

UNIT XI 448

Performance Test #1 448
Performance Test #2 449
Performance Test #3 450

UNIT XII 452

Performance Test #1 452
Performance Test #2 452
Performance Test #3 452

UNIT XIII 455

Performance Test #1 455
Performance Test #2 457
Performance Test #3 458

UNIT XIV 463

Performance Test #1 463
Performance Test #2 463

UNIT XV 464

Performance Test #1 464
Performance Test #2 465
Performance Test #3 466

UNIT XVI 468

Performance Test #1 468
Performance Test #2 469
Performance Test #3 471
Performance Test #4 471
Performance Test #5 473

SUPPLEMENTARY REFERENCES 475

Index   477

Foreword

As far as I know, this is the first self-paced (self-teaching) text in the mechanisms field. The author has put much effort into organizing the subject matter logically, presenting it clearly, and explaining it carefully.

I like particularly the introduction of a relatively high proportion of useful, modern material on kinematic analysis and design, in a simple and effective manner. This includes the complex-number method of kinematic analysis and synthesis, an extraordinary chapter on cam design (including envelope theory, as needed for numerically controlled machining, for instance), transmission-angle optimization in linkages, and an introduction to kinematic synthesis.

The use of self-teaching methods in engineering design is still in an early stage. Eventually their place and role will be more clearly perceived than at present. There seems little doubt, however, that such methods are destined to serve a useful educational purpose for large numbers of people.

The author has made an impressive contribution to self-teaching in the mechanisms field. His text will, I believe, be a valuable educational resource to those wishing to learn about the kinematics of mechanisms, either through the medium of self-paced instruction or through lecture courses.

F. Freudenstein

Columbia University

Preface

This textbook in undergraduate applied kinematics represents my personal experience as a student, an educator, a researcher, a consultant to industry designers, and a director of summer institutes for college teachers and industry designers. It is written to provide the most modern material in mechanisms synthesis and analysis and to present effective teaching techniques.

Technical content is presented in five parts. These five parts have been further divided into 16 units:

Mobility Analysis, Unit I
Kinetic Analysis, Units II–IV
Gears, Units V–VII
Cams, Units VII–X
Linkages, Units XI–XVI

Unit I on Mobility Analysis of Mechanisms presents an introduction to structural analysis and synthesis of kinematic chains and determination of degrees of freedom of a linkage system using Gruebler's mobility criteria. The material presented in this unit is unique, practical, and available at the textbook level for the first time in any language. This unit reflects our long-term involvement in a research program in structural synthesis and existence criteria. The material will prove to be an excellent approach in arriving at alternate designs of mechanisms. Furthermore, it shows to a student that a kinematic chain with a pin joint (revolute pair) is a basic mechanical system that contains kinematic elements such as sliders, cams, springs, belt-pulleys, chain-sprockets or gears.

Units II, III, and IV present methodologies to perform kinematic analysis of mechanisms. Unit II deals with the fundamentals of vector algebra, complex number algebra, and the application of these mathematical tools in studying the velocity of a particle and of a rigid body. The application of vector polygon technique in conducting velocity analyses of single and multi-loop mechanisms is demonstrated several times to provide working knowledge to students.

Units V, VI, and VII deal with the terminology, the analysis and the design of gear trains. These units provide a student with a working knowledge of design and analysis of gear trains. The vector approach is emphasized in deriving the equation of motion for different types of epicyclic gear trains.

Units VIII, IX, and X deal with the design of cams and the synthesis of motion programs of followers. Unit VIII presents the basic terminology of cams and followers and the types of motion programs more commonly used. Unit IX presents the types of motion programs already in existence and proved to be of practical importance. A student is presented with the criteria of selection of motion programs by providing their properties. This unit also presents analytical techniques which permit a student to synthesize a motion program according to his design requirements. All design procedures for motion programming are laid out in a stepwise manner for effective learning. Unit X deals with the sizing of cams. The student is presented with the most powerful theory of envelopes that is applied in obtaining cam profile for different types of followers. The importance of the pressure angle is demonstrated analytically, and some illustrative design charts are presented for obtaining quick design solutions. This material is expected to train a student for an advanced training in cam design.

Units XI through XVI deal with linkage design. The material presented in these units is quite extensive and practical for undergraduate students. In proportion, the amount of material on linkages occupies a large part of the book. This is primarily because linkages provide powerful tools in programming for non-uniform motion. Unit XI presents the fundamentals of a four-bar mechanism, types of four-bar mechanism, gross motion analysis and synthesis of a four-bar mechanism using an index that defines the quality of motion. Unit XII presents the graphical approach of synthesizing a four-bar mechanism. Techniques that have proven practical are presented in a stepwise manner for effective self-paced learning. These techniques deal with the property of poles and the application of the principle of inversion. The overlay technique which requires no previous knowledge in linkage synthesis is presented and three examples illustrate its simplicity and usefulness.

Unit XVI deals with the coupler cognate mechanisms of four-link, slider-crank, and geared five-bar mechanisms. The proof for the existence of coupler cognate mechanisms is presented using complex number algebra. Coupler cognate mechanisms have a practical importance. This is shown by demonstrating the synthesis of six-link mechanisms for generating parallel motion. I believe that the material presented in this unit will stimulate students as well as designers in proving the importance of linkages in machine design.

For the convenience of those students who are interested in learning views of other authors on the technical subjects of these units, references to existing textbooks are provided.

To make teaching more effective, to make learning more exciting, and to make our present education system perform more efficiently, new methods must be devised whereby students get the most of the learning package and the instructor is most effective in motivating students and contributing to the growth of his institution. The development of such methods inevitably requires adjustments by students, instructors, and educational institutions. The place where adjustment is greatest is in the style of presentation of textbooks and their technical contents. In so doing, not only the authors but also the publishers are expected to take on new challenges in writing and publishing textbooks based on such modern concepts.

The format of this book is based on modern teaching concepts that require a student to accept the responsibility for learning. Accordingly, each of the 16 units of this book are divided into subunits called "objectives." Each objective within a unit is a "mini-learning package" centered around some specific theme that the student must learn. At the end of each objective, there are competency items which the student must answer to check his performance in learning; it is the student's responsibility to seek counsel with his intructor in case of difficulties. At the end of each unit there are performance tests which serve as comprehensive tests for that unit.

Because of the format, an instructor may use this book either for a lecture-oriented or for a self-study-oriented teaching program. The technical content of the book may be covered in either one semester or two quarters.

This textbook can be used for a lecture-oriented teaching program and may prove advantageous over conventionally written textbooks:

- The instructor is in a position to select a course outline and prepare a complete reading assignment for each hour of the lecture for the entire semester.
- The instructor can prepare his lectures according to the objectives of each of the 16 units.
- The student can read *all* the instructional material prior to attending his instructor's formal presentation of his lecture.
- With proper audiovisual aids, the instructor can cover more ground in presenting his technical material.
- Students are more involved with the instructor's presentation than with taking notes.
- The instructor is not required to prepare any class notes for his lectures.

The author has the following reservations about lecture-oriented teaching programs:

- Due to the limitation on lecture time, the instructor is not able to devote much time to interacting with all his students and insuring that all students have learned everything that was presented.
- The program does not have enough feedback from the student to his instructor, frequently neither is aware of the need for remedial work.
- Student counseling is virtually impossible.
- No special efforts are made to provide more time for students who are slow in learning or to allow students who are quick in learning to progress faster.

An instructor-guided self-study program will prove a most effective teaching technique. The format of this textbook will permit the instructor to adopt such teaching methodologies. My experience in testing this format has shown that:

- Self-paced learning is effective when a student accepts the responsibility for learning.
- Instructional material presented in this format serves the need of all students.

- Because of the self-involvement of the student and the interactive role of the instructor, the student has been interested and motivated.
- The instructor is more organized in presenting the instructional and testing material.
- The self-study program provides the student with complete freedom to master all the instructional material at his own rate and to obtain a better knowledge of the subject.
- The instructor is forced to play the role of counselor and to prescribe remedial steps to the student having difficulty with a learning package.
- The self-study program permits better students to accomplish their learning objectives earlier than the other students.

At Oklahoma State University, I have tried the self-paced approach by week because of its disciplinary effect on both the instructor and the student. My experience in testing this format on two control groups (one group exposed to lectures and the other group exposed to self-study without lectures) showed that the self-study group had significantly higher test averages on identical tests. These results show that students exposed to the self-study program do acquire knowledge better than those exposed to the lecture system. In my experience, the self-study approach has motivated the students well and exposed them more frequently than the lecture method to my thinking.

No matter which of the two procedures is adopted, I sincerely hope that this textbook will prove significantly valuable to students of kinematics in achieving their learning objectives.

Acknowledgments

Writing a textbook is a mammoth project, seldom done without help or influence of other people. The men who have motivated me the most in my career are the Reverend Father M. M. Balaguer, Dr. Lee Harrisberger, and Dr. Robert E. Little. Their contributions were far beyond what anyone could ask. My dream to come to the United States was made a reality by the extensive efforts of Father Balaguer. Dr. Harrisberger's contribution in molding me into a full-time kinematician is known to all those who are practicing the science.

Dr. Michael Gaus of National Science Foundation, Prof. Ferdinand Freudenstein of Columbia University, Alcoa Professor G. N. Sandor of Rensselaer Polytechnic Institute, and Prof. Joseph E. Shigley of the University of Michigan encouraged me to undertake this project. I am especially grateful to Professor Freudenstein for so generously contributing his time in writing the foreword for this book.

The administration at Oklahoma State University has always urged faculty members to develop effective teaching techniques and I thank Dean M. R. Lohmann, Dr. Lee Harrisberger, and Dr. K. N. Reid for their leadership roles in such projects. Stimulating discussions with my colleagues Dr. Richard Lowery and Dr. Mike Mamoun are reflected all through the book.

I am grateful to my friends and colleagues J. Church, M. Huang, G. Dewey, Rao Dukkipati, D. Kohli, P. R. Pamidi, and L. Torfason for abetting this project.

My wife Ila and my children, Bina, Robert, and Lisa, suffered while I slaved on this book and I am happy for their kind cooperation and understanding. My brother Ramaniklal, his wife Rama, my sister Kantaben, and her husband Jayantilal were ever encouraging.

Finally, this project would not have become a book were it not for the initiative of Mr. B. J. Clark and Mrs. Harriet Malkin at McGraw-Hill and Mr. William Begell at Scripta Book Company. I am also grateful to the editorial and typesetting staff of Scripta Book Company who have done such an excellent job in producing this book.

How to Use This Book for Self-study Program

The technical content of this book is divided into 16 units which are distributed into five parts: Mobility Analysis, Kinematic Analysis, Gears, Cams, and Linkages. Each unit is subdivided into several objectives. Each objective is a mini-learning package. At the end of each objective there are competency items. At the end of each unit, there are performance tests. The answers to the performance tests are presented at the end of the textbook. For additional reading, other reading materials are cited by reference number and the appropriate section of that reference item in "Supplementary References." The complete list of references for the supplementary material is presented on page 475.

Table A below shows a hierarchy chart that should be followed by the student using the self-study program. It suggests that Unit I is the groundwork for the entire material in mechanism synthesis and analysis. After the first four objectives of Unit II, one could either learn the four objectives of Unit III or learn the remaining three objectives of Unit II. Once all the objectives of Units II and III are achieved, the student may study the units on cams, linkages, or gears. Unit IV, however, is a prerequisite to Unit XII. After XII, the student may go to either Unit XIII or Unit XV. Unit XVI may be studied after Unit XIV or Unit XV.

Table B shows that the student may pursue a self-paced approach by semester. In the self-paced approach by week, the student or the instructor sets a one-week deadline to learn all the objectives of one unit. The student checks his performance with his instructor by taking a performance test. In order to learn all the objectives of all the units, the student is thus required to have 16 weeks.

In the self-paced approach by semester, the student is left alone with the entire package of 16 units. He prepares his own study program with only one deadline—he must learn all the objectives of all the units by the end of a semester. According to this plan, the student may master the objectives of three to four units in one week and may not master any objectives the next week. He must, however, counsel with his intructor as often as he can to check his performance and to receive guidance for any remedial steps for the difficult subject.

TABLE A Learning Hierarchy for Mechanism Synthesis and Analysis

Δ Units
O Objectives

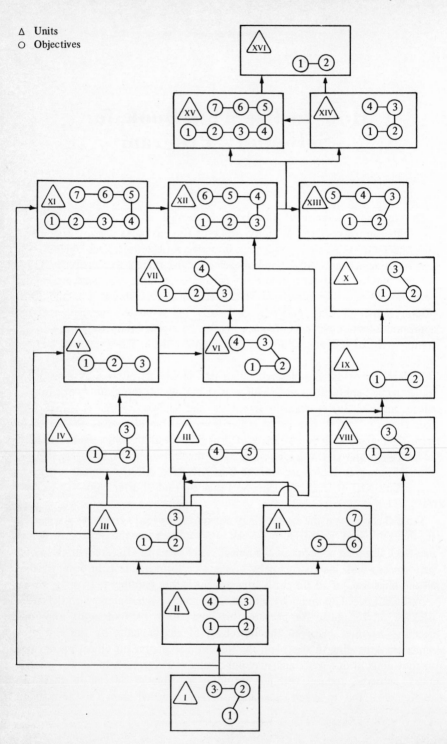

TABLE B Flow Chart for Individualized Approach

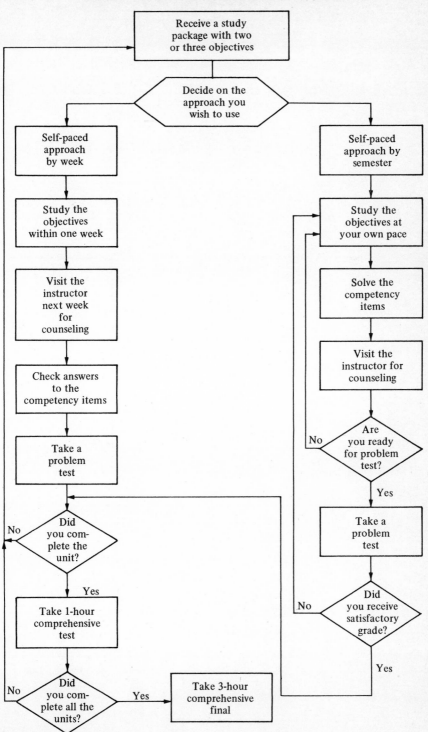

Unit I. A Systematic Approach
to Number Synthesis

Objectives
1. To name different types of motions displayed by the different types of joints, frequently called *kinematic pairs*
2. To build kinematic chains and calculate their degrees of freedom
3. To build kinematic chains and linkages with other types of kinematic pairs

Objective 1

To name different types of motions displayed by the different types of joints, frequently called *kinematic pairs*

Activities
- Read the material provided below and as you read:
  a. Sketch the form-closed and force-closed joints.
  b. Name the type of motion displayed by the joints.
  c. Calculate degrees of freedom of the joints.
  d. Calculate number of constraints placed on the movability of the joints.
  e. Classify the joints according to their degrees of freedom.
  f. Classify the joints according to their number of constraints.
- Check your answers to the competency items with your instructor.

Reading Material
A kinematic pair is a pair of elements, permanently kept in a contact, so that there exists a relative movement between these elements.

A rigid body which is free to move in the space has six degrees of freedom. These degrees of freedom are calculated on the basis that this rigid body is able to rotate about three independent axes and is also able to translate itself along these three independent axes.

Two such rigid bodies placed together so as to permit relative motion will build a kinematic pair. A kinematic pair can have a maximum of five degrees and a minimum of one degree of freedom. Table 1 presents all possible physically realizable kinematic pairs. The revolute pair has one degree of freedom and all it can do is rotate around one axis. The prism pair also has one degree of freedom;

TABLE 1 Classification of Kinematic Pairs

Class of kinematic pairs	Degrees of freedom	Number of point contacts	Name of kinematic pair and its symbol	Kinematic pairs, form-closed and force-closed
I	1	5	Revolute–R Prismatic–P Helical–H	
II	2	4	Slotted spheric–S_L Cylinder–C Cam–C_a	
III	3	3	Spheric pair–S Sphere-slotted cylinder–S_S Plane pair–P_L	
IV	4	2	Sphere-groove–S_g Cylinder plane pair–C_p	
V	5	1	Sphere-plane–S_P	

all it can do is to translate along one axis. The cylinder pair has two degrees of freedom which permit it to execute rotational and translational motions. The cylinder pair has two coincident axes. The axis about which the rotational motion takes place is also the axis about which the translational motion takes place. This is not, however, the case for the cam pair which also has two degrees of freedom.

Study carefully Table 1 and determine all possible motions these kinematic pairs can perform.

There are five different classes of kinematic pairs. The class of a kinematic pair depends on its degrees of freedom. The class I kinematic pairs have one degree of freedom; the class II kinematic pairs have two degrees of freedom; the class III kinematic pairs have three degrees of freedom, etc.

The class I kinematic pairs are revolute pair (pin joint), helical pair (screw joint), and prismatic pair (sliding pair). The class II kinematic pairs are cylinder pair (turn-slide pair), cam pair, and slotted spheric pair. The class III kinematic pairs include a spheric pair (Global pair), plane pair, and a sphere in slotted cylinder pair.

Reuleaux had classified the kinematic pairs into two groups: lower kinematic pairs and higher kinematic pairs. The male and female elements of the lower kinematic pairs make either area or surface contacts. For the higher kinematic pairs, the male and the female element make either line or point contact. In general, the lower kinematic pairs are employed in a linkage when large forces must be transmitted.

Competency Items
- List the five classes of kinematic pairs.
- Name the kinematic pairs that belong to class I.
- Name the kinematic pairs that belong to class II.
- Calculate the number of constraints on the motion of the cylinder, spherical, and prism pairs.
- Classify the five classes of kinematic pairs into higher and lower classes of kinematic pairs.
- State the difference in the motion executed by cylinder and cam pairs.

Objective 2

To build kinematic chains and calculate their degrees of freedom

Activities
- Read the material provided below, and as you read:
  - a. Name the different types of kinematic links.
  - b. Build kinematic chains.
  - c. Calculate the degrees of freedom of the linkages derived from the kinematic chains that you built.
- Check your answers to the competency items with your instructor.

Reading Material

A kinematic link is considered to be weightless and rigid. The type of a kinematic link is identified by the number of kinematic pairs that can be placed on it.

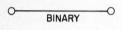

Fig. 1 Schematic representation of a binary link. It has two ends and a revolute pair, shown by a circle, is placed at each end.

Fig. 2 Schematic representation of a ternary link. It has three ends and a revolute pair is placed at each end.

Fig. 3 Schematic representation of a quaternary link. It has four ends and a revolute pair is placed at each end.

A kinematic link can have as many ends as one would like it to have. Note, however, that this link is a rigid link and there is no relative movement between the joints within a link. The polygonal-type links such as ternary, quaternary, etc., are normally shown as rigid links by drawing lines within the polygons.

A kinematic chain is built using kinematic pairs and kinematic links so as to form a closed circuit or loop.

Fig. 4 Kinematic chain with five binary links connected by five revolute pairs. This kinematic chain has one loop.

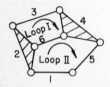

Fig. 5 Kinematic chain with four binary links and two ternary links connected by seven revolute pairs. This kinematic chain has two loops and a total of six links.

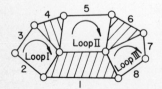

Fig. 6 Kinematic chain with five binary, two ternary, and one quaternary link, connected by ten revolute pairs. This kinematic chain has three loops and a total of eight links.

A linkage is obtained by fixing to the ground one of the links of the kinematic chain. Thus, the linkage shown in Fig. 7 is obtained when link 1 of Fig. 4 is fixed to the ground.

Once the linkage is obtained, it becomes necessary to determine the number of inputs that can be given to the linkage so that the linkage has a constrained motion. This information is obtained by calculating the degrees of freedom of the linkage.

The degrees of freedom of the linkage are determined using the mobility criterion proposed by Gruebler:

$$F = 3(N - 1) - 2 \times P_1 - 1 \times P_2$$

where

F = degrees of freedom of the linkage
N = total number of links
P_1 = total number of kinematic pairs having one degree of freedom
P_2 = total number of kinematic pairs having two degrees of freedom

NOTE: Gruebler's mobility criterion is valid for linkages that move in parallel planes.

Thus, for the linkage in Fig. 7

$$N = 5, \quad P_1 = 5, \quad P_2 = 0$$

$$F = 3 \times (5 - 1) - 2 \times 5 = 12 - 10 = 2$$

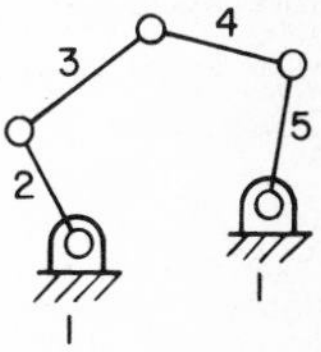

That is, the linkage has two degrees of freedom. Two input motions can be provided to the linkage in order that the linkage has a constrained motion.

Fig. 7

For the linkage in Fig. 8, derived from Fig. 5

$$N = 6$$
$$P_1 = 7$$
$$P_2 = 0$$
$$F = 3(6 - 1) - 7 \times 2 = 1$$

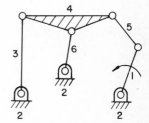

That is, the linkage has one degree of freedom. If link 1 is driven by an external force then the motions of links 3, 4, 5, and 6 are constrained. We can also say that the motions of links 3, 4, 5, and 6 are dependent on the motion of link 1, which is driven by an external force.

Fig. 8

A linkage with one or more degrees of freedom permits relative motions between its links. If a linkage has one degree of freedom, then the linkage is called a mechanism. A linkage with zero or negative degrees of freedom is a structure. A structure does not permit relative motion between its members. Examine the linkage in Fig. 9.

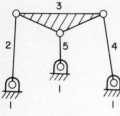

Fig. 9

$$N = 5, \quad P_1 = 6, \quad P_2 = 0$$

$$F = 3(5 - 1) - 6 \times 2 = 0$$

That is, the linkage in Fig. 9 is a structure. Because the linkage in Fig. 9 is a structure, links 3, 4, and 5 are not expected to move relative to link 1 when an external force is applied to link 2.

Competency Items

- Define a kinematic link and name the three different types of kinematic links.
- Define a kinematic chain.
- Define a loop in a kinematic chain.
- Define a linkage.
- State Gruebler's mobility criterion.
- Identify the notations F, N, P_1, and P_2.
- Calculate the degrees of freedom of the linkages shown below.

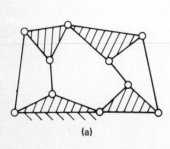

(a)

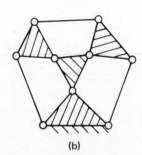

(b)

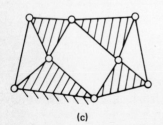

(c)

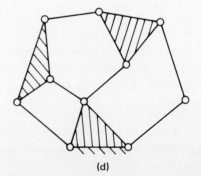

(d)

- Build a kinematic chain so that its linkage has $F = 3$.
- Build a kinematic chain so that its linkage has $F = -2$.
- Build a kinematic chain having four binary and four ternary links so that its resultant linkage has $F = 1$.

Objective 3

To build kinematic chains and linkages with other types of kinematic pairs

Activities
- Read the material provided below, and as you read:
  a. Build linkages with revolute and prism pairs.
  b. Build linkages with cam pairs.
  c. Build linkages which are spring loaded.
  d. Build linkages with belts and pulley drives.
- Check your answers to the competency items with your instructor.

Reading Material
We are concerned here with the linkages that move in a plane or in parallel planes. Note that in kinematic chains with revolute pairs, the axes of the revolute pairs are all parallel to one another and are perpendicular to the plane of this paper. In order to maintain this planar motion of the linkages with other types of joints, we are required to observe certain rules.

Rules for Substituting Prism Pairs in Place of Revolute Pairs
- If a revolute pair is to be replaced by a prism pair, then the axis of the prism pair must be either in the plane of the linkage or in a plane parallel to the plane of the linkage.
- A maximum of two revolute pairs may be replaced by two prism pairs in a given loop of a linkage. The axes of the substituted prism pairs must intersect one another.

Let us consider some examples demonstrating the application of this rule. Examine Fig. 10a–i. Figure 10a is a four-link mechanism with four revolute pairs. The revolute pair at the joint 4 is replaced by a prism pair with its axis coincident with the link 1. The resultant mechanism is a slider-crank mechanism shown in Fig. 10b. When the revolute pair at the joint 3 is replaced by a slider pair with its axis coincident with link 3, an inverted slider-crank mechanism, shown in Fig. 10c, is obtained. Now, examine how the mechanism in Fig. 10d is obtained.

There are three possible ways in which two of the four revolute pairs of the mechanism of Fig. 10a can be replaced by two prism pairs. These cases are shown in Figure 10e–g. The axes of the prism pairs in all the cases are intersecting.

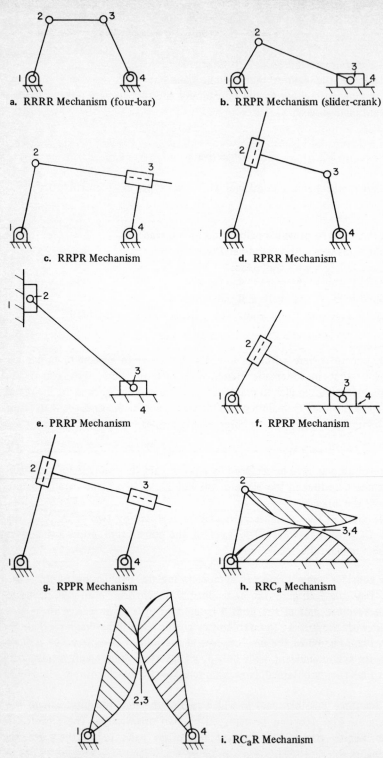

a. RRRR Mechanism (four-bar)

b. RRPR Mechanism (slider-crank)

c. RRPR Mechanism

d. RPRR Mechanism

e. PRRP Mechanism

f. RPRP Mechanism

g. RPPR Mechanism

h. RRC_a Mechanism

i. RC_aR Mechanism

Fig. 10 Different four-link mechanisms with revolute, prism, and cam pairs.

Rules for Substituting Cam Pairs in Place of Revolute Pairs

A cam pair has two degrees of freedom. Examination of Gruebler's mobility criterion reveals that a binary link with revolute pairs at each end can be replaced by a cam pair. For example, the binary link 4 (Fig. 10a) with its two revolute pairs at joints 3 and 4 can be replaced by a cam pair to obtain the resultant mechanism in Fig. 10h. Let us now check if $F = 1$ for this mechanism. Since after the substitution

$$N = 3$$
$$P_1 = 2$$
$$P_2 = 1$$
$$F = 3(N - 1) - 2P_1 - P_2$$
$$= 3(3 - 1) - 2 \times 2 - 1$$
$$= 1$$

So, the linkage in Fig. 10h has one degree of freedom. The mechanism in Fig. 10i is obtained by substituting a cam pair for the binary link 3 and its two revolute pairs.

*Rules for Substituting a Spring in Place of a Revolute Pair
and Two Binary Links*

In order to simulate the motion produced by a spring, examine the motion generated by two binary links connected by a revolute pair, as shown in Fig. 11.

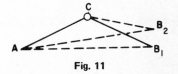

Fig. 11

Note that as AC is held fixed and BC is moved away or towards the point A, the length AB varies. This variation in the length AB can be accomplished by a spring which is either in tension or in compression. For this reason, one is able to replace a pair of binary links connected with a revolute pair by a spring.

Let us consider some examples. Examine the Watt six-link chain shown in Fig. 12. Examine the connection for links 2 and 3. This connection is satisfactory for

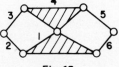

Fig. 12

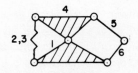

Fig. 13

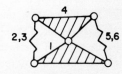

Fig. 14

substituting a spring. The resultant kinematic chain is shown in Fig. 13. Also, examine the connections of links 5 and 6. Note, here also we can substitute a spring in place of two binary links 5 and 6. The resultant Watt chain with two springs is shown in Fig. 14.

Figures 12, 13, and 14 are kinematic chains. When link 4 or 1 is fixed with ground, we will obtain a spring-loaded mechanism.

Rule for Substituting Belt and Pulley Combination in Place of Revolute Pairs, Two Binary Links and One Ternary Link

We can simulate the motion of the belt and pulley combination by two binary links and one ternary link which are connected by two revolute pairs. The motion equivalence is described in Fig. 15a and b. Figure 15a describes the link and joint combination which can be substituted by a belt-pulley combination shown in Fig. 15b.

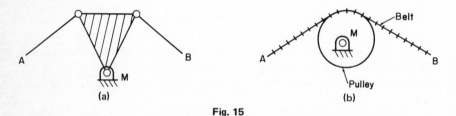

(a)   (b)

Fig. 15

Let us consider an example. Examine the Watt chain shown in Fig. 12. Examine the connection of links 3, 4, 5, or 2, 1, 6. For the links 3, 4, 5 note that links 3 and 5 are binary links connected to the ternary link 4. This combination can be replaced by a belt-pulley combination to obtain the chain shown in Fig. 16. A similar chain is obtained if we substitute a belt-pulley for the combination consisting of links 2, 1, 6.

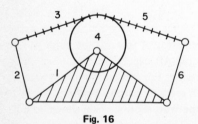

Fig. 16

Revolute, prism, cam, spring, and belt-pulley combinations are normally used in designing planar mechanisms. A kinematic pair such as a cylinder pair may be used. However, it will either act as a slider pair if the axis of the cylinder pair lies in the plane of the mechanism, or act as a revolute pair if the axis of the cylinder pair lies normal to the plane of the mechanism.

Competency Items

- State the rules for substituting:
  a. prism pairs in place of revolute pairs
  b. cam pairs in place of revolute pairs
  c. spring joints in place of revolute pairs
  d. belt and chain combination in place of revolute pairs
- Draw the equivalent kinematic chains with revolute pairs for the chains shown in the figures below.

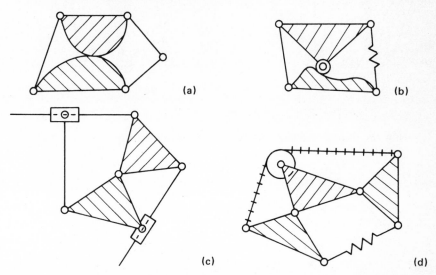

(a)　(b)

(c)　(d)

Performance Test #1

1. Describe the motion of a kinematic pair consisting of a spherical ball forced to move in a slotted cylinder.
   a. How many degrees of freedom does it have?
   b. What are the types of motion this kinematic pair can perform?
2. Write three examples of kinematic pairs that have four degrees of constraint on their motions.
3. Calculate the number of binary, ternary, etc., links, number of joints, and number of loops in the following kinematic chains.

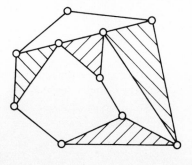

(a)

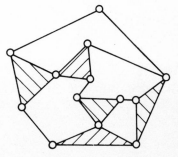

(b)

4. Compute the degrees of freedom of the following linkages.

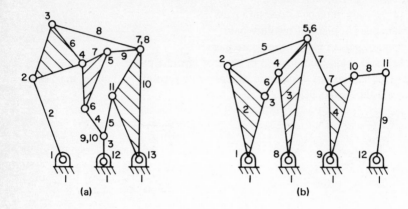

(a)   (b)

5. Draw the equivalent kinematic chains with revolute pairs.

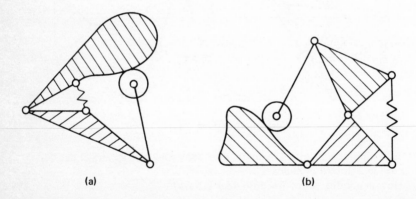

(a)   (b)

6. Draw all unique mechanisms from the six-link kinematic chain shown below.

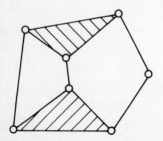

Performance Test #2

1. State the type of motion found in a helical pair.
2. Name a kinematic pair that has the following motions:
    a.   two translations and one rotation
    b.   two rotations
    c.   three rotations and one translation
3. Calculate the number of constraints each of the pairs in question 2 has.
4. Name the classes of pairs that have *only* higher kinematic pairs.
5. For the kinematic linkage shown below calculate the following:
    a.   number of loops
    b.   number of binary links
    c.   number of ternary links
    d.   number of other links
    e.   number of joints
    f.   number of degrees of
         freedom

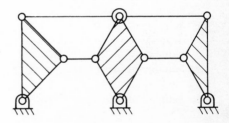

6. State the difference between a linkage and a mechanism.
7. State whether or not the following linkage is a structure. Give reasons.

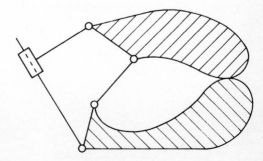

8. Draw the equivalent kinematic chain with revolute pairs.

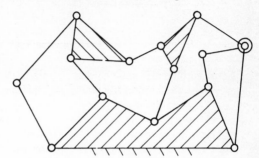

9. Replace appropriate elements of the kinematic chain so that it contains one spring and one belt-pulley combination.

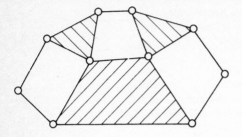

Performance Test #3

1. Name a kinematic pair that has two translations and two rotations.
2. a. Name the kind of kinematic pair that has the same motions as:
   (1) your elbow
   (2) the piston and cylinder as it acts in an internal combustion engine
   b. How many rotations and translations does each pair have?
3. State the basic difference between a cylinder pair and a cam pair.
4. Name the lower kinematic pairs of class III.
5. Build a kinematic chain with four loops, one quaternary link, and as many ternary and binary links as you need.
6. For the kinematic linkage shown calculate the following:
   a. number of binary links
   b. number of ternary links
   c. number of quaternary links
   d. number of other links
   e. number of joints
   f. number of degrees of freedom

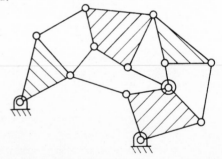

7. Compute the degrees of freedom of the following linkage.

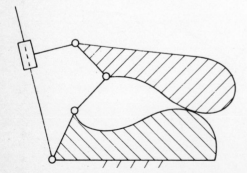

8. Draw the equivalent kinematic chain with revolute pairs.

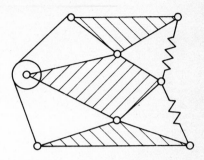

9. Draw all *unique* mechanisms from the seven-link kinematic chain shown.

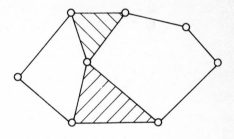

Supplementary References

2(2.1–2.7, 2.12–2.16); 3(14.1); 7(6.3); 8(1.1–1.5, 2.1, 2.4, 2.6, 2.7, 3.1–3.4)

Unit II. Velocity Analysis
of Linkages

Objectives
  1. To define vectors and vector properties
  2. To define velocity and its components for a particle moving on a path
  3. To describe the velocity of a particle moving on a path using complex number notation
  4. To derive a velocity relationship for a rigid body displacement
  5. To demonstrate the application of vector polygon technique in performing velocity analysis of single-loop mechanisms
  6. To perform velocity analysis of six-link mechanisms using the vector polygon technique
  7. To state the properties of instant center and demonstrate their application in performing velocity analysis of mechanisms

Objective 1

To define vectors and vector properties

Activities
  • Read the material provided below, and as you read:
    a. Define a position vector.
    b. State the properties of vectors.
    c. Define unit vectors and state their properties.
    d. Define dot product and cross product.
    e. Define vector differentiation.
  • Check your answers to the competency items with your instructor.

Reading Material
   Before one can proceed to learn to perform velocity and acceleration analysis of a mechanism, it is necessary to get well acquainted with the basic concepts of vectors, their properties, and how vectors are utilized in performing kinematic analysis.

A particle P is moving on path S which is located in an xyz frame of reference as shown in Fig. 1. The different positions of P can be located from origin point O by drawing lines from O to connect positions $P_1, P_2, \ldots, P_7$. In performing this operation, we are letting the point of a pencil travel along line $OP_1, OP_2, \ldots, OP_7$ from O to $P_1, P_2, \ldots, P_7$. In addition, we observe that the lines $OP_1, OP_2, \ldots, OP_7$ have length, direction (since we are traveling along lines $OP_1, OP_2, \ldots, OP_7$), and sense (since we are moving from O to P_1, $P_2, \ldots, P_7$).

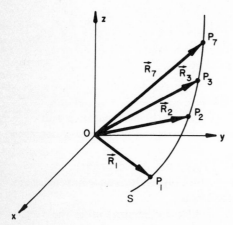

Fig. 1

The three properties listed above for lines $OP_1, OP_2, \ldots, OP_7$ define a vector. Thus, a vector is defined as a quantity which has magnitude, direction, and sense. A vector such as OP_1 can be symbolically represented by

$$\mathbf{R}_1 = \mathbf{OP}_1$$

The magnitude of the vector is usually represented by $|\mathbf{R}_1|$, R_1, or r_1.

Examples of vector quantities are velocity, acceleration, force, etc.

By definition a *scalar* is a quantity that has magnitude, but does not have direction and sense. Examples of scalars are length, volume, mass, etc.

The important properties of vectors are listed below.

1. Two vectors, $\mathbf{R}_1$ and $\mathbf{R}_2$, are equal provided they have equal magnitude, parallel or coaxial directions, and same senses. Figure 2 shows two equal vectors $\mathbf{R}_1$ and $\mathbf{R}_2$.

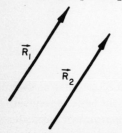

Fig. 2

2. When two vectors R_1 and R_2 have equal magnitudes, parallel directions, but opposite sense, $R_1 = -R_2$, or $R_2 = -R_1$. Figure 3 shows two such vectors.

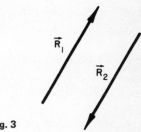

Fig. 3

3. The algebraic sum of two vectors R_1 and R_2 is a vector. Thus, $R_1 + R_2 = R_3$ and $R_1 - R_2 = R_4$ where R_3 and R_4 are the resultant vectors. The resultant vector R_3 is the same whether R_1 is added to R_2 or R_2 is added to R_1. See Fig. 4.

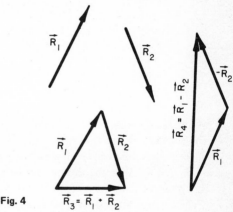

Fig. 4

4. Multiplication by a scalar quantity to a vector is a vector.
5. A unit vector U is a vector with unit magnitude. Three unit vectors associated with xyz coordinate axes are U_x, U_y, and U_z.
6. If xp, yp, zp are coordinates of point P then, the vector Rp can be described mathematically using the unit vectors U_x, U_y, U_z in the following manner:

$$Rp = xpU_x + ypU_y + zpU_z$$

The magnitude $|Rp|$ is

$$r_p = (xp^2 + yp^2 + zp^2)^{1/2}$$

and direction of $\mathbf{R}p$ is defined by the unit vector $\mathbf{U}p$ associated with $\mathbf{R}p$ where

$$\mathbf{U}p = \frac{xp}{r_p}\mathbf{U}_x + \frac{yp}{r_p}\mathbf{U}_y + \frac{zp}{r_p}\mathbf{U}_z$$

7. The *dot product* of two vectors $\mathbf{R}_1$ and $\mathbf{R}_2$ is defined as

$$\mathbf{R}_1 \cdot \mathbf{R}_2 = r_1 r_2 \cos\theta$$

where θ is the angle between the two vectors $\mathbf{R}_1$ and $\mathbf{R}_2$. The dot product of two vectors is a scalar quantity.

8. Since the unit vectors $\mathbf{U}_x, \mathbf{U}_y,$ and $\mathbf{U}_z$ are perpendicular to one another

$$\mathbf{U}_x \cdot \mathbf{U}_y = \mathbf{U}_y \cdot \mathbf{U}_z = \mathbf{U}_x \cdot \mathbf{U}_z = 0$$

and

$$\mathbf{U}_x \cdot \mathbf{U}_x = \mathbf{U}_y \cdot \mathbf{U}_y = \mathbf{U}_z \cdot \mathbf{U}_z = 1$$

9. The cross product of two vectors $\mathbf{R}_1$ and $\mathbf{R}_2$ is defined by

$$\mathbf{R}_1 \times \mathbf{R}_2 = (r_1 r_2 \sin\theta)\mathbf{U}$$

where $\mathbf{U}$ is a unit vector perpendicular to the plane of vectors $\mathbf{R}_1$ and $\mathbf{R}_2$. The sense of vector $\mathbf{U}$ is determined using the right-hand rule. If the index finger and middle finger of your right hand represent directions and senses of $\mathbf{R}_1$ and $\mathbf{R}_2$, then the thumb of your right hand represents the direction of $\mathbf{U}$. Figure 5 shows the cross product operation of vectors $\mathbf{R}_1$ and $\mathbf{R}_2$.

The cross product of two vectors is a vector quantity

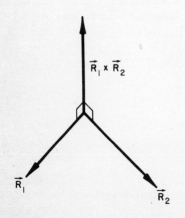

Fig. 5

10. The cross product of two vectors R_1 and R_2 can be obtained by using a determinant. Let

$$R_1 = r_{1x}U_x + r_{1y}U_y + r_{1z}U_z$$
$$R_2 = r_{2x}U_x + r_{2y}U_y + r_{2z}U_z$$

then the cross product of R_1 and R_2 can be obtained from the determinant

$$R_1 \times R_2 = \begin{vmatrix} U_x & U_y & U_z \\ r_{1x} & r_{1y} & r_{1z} \\ r_{2x} & r_{2y} & r_{2z} \end{vmatrix}$$

$$= U_x(r_{1y}r_{2z} - r_{1z}r_{2y}) + U_y(r_{1z}r_{2x} - r_{1x}r_{2z})$$
$$+ U_z(r_{1x}r_{2y} - r_{2x}r_{1y})$$

11. The cross products of the three unit vectors U_x, U_y, U_z are

$$U_x \times U_x = U_y \times U_y = U_z \times U_z = 0$$

and

$$U_x \times U_y = U_z, \quad U_y \times U_z = U_x, \quad U_z \times U_x = U_y$$

12. Differentiation of vectors with respect to some scalar quantity (time t, as an example), can be carried out using the rules of the limiting process. For example, if a vector $R(t)$ becomes $R(t + \Delta t)$ at some time $t + \Delta t$, such that $R(t + \Delta t) = R(t) + \Delta R$, then the ordinary derivative of R with respect to t is defined as

$$\lim_{\Delta t \to 0} \frac{R(t + \Delta t) - R(t)}{\Delta t} = \dot{R}$$

When R is defined in an x, y, z frame of reference as

$$R = r_x U_x + r_y U_z + r_z U_z$$

then time derivative of R will yield

$$\dot{R} = \dot{r}_x U_x + \dot{r}_y U_y + \dot{r}_z U_z$$

The following are the important results in vector differentiations:

(1) $\dfrac{d}{dt} (R_1 \pm R_2) = \dfrac{dR_1}{dt} \pm \dfrac{dR_2}{dt}$

(2) $\dfrac{d}{dt} (R_1 \cdot R_2) = \dfrac{dR_1}{dt} \cdot R_2 + R_1 \cdot \dfrac{dR_2}{dt}$

(3) $\dfrac{d}{dt} (R_1 \times R_2) = \dfrac{dR_1}{dt} \times R_2 + R_1 \times \dfrac{dR_2}{dt}$

Competency Items
- Write the position vector Rp for a particle P with coordinates $xp = 1, yp = 2$, $zp = 3$.
- Calculate the magnitude of the vector Rp.
- Calculate the unit vector Up associated with Rp.
- Show graphically the vector summation of the two vectors:

$R_1 = 2U_x + 3U_y$

$R_2 = 3U_x - 2U_y$

- Calculate the dot product $R_1 \cdot R_2$.
- Calculate the cross product $R_1 \times R_2$ and sketch the resultant vector.

Objective 2

To define velocity and its components for a particle moving on a path

Activities
- Read the material provided below, and as you read:
  a. Define the velocity of a particle moving on a path.
  b. Calculate the radial and transverse components of the velocity of the moving particle using vector relationships.
  c. Calculate vertical and horizontal components of the velocity of the moving particle.
- Check your answers to the competency items with your instructor.

Reading Material

Part I: Definition of velocity of a particle moving on a path

Let us examine the motion of a point in fixed coordinate system xyz. The point P is traveling along the continuous paths shown in Fig. 6. The positions P_1, P_2, etc., of the point P can be described by using the position vectors R_1, R_2, etc. Note that the magnitude and direction of the position vector R are function of time. The point P travels through ΔR during the time Δt where

$\Delta R = R_2 - R_1$

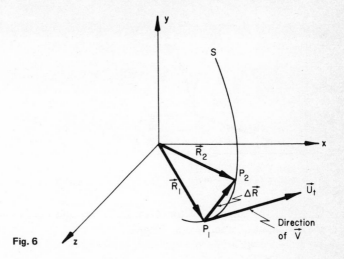

Fig. 6

The velocity V of the point P is given by the time derivative. That is,

$$V = \lim_{\Delta t \to 0} \frac{\Delta R}{\Delta t} = \frac{dR}{dt}$$

The direction of V is obtained as the limiting direction of ΔR as it approached zero. Thus

$$V = \lim_{\Delta t \to 0} \frac{\Delta R}{\Delta s} \frac{\Delta s}{\Delta t}$$

But the limiting value of $\Delta R/\Delta S$ is a unit vector U_t along the tangent to the curve S. Thus, the velocity can be written as

$$V = \frac{dS}{dt} U_t = \dot{S} U_t$$

We observe from the above relationship, that for a particle moving on path S, the magnitude of the velocity is given by dS/dt, and its direction is given by U_t.

Part II: Derivation to calculate horizontal and vertical
components of velocity

Consider an xy frame of reference shown in Fig. 7. The particle P is located using the position vector R which has magnitude r and is inclined at an angle θ with the x axis. Let $R_x = r \cos \theta$ be the magnitude of the component R along x axis, $Ry = r \sin \theta$ be the magnitude of the component of R along y axis and

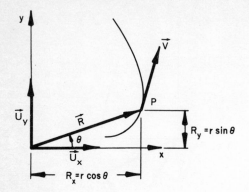

Fig. 7

U_x and U_y be the unit vectors (vectors having unit magnitude), associated with the directions of x and y axes. Then

$$R = R_x U_x + R_y U_y$$

The velocity V of the particle P is obtained by taking the time derivative of R.

$$V = \frac{dR}{dt} = \frac{d}{dt}(R_x U_x + R_y U_y)$$

Since U_x and U_y are associated with the x and y axes which are fixed with the ground, the time derivatives of the unit vector U_x and U_y are zero. Hence,

$$V = \dot{R}_x U_x + \dot{R}_y U_y$$

but

$$\dot{R}_x = \dot{r} \cos\theta - r\frac{d\theta}{dt}\sin\theta$$

and

$$\dot{R}_y = \dot{r} \sin\theta + r\frac{d\theta}{dt}\cos\theta$$

Hence

$$V = \left(\dot{r} \cos\theta - r\frac{d\theta}{dt}\sin\theta\right)U_x + \left(\dot{r} \sin\theta + r\frac{d\theta}{dt}\cos\theta\right)U_y$$

The components $\dot{R}_x U_x$ and $\dot{R}_y U_y$ are the horizontal and vertical components of the velocity V.

The quantity $d\theta/dt$ is the time rate of change of angular position of vector R. It defines the angular velocity. For Fig. 7, the vector quantity for the angular velocity is

$$\omega = \frac{d\theta}{dt} U_z$$

since the particle is rotating in the plane of the xy frame of reference. The magnitude of ω is $\omega = d\theta/dt$

Part III: Derivation to calculate radial and transverse
components of velocity

Let us now obtain the radial and transverse components of the velocity of the particle P moving along the path S.

If U_r is the unit vector associated with the direction of the position vector R, then

$$R = rU_r$$

where r is the magnitude of R.

The velocity of the particle P is obtained by taking the time derivative of R.

$$V = \frac{dR}{dt} = \frac{d}{dt} rU_r$$

Since the particle is moving on the path S so that both r and U_r are changing with respect to time,

$$V = \dot{r}U_r + r\dot{U}_r$$

As the particle P moves along the path S, the position vector R and its associated unit vector U_r are both rotating about the origin through varying values of θ. Since U_r is of unit magnitude, the velocity of the particle placed at the end of U_r is $d\theta/dt = \omega$. The direction of the velocity ω is perpendicular to the unit vector U_r. Thus

$$\dot{U}_r = \omega U_\theta$$

where U_θ is the unit vector perpendicular to U_r. Hence, the velocity of the particle P is

$$V = \dot{r}U_r + r\omega U_\theta$$

The components $\dot{r}$ and $r\omega$ are called the radial and transverse components of V. These are shown in Fig. 8.

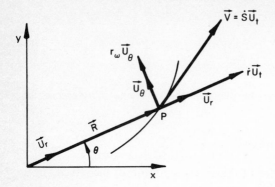

Fig. 8

Competency Items

- Define velocity of a particle moving in a circular path.
- Calculate the velocity of a particle moving in a path $y = 1/3 \, at^3$ at time $t = 0.5$ sec.
- Plot velocity versus time for the motion of a particle moving on a path $y = a \cos 2\pi t$.
- Plot the velocity of a particle moving on a path described by $y = 1/2 \, x^2$ when $\mathbf{R} = 4\mathbf{U}_x + 8\mathbf{U}_y$ if $dx/dt = 2$ in./sec.
- Plot the radial and the transverse components of the velocity of that particle.
- Calculate the rate of change of length of the position vector $\mathbf{R}$ above. Calculate the angular velocity of the position vector in this position.
- A particle is moving on a path described by $y = 4x^2 + x/2$ and has $\dot{x} = 2$ in./sec, $\dot{y} = 17$ in./sec. Locate the position vector $\mathbf{R}$.

Objective 3

To describe the velocity of a particle moving on a path using complex number notation

Activities

- Read the material provided below, and as you read:
  a. Describe the position of a particle using complex number notation.
  b. Derive horizontal and vertical components of the velocity of a particle moving on a path.
  c. Derive transverse and radial components of the velocity of a particle moving on a path.
- Check your answers to the competency items with your instructor.

Reading Material

The position of a particle P is described using the position vector $\mathbf{R}$ which is inclined at an angle θ with the x axis, as shown in Fig. 9.

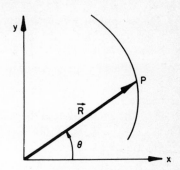

Fig. 9

The vector $\mathbf{R}$ can be expressed using complex number notation as

$$\mathbf{R} = re^{j\theta}$$

where r is the magnitude (the length) of $\mathbf{R}$, and θ describes the inclination of $\mathbf{R}$ with respect to the x axis. The angle θ is measured with reference to x axis in a counterclockwise direction. The complex operator j is defined by the relationship $j^2 = -1$.

The following are the important properties of complex numbers:

1. $\mathbf{R}_1 + \mathbf{R}_2 = r_1 e^{j\theta_1} + r_2 e^{j\theta_2}$

2. $\mathbf{R}_1 - \mathbf{R}_2 = r_1 e^{j\theta_1} - r_2 e^{j\theta_2}$

3. $\mathbf{R}_1 \mathbf{R}_2 = r_1 e^{j\theta_1} \times r_2 e^{j\theta_2} = r_1 r_2 e^{j(\theta_1 + \theta_2)}$
   if $r_1 r_2 = r_3$ and $\theta_1 + \theta_2 = \theta_3$
   Then $\mathbf{R}_1 \mathbf{R}_2 = r_3 e^{j\theta_3} = \mathbf{R}_3$

NOTE: The product of two vectors is a third vector. The result of the simple product of two vectors is different from the result of the vector cross product.

4. $\dfrac{\mathbf{R}_1}{\mathbf{R}_2} = \dfrac{r_1 e^{j\theta_1}}{r_2 e^{j\theta_2}} = \dfrac{r_1}{r_2} e^{j(\theta_1 - \theta_2)}$

5. $\dfrac{d\mathbf{R}_1}{dt} = \dfrac{d}{dt}(r_1 e^{j\theta_1}) = \dot{r}_1 e^{j\theta_1} + jr_1 \dfrac{d\theta_1}{dt} e^{j\theta_1}$

6. $e^{j\theta} = \cos\theta + j\sin\theta$

7. $\mathbf{R}_1 = r_1 e^{j\theta_1} = r_1(\cos\theta + j\sin\theta)$

8. $\mathbf{R}_1 + \mathbf{R}_2 = (r_1 \cos\theta_1 + jr_1 \sin\theta_1) + (r_2 \cos\theta_2 + jr_2 \sin\theta_2)$
   $\qquad\qquad = (r_1 \cos\theta_1 + r_2 \cos\theta_2) + j(r_1 \sin\theta_1 + r_2 \sin\theta_2)$

9. In the property 8, a component similar to $(r_1 \cos\theta_1 + r_2 \cos\theta_2)$ is called the *real part* of the complex number. A component similar to $(r_1 \sin\theta_1 + r_2 \sin\theta_2)$ is called the *complex part*. The complex part is identified by the presence of the complex operator j.

Example 1.

Find the summation of the vectors $\mathbf{R}_1$ and $\mathbf{R}_2$ using complex number notation $r_1 = 2, r_2 = 1, \theta_1 = 60°, \theta_2 = 30°$.

Solution.

$$\mathbf{R}_1 + \mathbf{R}_2 = r_1 e^{j\theta_1} + r_2 e^{j\theta_2}$$
$$= (r_1 \cos\theta_1 + r_2 \cos\theta_2) + j(r_1 \sin\theta_1 + r_2 \sin\theta_2)$$
$$= (2\cos 60° + 1\cos 30°) + j(2\sin 60° + 1\sin 30°)$$
$$= 1.866 + j2.232$$
$$\mathbf{R}_1 + \mathbf{R}_2 = \mathbf{R}_3 = r_3 e^{j\theta_3}$$

where $\theta_3 = \tan^{-1}\left(\dfrac{2.232}{1.866}\right) = \tan^{-1}(1.196)$
$$= 50.0°$$
$$r_3 = (1.866^2 + 2.232^2)^{1/2}$$
$$= 2.9428$$

Figure 10 shows $\mathbf{R}_1$, $\mathbf{R}_2$, and $\mathbf{R}_3 = \mathbf{R}_1 + \mathbf{R}_2$.

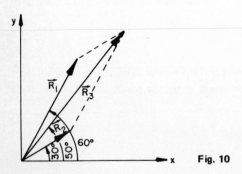

Fig. 10

Example 2.

Using the same vectors $\mathbf{R}_1$ and $\mathbf{R}_2$ as in the previous example find $\mathbf{R}_1 - \mathbf{R}_2$.

Solution.

$$\mathbf{R}_1 - \mathbf{R}_2 = r_1 e^{j\theta_1} - r_2 e^{j\theta_2}$$
$$= (r_1 \cos\theta_1 - r_2 \cos\theta_2) + j(r_1 \sin\theta_1 - r_2 \sin\theta_2)$$
$$= (2\cos 60° - \cos 30°) + j(2\sin 60° - \sin 30°)$$
$$= 0.134 + j1.232$$

Example 3.

Find $R_1 \times R_2$.

Solution.

$$
\begin{aligned}
R_1 \times R_2 &= r_1 e^{j\theta_1} \times r_2 e^{j\theta_2} \\
&= r_1 r_2 e^{j(\theta_1 + \theta_2)} \\
&= 2 e^{j(90°)} \\
&= 2 \cos 90° + 2j \sin 90° \\
&= 2j
\end{aligned}
$$

NOTE: This product of complex numbers is *not* the cross product of the vectors.

Example 4.

Find R_1/R_2.

Solution.

$$
\begin{aligned}
\frac{R_1}{R_2} &= \frac{r_1 e^{j\theta_1}}{r_2 e^{j\theta_2}} \\
&= \frac{r_1}{r_2} e^{j(\theta_1 - \theta_2)} \\
&= \frac{2}{1} e^{j(60° - 30°)} \\
&= 2 \cos 30° + 2j \sin 30° \\
&= 1.732 + j(1)
\end{aligned}
$$

Example 5.

Calculate the x and y components of the vectors R_1 and R_2 of example 1.

Solution.

$$
R_1 = r_1 e^{j\theta_1} = 2 e^{j60°} = 2 \cos 60° + j2 \sin 60°
$$

$$
R_2 = r_2 e^{j\theta_2} = e^{j30°} = \cos 30° + j \sin 30°
$$

Since property 9 pointed out that the real parts are the x components and that the complex parts are the y components, we can readily see that the x components

of $R_1 = 2 \cos 60° = 1$

of $R_2 = \cos 30° = 0.866$

The y components

of $R_1 = 2 \sin 60° = 1.732$

of $R_2 = \sin 30° = 0.500$

Let us now obtain the velocity of a particle moving on a path. The position of the particle is described by

$$\mathbf{R} = re^{j\theta}$$

By definition,

$$\mathbf{V} = \frac{d\mathbf{R}}{dt} = \frac{d}{dt}(re^{j\theta})$$

$$= \dot{r}e^{j\theta} + jr\frac{d\theta}{dt}e^{j\theta}$$

Let $d\theta/dt = \omega$, the angular velocity of the particle. (The angular velocity helps us to calculate the amount of rotational motion experienced by the moving particle.)

Hence,

$$\mathbf{V} = \dot{r}e^{j\theta} + jr\omega e^{j\theta}$$

Notice that $\dot{r}$ describes the radial component while $r\omega$ describes the transverse component of the velocity $\mathbf{V}$ of the particle P.

Substituting for $e^{j\theta} = \cos\theta + j\sin\theta$, we get

$$\mathbf{V} = \dot{r}(\cos\theta + j\sin\theta) + jr\omega(\cos\theta + j\sin\theta)$$

Rearranging the above relationship, we get

$$\mathbf{V} = (\dot{r}\cos\theta - r\omega\sin\theta) + j(\dot{r}\sin\theta + r\omega\cos\theta)$$

NOTE: $(\dot{r}\cos\theta - r\omega\sin\theta) = x$ component of the velocity of particle P.

$(\dot{r}\sin\theta + r\omega\cos\theta) = y$ component of the velocity of particle P.

Figure 11 shows x, y, transverse and radial components of the velocity of the particle P.

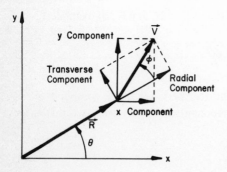

Fig. 11

Example 6.

Refer to Fig. 11. If $r = 2$ in., $\phi = 30°$, $\theta = 30°$, and the magnitude of $\mathbf{V} = 10$ in./sec., find the x, y, transverse, and radial components of velocity $\mathbf{V}$.

Solution.

Using the notation

V_x = the x component of velocity
V_y = the y component of velocity
V_r = the radial component of velocity
V_t = the transverse component of velocity

$V_x = V \cos(\phi + \theta) = 10\,\text{in./sec}\cos 60° = 5.00\,\text{in./sec}$
$V_y = V \sin(\phi + \theta) = 10\,\text{in./sec}\sin 60° = 8.66\,\text{in./sec}$
$V_r = V \cos\phi = 10\,\text{in./sec}\cos 30° = 8.66\,\text{in./sec}$
$V_t = V \sin\phi = 10\,\text{in./sec}\sin 30° = 5.00\,\text{in./sec}$

Competency Items
- Calculate the horizontal and vertical components of the velocity of a particle which has velocity $25\,e^{j0°}$ in./sec and position $5\,e^{j60°}$ in.
- Calculate the radial and transverse components of that particle.
- Calculate ω_2 if $\omega_4 = 10$ rad/sec for the crank MA of the mechanism shown in the figure. $MQ = 3.6$ in.

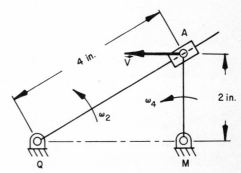

- If a particle whose position is described by $\mathbf{R} = 2e^{j45°}$ in. has a velocity of $20e^{j60°}$ in./sec, calculate the radial component of velocity V and the angular velocity of the position vector $\mathbf{R}$.

Objective 4

To derive a velocity relationship for a rigid body displacement

Activities
- Read the material provided, and as you read:
  a. Write the vector equation describing the velocity relationship for a rigid body motion.
  b. Construct a vector polygon from the vector equation.
- Check your answers to the competency items with your instructor.

Reading Material

A rigid body is shown in its two positions in Fig. 12. In order to study the motion of a rigid body W, it is necessary to examine the motion of two fixed points A and B in the rigid body.

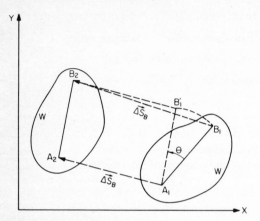

Fig. 12

Since the motion of the rigid body can be decomposed into pure translation along the line $A_1 A_2$ and pure rotation of the point B_1 through angle θ^* about A_1, the distance traveled by the point B from position B_1 to B_2 is given by

$$\Delta S_B = \Delta S_A + A_1 B_1 (\theta U_z) \times (A_1 B_1)$$

The time derivative of the above relation will give the velocity of point B in the rigid body

$$\frac{dS_B}{dt} = \frac{dS_A}{dt} + \frac{d\theta}{dt} U_z \times A_1 B_1$$

Since $\omega = (d\theta/dt) U_z$,

$$V_B = V_A + A_1 B_1 \omega \times A_1 B_1$$

$$V_B = V_A + V_{B/A} \tag{1}$$

The quantity $\omega \times A_1 B_1$ is the relative velocity of point B with respect to point A. Let $V_{B/A} = \omega \times A_1 B_1$

The magnitude of the velocity $V_{B/A}$ is $AB \cdot \omega$ and its direction is normal to the line AB. The sense of $V_{B/A}$ is determined by the sense of rotation of the rigid body.

Equation (1) describes one of the most useful relationships in conducting the velocity analysis of mechanisms. We shall now study the steps required to construct the vector polygon from the information given by Eq. (1).

*θ is infinitesimally small.

Equation (1) has three vector quantities V_B, V_A, $V_{B/A}$. Each of these vector quantities have magnitudes and directions. In order to know V_B completely, we must know its magnitude and direction. In general, the direction of V_B is known, and its magnitude must be determined from Eq. (1). In order to solve this vector equation, we must know the magnitude and direction of V_A and direction of $V_{B/A}$. Work the example below to learn the steps to construct the vector polygon and the graphical method of solving Eq. (1).

Example 7.

V_A = 20 in./sec parallel to the x axis.

V_B is moving parallel to the y axis.

$V_{B/A}$ is moving along the line inclined at angle $135°$ with the x axis.

Draw a vector polygon to determine the magnitude and direction of V_B and $V_{B/A}$.

Solution.

Repeat the following steps and examine Fig. 13.

1. Select a scale 1 in. = 10 in./sec.
2. Select the origin point O_v and draw a line $O_v X$ parallel to x axis. Measure length $O_v A$ = 2 in. to represent magnitude of the velocity V_A. Place a "tail" mark at O and a "head" mark at a'. Then $O_v A$ is V_A. See Fig. 13.

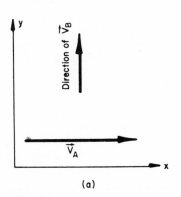
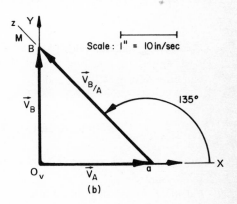

(a)   (b)

Fig. 13

3. At 'a' draw a line z inclined at angle $135°$ with the line $O_v X$.
4. At O_v draw a line $O_v Y$ parallel to y axis.
5. Let the line $O_v z$ intersect line $O_v Y$ in b.
6. Place tail at O_v and head at b so that $O_v b$ is V_B.
7. Place tail at a and head at b on line az so that AB is $V_{B/A}$.
8. Measure the length $O_v b$, convert it to the scale in in./sec. This is the magnitude of V_B.

9. Measure the length ab, convert it to the scale in 1 in. = 10 m/sec. This is the magnitude of $V_{B/A}$.

10. The polygon $O_v ab$ is called the *vector polygon* solving the vector equation

$$V_B = V_A + V_{B/A}$$

Competency Items

- In the figure below the velocities of points A and B of a rigid body AB are given. Draw a vector polygon to determine $V_{B/A}$.

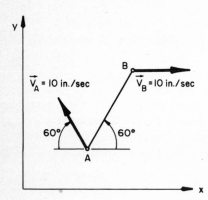

- Calculate the angular velocity of the rigid body if length AB = 1 in.
- Write the vector equation relating V_A, V_B, and $V_{A/B}$. Also write the vector equation relating V_A, V_B, and $V_{B/A}$.

Performance Test #1—Objectives 1, 2, 3, 4

1. Calculate the velocity of a particle moving on a path $y = 1/2\, a^2 t^2 + at$ at t = 0.5 sec and a = 1.

2. Calculate the velocity of a particle moving on a path $y = a \sin 2\pi t$ when a = 1, t = 0.5 sec.

3. Calculate

   a. $R_1 + R_2$

   b. $R_1 \times R_2$

   c. R_1/R_2

   when

   $R_1 = 3\,e^{j(30°)}$, $R_2 = 4\,e^{j(60°)}$

4. A particle is tied to a string of length l = 5 in. This particle is moving with an angular velocity ω = 5 rad/sec. Calculate the magnitude of the velocity V of the particle.

5. The position of a particle P is described using the position vector $3\,e^{j(30°)}$. Its velocity is given by $\mathbf{V} = 10\,e^{j(90°)}$.
   a. Plot this position vector.
   b. Show the radial and transverse components of its velocity $\mathbf{V}$.
   c. Show the x and y components of its velocity $\mathbf{V}$.

6. Determine the velocity of a particle moving on a path described by $y = 1/2\,x^2$ when $\mathbf{R} = 4\mathbf{U}_x + 8\mathbf{U}_y$ and if $dx/dt = 2$ in./sec.

Performance Test #2—Objectives 1, 2, 3, 4

1. Calculate the velocity of a particle moving on a path $y = 2a^3t^2 - a^2 \sin 2\pi t$ when $a = 2$ and $t = 1$ sec.

2. Calculate
   a. $\mathbf{R}_1 - \mathbf{R}_2$
   b. $\mathbf{R}_2/\mathbf{R}_1$
   when
   $$\mathbf{R}_1 = 2e^{j(45°)} \qquad \mathbf{R}_2 = 5e^{j(30°)}$$

3. The position of a particle is given by the vector $\mathbf{R} = 2e^{j(60°)}$ in. The magnitude of the velocity vector is 6 in./sec and the radial component is -3 in./sec. The particle is moving in a clockwise direction about the origin. Calculate
   a. the transverse component of the velocity
   b. the values of V and θ of the velocity vector $\mathbf{V} = Ve^{j\theta}$

4. A particle is moving on a path described by $y = 4x^2 - x/2$ and has $\dot{x} = 4$ in./sec when $dy/dx = 31/2$. Calculate the velocity vector $\mathbf{V}$ and the position vector $\mathbf{R}$; i.e., calculate R_x, R_y, $\dot{R}_x$, and $\dot{R}_y$.

Performance Test #3—Objectives 1, 2, 3, 4

1. Determine the x and y components of the product of the two vectors $-6e^{j25°}$ and $4e^{-j45°}$.

2. A particle has a velocity described by $\mathbf{V} = 10e^{j(330°)}$ in./sec. If the radial component of the velocity is $\dot{r} = +5$ in./sec and the angular velocity is $\omega = -3$ rad/sec, find the position vector of the particle in the form $\mathbf{R} = re^{j\theta}$; i.e., find r and θ.

3. A particle is traveling on a path described by $y = 2x^3 - 54$. The vertical component of the velocity is $\dot{R}_y = 27$ in./sec, and the horizontal position component is $R_x = 3$ in. Find the horizontal component velocity $\dot{R}_x$ and the angular velocity ω of the particle.

4. The two vectors shown represent velocities (one absolute, the other relative) of two points (A and B) of a rigid body. They are drawn to scale.

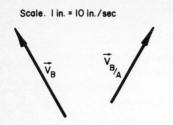

Scale. I in. = 10 in./sec

a. Write a vector equation to relate the velocities of the two points.
b. Construct the vector polygon.
c. Find the magnitude of the velocity of point A.

Objective 5

To demonstrate the application of a vector polygon technique in performing velocity analysis of single-loop mechanisms

Activities

- Read the material provided below, and as you read demonstrate the application of the vector polygon technique in performing velocity analysis of:
  a. a four-link mechanism
  b. a slider-crank mechanism
  c. an inverted slider-crank mechanism
- Check your answers to the competency items with your instructor.

Reading Material

We will demonstrate the application of the velocity-vector relationship derived in objective 4. The velocity-vector relationship

$$\mathbf{V}_B = \mathbf{V}_A + \mathbf{V}_{B/A}$$

describes vectorally the velocity of point B, moving relative to point A, which, in turn, may be either moving or stationary. The vector quantities $\mathbf{V}_B$, $\mathbf{V}_A$, $\mathbf{V}_{B/A}$ have magnitudes and directions. In general, in solving the velocity-vector relationship, we will know the magnitude and direction of $\mathbf{V}_A$. The directions of $\mathbf{V}_B$ and $\mathbf{V}_{B/A}$ are also known. The velocity-vector relationship is applied to determine the magnitudes of velocities $\mathbf{V}_B$ and $\mathbf{V}_{B/A}$. The velocity-vector relationship can be solved analytically or graphically by constructing a velocity polygon. This technique is demonstrated by considering three examples of single loop planar mechanisms.

Example 8.

A four-link mechanism $MABQ$ is shown in Fig. 14a with link lengths $MA = 4$ in., $AB = 8$ in., $BQ = 5.5$ in. The input link MA has an angular velocity $\omega_2 = 94.2$ in./sec in a clockwise direction. Draw, in a stepwise manner, the vector polygon and calculate magnitudes and directions of angular velocities ω_3 and ω_4 of links AB and QB.

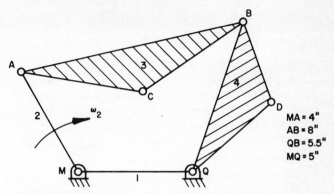

Fig. 14a A four-link mechanism.

Solution.

Repeat the following steps that lead to the determination of ω_3 and ω_4.

1. Mark links MQ, MA, AB, and BQ as link number $1, 2, 3$, and 4.

2. Compute the velocity of point A using

$$V_A = MA \cdot \omega_2 = \left(\frac{4}{12}\right)(94.2) - 31.4 \text{ ft/sec}$$

Note that the direction of $\mathbf{V}_A$ is perpendicular to link MA.

3. Select an origin point O_v. Also select a scale, say, 1 in. = 10 ft/sec. Now at O_v draw a line $O_v a$ of length 3.14 in. (since $V = 31.4$ ft/sec, and scale 1 in. = 10 ft/sec), perpendicular to link MA. See Fig. 14b.

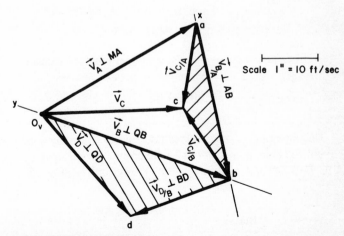

Fig. 14b Velocity polygon.

4. The velocity of point B can be computed using two paths.

a. Equation for the path MAB yields

$$V_B = V_A + V_{B/A}$$

The magnitude and direction of velocity V_A are completely known. The magnitude of the velocity $V_{B/A}$ is not known. However, since the point B moves on a circular path relative to the point A, the direction of the velocity $V_{B/A}$ must be along the line perpendicular to link AB.

Since the velocity $V_{B/A}$ is vectorally added to the velocity V_A, at the point "a" draw line x perpendicular to AB.

5. Since the point B also moves on a circular path about the fixed center Q, equation for V_B yields

$$V_B = QB \cdot \omega_4$$

where ω_4 is the angular velocity of link 4. Since ω_4 is not known, the magnitude of velocity V_B is not known. However, since the point B moves on a circle with Q as center, the direction of V_B is normal to the link QB. Therefore, draw line y passing through the origin O_v and perpendicular to link QB. Let the line x and y intersect in point b.

6. Steps 4 and 5 complete the construction to obtain the velocity polygon $O_v ab$ in Fig. 14b. Note that in adding the velocity vectors, the head of the vector V_A must meet the tail of the vector $V_{B/A}$.

7. The magnitudes of the velocities $V_{B/A}$ and V_B are obtained by measuring the length of the $V_{B/A}$ and V_B. Thus, after converting to the actual units

$$V_{B/A} = 10 \text{ ft/sec/in.} \times 2 \text{ in.} = 20 \text{ ft/sec}$$
$$V_B = 10 \times 3.4 = 34.0 \text{ ft/sec}$$

8. The angular velocity ω_3 of the coupler link is obtained as

$$\omega_3 = \frac{V_{B/A}}{AB} = 30 \text{ rad/sec}$$

The direction of ω_3 is clockwise since $V_{B/A}$ makes the point B to move about the point A in a clockwise direction.

9. The angular velocity ω_4 of the follower link is obtained as

$$\omega_4 = \frac{V_B}{BQ} = 74.2 \text{ rad/sec (clockwise)}$$

The absolute velocities of points C and D lying in the plane of links AB and QB can be determined using the following steps:

1. Velocity of point C is obtained by constructing the velocity polygon using

the following vector relationships

$$V_C = V_A + V_{C/A}$$
$$V_C = V_B + V_{C/B}$$

Note that the directions of $V_{C/A}$ and $V_{C/B}$ are perpendicular to AC and BC. Also, the magnitude and direction V_A and V_B are known. Hence, the vector polygon constructed using the above two relationships yields the magnitude of the velocities of the points C and B.

Thus, from the velocity polygon, Fig. 14b,

$$V_C = \text{ft/sec/in.} \times 2.7 \text{ in} = 27.0 \text{ ft/sec}$$

Note that there is an alternate way of obtaining the velocity of point C.

Since the body ABC is moving with angular velocity ω_3, the magnitude of velocity is given by

$$V_{C/A} = AC \cdot \omega_3 = \frac{4.5 \times 30}{12} = 11.25 \text{ ft/sec}$$

Since the magnitude and direction of $V_{C/A}$ are available, the velocity of point C can be obtained by constructing the velocity polygon using the relationship

$$V_C = V_A + V_{C/A}$$

From Fig. 14b, V_c = 27 ft/sec.

2. Velocity of the point D can be obtained in a following manner:

$$V_D = \omega_4 \cdot QD = \frac{74.2 \times 4}{12} = 24.73 \text{ ft/sec}$$

The direction of V_D is perpendicular to QD.

Example 9.

A slider-crank mechanism shown in Fig. 15a is driven by crank 2 at 600 rpm, in a counterclockwise direction. Calculate the velocities of points A, B, C, and the angular velocity of link 3.

Solution.

Here again we shall consider the graphical approach in which the velocity polygon will be constructed using the results of the velocity-vector relationship. Let n denote rpm of crank 2.

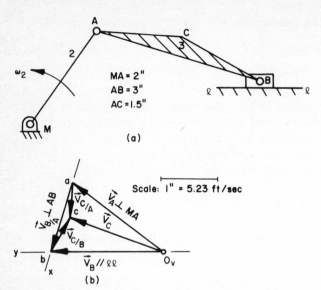

Fig. 15 (a) A slider-crank mechanism; (b) velocity polygon.

1. Compute the angular velocity ω_2

$$\omega_2 = \frac{2\pi n}{60} = \frac{2\pi\,(600)}{60} = 62.8 \text{ rad/sec}$$

2. Compute the velocity V_A

$$V_A = MA \cdot \omega_2 = \tfrac{2}{12} \times 62.8 = 10.46 \text{ ft/sec}$$

The direction of the velocity V_A is perpendicular to MA.

3. Construct the velocity polygon using

$$V_B = V_A + V_{B/A}$$

Note that the direction of the velocity V_B is along the axis of slider. The direction of the velocity $V_{B/A}$ is perpendicular to the link AB. It is required to find the magnitudes of the velocities V_B and $V_{B/A}$. These magnitudes can be found by constructing the velocity polygon. See Fig. 15b.

4. Select an origin point O_v. At O_v draw line $O_v a$ perpendicular to link 2. Measure distance $O_v a$ to scale velocity V_A.

5. At "a" draw line x perpendicular to line AB.

6. Draw line y parallel to the axis of the slider and pass through O_v. Let line y intersect line x in point b.

7. $O_v ab$ is the velocity polygon. Measure distances $O_v b$ and ab and convert it to obtain

$$V_B = 5.23 \text{ ft/sec/in.} \times 2 \text{ in.} = 10.46 \text{ ft/sec}$$
$$V_{B/A} = 5.23 \times 1.25 = 6.54 \text{ ft/sec}$$

8. The angular velocity of the link 3 can be obtained as

$$\omega_3 = \frac{V_{B/A}}{AB} = \frac{6.54 \times 12}{3} = 26.16 \text{ rad/sec (clockwise)}$$

9. The velocity of the point C can be obtained in two ways:
   a. Construct the velocity polygon using the following vector relationships

$$\mathbf{V}_C = \mathbf{V}_A + \mathbf{V}_{C/A}$$
$$\mathbf{V}_C = \mathbf{V}_B + \mathbf{V}_{C/B}$$

   The directions of the velocities $\mathbf{V}_{C/A}$ and $\mathbf{V}_{C/B}$ are perpendicular to AC and BC.
   b. Since ω_3 is known, the velocity $V_{C/A}$ can be computed.

$$V_{C/A} = AC \cdot \omega_3 = \frac{1.5}{12} \times 26.16 = 3.27 \text{ ft/sec}$$

Example 10.
The inverted slider-crank mechanism shown in Fig 16a is driven by crank 2 at 600 rpm counterclockwise direction. Calculate the velocities of points A_2 and A_3 and D. Also find the angular velocities of links 2 and 3 $GA = 3$ in.
Solution.
Once more, we shall consider the graphical approach. Let n denote rpm of crank 2.

1. Compute the angular velocity ω_2

$$\omega_2 = \frac{2\pi n}{60} = \frac{2\pi \times 600}{60} = 62.8 \text{ rad/sec}$$

2. Compute the velocity V_A

$$V_A = MA \cdot \omega_2 = \frac{2}{12} \times 62.8 = 10.46 \text{ ft/sec}$$

The direction of V_A is perpendicular to MA.

3. Construct the velocity polygon using the following relationship:

$$\mathbf{V}_{A_3} = \mathbf{V}_{A_2} + \mathbf{V}_{A_3/A_2}$$

Note that the point A is common in both link 2 and link 3. The direction of velocity V_{A_2} of point A when it is considered to lie on link 2 is perpendicular to link 2. The direction of the velocity V_{A_3} when the point A is considered to lie on link 3 is perpendicular to link 3. Also note that there is a relative motion between the points A_2 and A_3. This motion is the motion of the slider along the direction of the link 3. Thus the direction of the velocity V_{A_3/A_2} is along the direction of the link 3.

4. Select origin point O_v. Select scale for the velocity polygon. At O_v draw a line $O_v a$ perpendicular to MA to the appropriate scale to represent velocity V_{A_2}. See Fig. 16b.
5. At "a" draw line x parallel to link 3.
6. Draw line y passing through O_v and perpendicular to link 3. Let line x intersect line y in point b. The constructed velocity polygon $O_v ab$ is shown in Fig. 16b.
7. Measure distance $O_v b$ and convert it to scale to obtain velocity V_{A_3}. The angular velocity ω_3 is computed using

$$\omega_3 = \frac{V_{A_3}}{QA}$$

$$= \frac{5.23 \times 0.6 \times 12}{3} = 12.55 \text{ rad/sec (clockwise)}$$

The direction of the angular velocity ω_3 is determined from the direction of V_{A_3}.

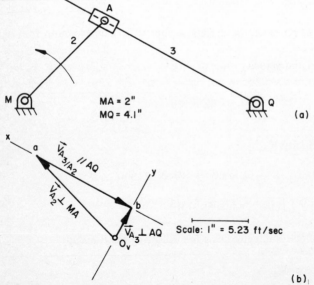

MA = 2"
MQ = 4.1"   (a)

Scale: 1" = 5.23 ft/sec

(b)

Fig. 16 (a) Inverted slider-crank mechanism; (b) velocity polygon.

Competency Items

- Demonstrate the application of vector polygon technique in performing velocity analysis of a four-link mechanism whose input link MA has an angular velocity $\omega_2 = 200$ rad/sec in a counterclockwise direction. For the four-link mechanism shown below:

    a. Calculate the velocity V_A of the point A.
    b. Write the velocity-vector relationship to determine velocity V_B of the point B.
    c. Write three steps to solve the vector relationship written in item b.
    d. Show how you will compute ω_3, ω_4, the velocities of points C and D.
    e. State the properties of triangles ACB and acb.
    f. State the properties of triangles QBD and O_vbd.
    g. Write the procedure to determine the directions of ω_3 and ω_4.
    h. Perform the velocity analysis of the four-link mechanism to obtain numerical values of ω_3 and ω_4.

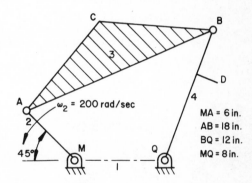

- Demonstrate the application of vector polygon technique in performing velocity analysis of a slider-crank mechanism whose input link MA rotates at 600 revolutions per minute in a clockwise direction. For the slider-crank mechanism shown below:

    a. Calculate ω_2.
    b. Calculate V_A.

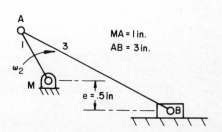

   c. Write the velocity-vector relationship for moving points A and B.
   d. Draw the vector polygon.
   e. Calculate ω_3.
   f. Calculate V_B.

- For the slider-crank mechanism shown below with the slider block moving along BM with velocity $V_B = 5$ in./sec,

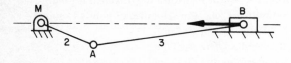

   a. Write the velocity-vector relationship for the moving points B and A.
   b. Draw the vector polygon.
   c. Calculate ω_2 and ω_3.

- For the inverted slider-crank mechanism shown below with link 3 having an angular velocity $\omega_3 = 100$ rad/sec,

   a. Write the velocity-vector relationship for points A_2 and A_3.
   b. Calculate ω_2.
   c. Calculate V_{A_3/A_2}.

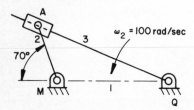

Objective 6

To perform velocity analysis of six-link mechanisms using the vector polygon technique

Activities

- Read the material provided below and as you read:
   a. List the six-link mechanisms with revolute and prism pairs.
   b. Extend the application of the vector polygon technique to perform the velocity analysis of six-link mechanisms.
   c. Demonstrate the application of velocity-vector relationship in constructing the vector polygon.

d. Calculate the angular and linear velocities of the different points in any given linkage system.

• Check your answers to the competency items with your instructor.

Reading Material

There are 42 different types of six-link mechanisms with three fixed pivots and are built using revolute and slider pairs. These different types of six-link mechanisms are shown in Figs. 17–22. Once the reader has familiarized himself with the vector polygon technique to perform the velocity analysis of four-link, slider-crank, and inverted slider-crank mechanisms, he will find it relatively simple to perform the velocity analysis of these six-link mechanisms. The necessary steps required in performing the velocity analysis involve repeated application of the velocity-vector relationship derived in Unit II, Objective 4. In this unit, we will study such a repeated procedure while performing the velocity analysis of a six-link mechanism with seven revolute pairs.

Example 11.

The six-link mechanism shown in Fig. 23a is driven by crank 2 at 900 rpm in a clockwise direction. Calculate the velocities of points *A*, *B*, *C*, and *D*, and the angular velocities of links 3, 4, 5, and 6.

Solution.

The following steps will lead to the desired answers.

1. Compute angular velocity ω_2.

$$\omega_2 = \frac{2\pi n}{60} = \frac{2\pi \times 900}{60} = 94.2 \text{ rad/sec}$$

2. Compute velocity V_A.

$$V_A = MA \cdot \omega_2 = 157 \text{ ft/sec}$$

3. Using the appropriate scale, draw velocity polygon to find the velocity $\mathbf{V}_B$ where

$$\mathbf{V}_B = \mathbf{V}_A + \mathbf{V}_{B/A}$$

NOTE: Direction of $\mathbf{V}_A$ is perpendicular to MA. Direction of $\mathbf{V}_B$ is perpendicular to QB. Direction of $\mathbf{V}_{B/A}$ is perpendicular to AB.

4. Determine the velocity of point *C* by constructing velocity polygon using the following vector equations:

$$\mathbf{V}_C = \mathbf{V}_A + \mathbf{V}_{C/A}$$
$$\mathbf{V}_C = \mathbf{V}_B + \mathbf{V}_{C/B}$$

NOTE: Direction of $\mathbf{V}_{C/A}$ is perpendicular to AC. Direction of $\mathbf{V}_{C/B}$ is perpendicular to BC.

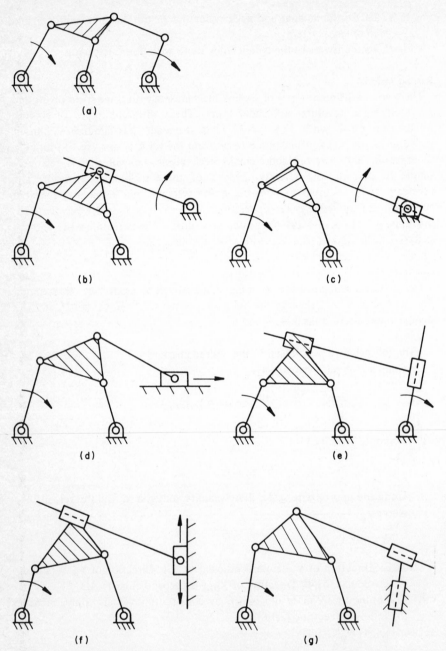

Fig. 17 Coupler drive six-link mechanisms derived from a four-revolute, four-link mechanism.

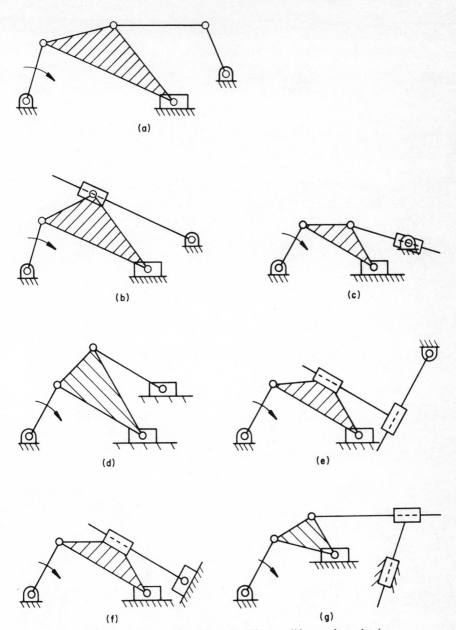

Fig. 18 Coupler drive six-link mechanisms derived from a slider-crank mechanism.

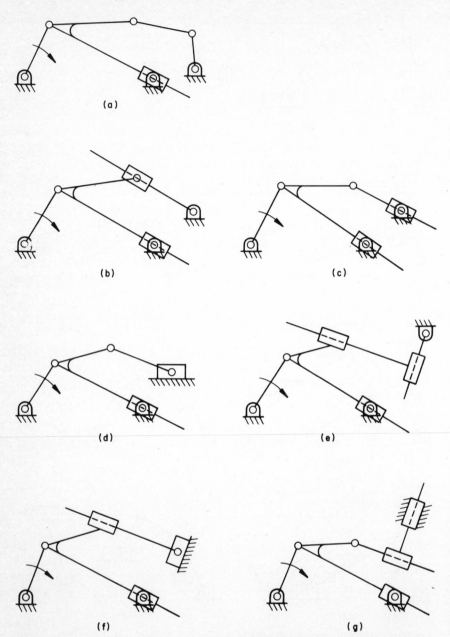

(a)

(b)

(c)

(d)

(e)

(f)

(g)

Fig. 19 Coupler drive six-link mechanisms derived from inverted slider-crank mechanism, type I.

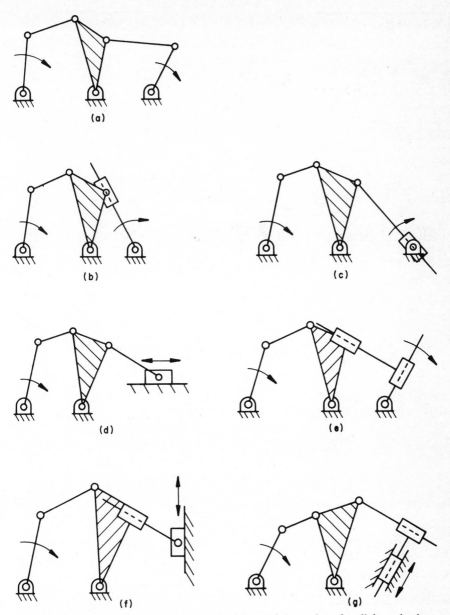

Fig. 20 Series drive six-link mechanisms derived from a four-revolute, four-link mechanism.

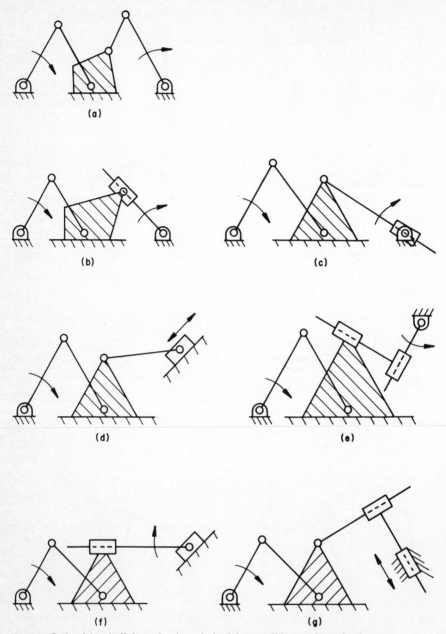

Fig. 21   Series drive six-link mechanisms derived from a slider-crank mechanism.

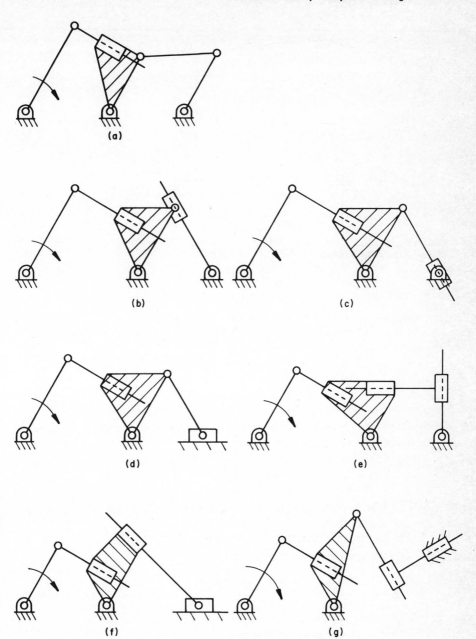

Fig. 22 Series drive six-link mechanisms derived from inverted slider-crank mechanism, type I.

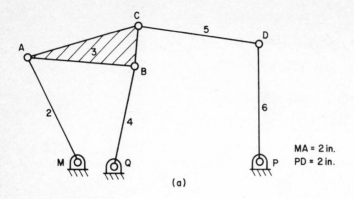

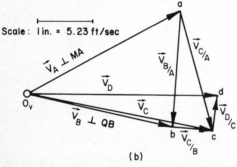

Fig. 23 (a) A six-link mechanism; (b) velocity polygon.

5. Determine the velocity of point D by constructing velocity polygon using the vector equation:

$$\mathbf{V}_D = \mathbf{V}_C + \mathbf{V}_{D/C}$$

NOTE: Direction of $\mathbf{V}_{D/C}$ is perpendicular to CD. Direction of $\mathbf{V}_D$ is perpendicular to PD.

The constructed velocity polygon is shown in Fig. 23b.

6. Determine ω_3 from the relationship

$$\omega_3 = \frac{V_{B/A}}{AB} = \frac{5.23 \times 1.9 \times 12}{1.9} = 62.8 \text{ rad/sec} \quad \text{(clockwise)}$$

7. Determine ω_4 from the relationship

$$\omega_4 = \frac{V_B}{QB} = \frac{5.23 \times 2.6 \times 12}{1.75} = 93.6 \text{ rad/sec}$$

8. Determine ω_5 from the relationship

$$\omega_5 = \frac{V_{D/C}}{CD} = \frac{5.23 \times 0.6 \times 12}{2.1} = 17.9 \text{ rad/sec} \quad \text{(counterclockwise)}$$

9. Determine ω_6 from the relationship

$$\omega_6 = \frac{V_D}{PD} = \frac{5.23 \times 3.3 \times 12}{2} = 103.55 \text{ rad/sec} \quad \text{(clockwise)}$$

Note that in steps 6 through 9, the directions of the angular velocities ω_3, ω_4, ω_5, and ω_6 must correspond to the direction of the velocities $V_{B/A}$, V_B, $V_{D/C}$, and V_D.

Competency Item
Perform the velocity analysis of the six-link mechanisms shown in Figs. 17–22 when the angular velocity of the input crank of the driving mechanism is 10 rad/sec. Measure link lengths and angular positions from the figures.

Objective 7

To state the properties of instant center and demonstrate their application in performing velocity analysis of mechanisms

Activities
- Read the material provided below, and as you read:
  a. Define an instant center for two bodies having relative motion.
  b. State the properties of instant center.
  c. Enumerate the instant centers of a mechanism.
  d. State Kennedy's theorem for three instant centers.
  e. State Freudenstein's theorem for the extreme value of velocity ratio.
  f. Demonstrate the applications of instant-center technique for performing velocity analysis of mechanisms.
- Check your answers to the competency items with your instructor.

Reading Material
The reading material is presented in the same order as you are required to do your activities.

Part I: Definition of an instant center
Let us consider that a body is executing planar motion with reference to an xy frame of reference as shown in Fig. 24a. Locate point I in xy plane such that

$$V_A = \omega \times IA$$

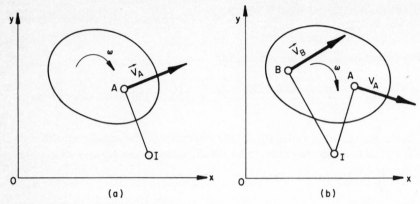

Fig. 24 Definition of instant center.

where $\boldsymbol{\omega} = \omega\mathbf{U}_z$ is the angular velocity of the rigid body, $\mathbf{U}_z$ is a unit vector parallel to $\mathbf{U}_Z$, and $\mathbf{V}_A$ is the velocity of some point A within the rigid body. We can now examine the velocity of the point I using

$$\mathbf{V}_I = \mathbf{V}_A + \mathbf{V}_{I/A}$$

Since

$$\mathbf{V}_{I/A} = -\boldsymbol{\omega} \times \mathbf{I}A$$
$$\mathbf{V}_I = 0$$

Hence, the velocity of the point I is zero.

For two points A and B in the rigid body, as shown in Fig. 24a, the point I is located at a point of intersection of two perpendiculars to $\mathbf{V}_A$ and $\mathbf{V}_B$.

The point I located in the xy plane is called the instant center. This point is often called "instantaneous center," "centro," and "pole." We will refer to it as instant center.

From the above discussion, we observe that there exists an instant center when two bodies are moving relative to one another. The instant center is a point common to both bodies and it has the same velocity in each of the two bodies.

Hence, an instant center can be defined as a point common to two bodies and having equal velocity in each.

Part II: Procedure to enumerate instant centers
of a mechanism

For two bodies moving relative to one another, there is one instant center. For three bodies, moving relative to one another, there are three instant centers. The number N of instant centers for n bodies moving relative to one another can be calculated from

$$N = \frac{n(n - 1)}{2}$$

Thus, for a four-link mechanism, there are four links (rigid bodies) moving relative to one another, and therefore it will have six instant centers. If these links are numbered as shown in Fig. 25, the instant centers obtained due to the relative motion of these links are

$$I_{12}\ I_{13}\ I_{14}$$
$$I_{23}\ I_{24}$$
$$I_{34}$$

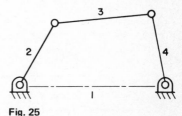

Where I_{ij} denotes the instant centers between ith and jth links.

Fig. 25

Part III: Kennedy's theorem for three instant centers

According to Kennedy's theorem, the three instant centers obtained as a result of three bodies moving relative to one another, lie on a straight line.

For example consider bodies 2 and 3 moving relative to another and relative to fixed link MQ as shown in Fig. 26.

The two bodies 2 and 3 make a contact at C. The instant centers I_{12} and I_{13} which are obtained as a result of relative motion between bodies 1 and 2, and bodies 1 and 3 are coincident with the fixed pivots M and Q. Then, according to Kennedy's theorem, the third instant center I_{23} due to the relative motion between bodies 1 and 3 must lie on MQ.

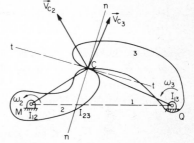

Fig. 26

We can prove the Kennedy theorem by assuming that I_{23} does not lie on MQ. Instead it coincides with point C. If I_{23} coincides with the point C, then $\mathbf{V}_{c_2}$ and $\mathbf{V}_{c_3}$, the velocities of the point C as it lies on body 2 and 3, must be equal. As we observe from Fig. 26, since $\mathbf{V}_{c_2} \neq \mathbf{V}_{c_3}$, C must not be I_{23}. The only time when $\mathbf{V}_{c_2}$ becomes equal to $\mathbf{V}_{c_3}$ is when the instant center I_{23} lies on MQ or I_{12} I_{13}.

Let

$$I_{12}I_{23}\ =\ l_2$$
$$I_{13}I_{23}\ =\ l_3$$

then since the velocities of the two bodies are equal at the instant center,

$$l_2\omega_2\ =\ l_3\omega_3$$
or
$$\frac{\omega_2}{\omega_3}\ =\ \frac{l_3}{l_2}$$

Part IV: Freudenstein's theorem for extreme value of the angular velocity ratio

A four-link mechanism is shown in Fig. 27.

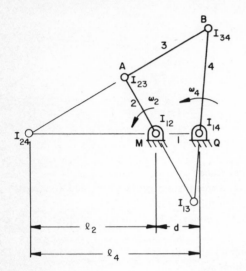

Fig. 27 Instant centers of a four-link mechanism.

Since three links MA, AB, and BQ are moving relative to the fourth link MQ, there are six instant centers, I_{12}, I_{13}, I_{14}, I_{23}, I_{24}, and I_{34}. Since links MQ and QB are rotating about M and Q, the instant centers I_{12} and I_{14} are coincident with the fixed pivots M and Q. Since link 3 rotates about pivot points A and B, the instant centers I_{23} and I_{34} coincide with A and B. The other two instant centers I_{24} and I_{13} can be located using the Kennedy's theorem. The instant center I_{24} is located using the following arguments:

1. Since links 2 and 4 are rotating about instant centers I_{12} and I_{14}, the instant center I_{24} must lie on the line joining I_{14} and I_{12}. (Note that the noncommon subscript in I_{12} and I_{14} is 24.)
2. Since links 2 and 4 are also rotating about instant centers I_{23} and I_{34}, the instant center I_{24} must lie on the line joining I_{23} and I_{34}. (Note that the noncommon subscript in I_{23} and I_{34} is 24.)
3. From the above arguments, we conclude that I_{24} must be the point of intersection of lines $I_{23}I_{34}$ and $I_{12}I_{14}$. Similarly, the instant center I_{13} is located as a point of intersection of lines $I_{12}I_{23}$ and $I_{14}I_{34}$. Let

$$I_{24}I_{12} = l_2$$
$$I_{24}I_{14} = l_4$$

then since I_{24} is the instant center for the links 2 and 4,

$$l_2\omega_2 = l_4\omega_4$$

where ω_2 and ω_4 are the angular velocities of links 2 and 4. The angular velocity ratio

$$\frac{\omega_4}{\omega_2} = \frac{l_2}{l_4}$$

$$= \frac{l_2}{l_2 + d} = \frac{1}{1 + (d/l_2)}$$

where $MQ = d$

The velocity ratio ω_4/ω_2 becomes maximum when (d/l_2) is minimum or l_2 is maximum. Since the instant center I_{24} travels along line MQ extended, l_2 is maximum when the velocity of I_{24} as a point on link AB is zero. This will happen when the velocity of I_{24} is along AB. Since I_{13} is the instant center for links 3 and 1, I_{24} when considered as moving with link 3 will be at rest when the line $I_{13}I_{24}$ is perpendicular to link AB. The line $I_{13}I_{24}$ is called the *collineation axis*.

The above results are summarized using Freudenstein's theorem which states that the velocity ratio ω_4/ω_2 of a four-link mechanism is maximum when the collineation axis is perpendicular to its coupler link AB.

Part V: Velocity analysis of mechanisms

In this section the properties of instant centers will be applied to demonstrate the velocity analysis of a four-link mechanism shown in Fig. 28. The input link MA rotates at 600 rpm in a counterclockwise direction. $MA = 0.55$ in., $AB = 2.85$ in., $QB = 1.55$ in., and $MQ = 1.90$ in. We are required to determine ω_3 and ω_4, the angular velocities of links AB and QB. We may proceed in a following stepwise manner:

1. Since MA rotates at 600 rpm

$$\omega_2 = \frac{2 \times \pi \times 600}{60}$$

$$= 62.8 \text{ rad/sec}$$

2. $V_A = MA \cdot \omega_2 = 0.55 \times 62.8$
$$= 34.54 \text{ in./sec}$$

3. Since the point A also lies on link 3,

$$V_A = \omega_3 \cdot I_{13} \; I_{23}$$

Hence,

$$\omega_3 = \frac{V_A}{I_{13}I_{23}} = \frac{34.54}{3.0}$$

$$= 11.51 \text{ rad/sec (counterclockwise)}$$

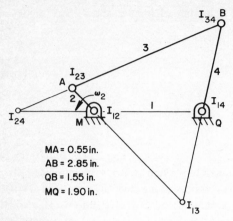

Fig. 28

4. Since the point B lies on link 3

$$V_B = \omega_3 \cdot I_{13}I_{34}$$
$$= 11.51 \times 3.50$$
$$= 40.28 \text{ in./sec}$$

5. Since the point B also lies on link 4

$$V_B = \omega_4 \cdot QB$$

Hence,

$$\omega_4 = \frac{V_B}{QB} = \frac{40.28}{1.55}$$
$$= 26.00 \text{ rad/sec (counterclockwise)}$$

This procedure provides an alternate approach to performing velocity analysis of mechanisms.

Competency Items
- Locate the instant centers of the mechanisms shown below:

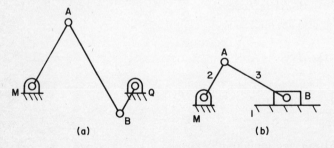

- List all the instant centers of a six-link mechanism using subscripts $1, 2, \ldots,$ 6 for its links.
- Calculate V_B and ω_3 of the slider-crank mechanism in the figures above, if the ω_2 = 92.4 rad/sec in a counterclockwise direction.

Performance Test #1—Objectives 5, 6, 7

1. If *MA* rotates with a constant angular velocity of 20 rad/sec in a clockwise direction, calculate the angular velocities of links *AB* and *BQ* using the polygon method. Compare your answers with those obtained using the instant-center technique.

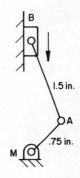

2. If the velocity of a piston at *B* is 25 in./sec at a particular instant, calculate the angular velocity of *BA* and *AM* using the polygon method. Compare your answers with those obtained using the instant-center technique.

3. Calculate the angular velocity of *BQ* and the velocity at which the sliding block moves on the link *BQ* using the polygon method. $\omega_2 = 50$ rad/sec.

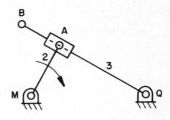

4. Construct a complete velocity polygon for the six-link mechanism shown. $\omega_2 = 20$ rad/sec in a counterclockwise direction.

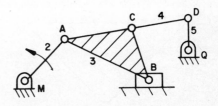

Performance Test #2—Objectives 5, 6, 7

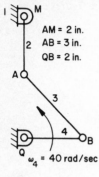

AM = 2 in.
AB = 3 in.
QB = 2 in.

ω_4 = 40 rad/sec

1. For the position of the mechanism shown calculate the angular velocity of link 3 using the polygon method. Show on the velocity polygon the vector representing the velocity of the midpoint of link *AB*. *MG* = 4.25 in.

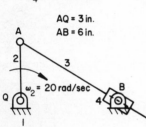

AQ = 3 in.
AB = 6 in.

ω_2 = 20 rad/sec

2. For the position of the mechanism shown calculate the angular velocity of link 3 using the polygon method. BQ = 6.75 in.

3. Construct a complete velocity polygon for the six-link mechanism shown. Find the angular velocity of links 3, 4, 5, and 6.

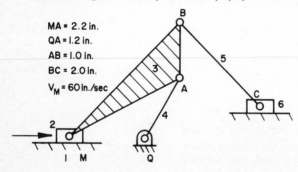

MA = 2.2 in.
QA = 1.2 in.
AB = 1.0 in.
BC = 2.0 in.
V_M = 60 in./sec

Performance Test #3—Objectives 5, 6, 7

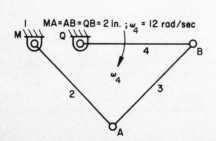

MA = AB = QB = 2 in. ; ω_4 = 12 rad/sec

1. For the position of the mechanism shown calculate the angular velocity of link 2 using the vector polygon method. Show *neatly* the complete polygon, and show all equations used in your solution. Compare your answers with those obtained using the instant-center technique. MQ = 0.85 in.

2. For the position of the mechanism shown calculate the velocity of point *B* using the vector polygon method. Show *neatly* the complete polygon, and show all equations used in your solution. Compare your answers with those obtained using the instant-center technique.

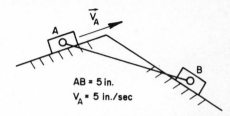

AB = 5 in.
V_A = 5 in./sec

3. The four-bar linkage *MABQ* is the same as in problem 1. Calculate the angular velocity of link 5 and the velocity at which the block at *C* slides along bar *DE*. Repeat the vector polygon of problem 1 and complete it *neatly* for this problem. Show all equations in addition to those of problem 1 used in your solution.

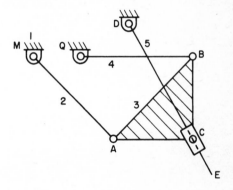

Supplementary References

3(2.1–2.3 [parts], 2.7, 3.1, 4.1–4.2); 5(1.15–1.17, 2.1–2.4, 2.8–2.9); 7(2.1–2.10, 3.1–3.6, 4.1–4.7); 8(2.2–2.3 [parts], 5.1–5.2, 5.7)

Unit III. Acceleration Analysis of Linkages

Objectives

1. To define acceleration and its components for a particle executing planar and space motion in stationary and moving coordinates
2. To calculate, using complex number notation, the acceleration of a particle moving on a path
3. To describe the acceleration-vector relationship of a rigid body moving in a plane
4. To demonstrate the application of the vector polygon technique in order to perform acceleration analysis of mechanisms
5. To perform acceleration analysis of a six-link mechanism using the vector polygon technique

Objective 1

To define acceleration and its components for a particle executing planar and space motion in stationary and moving coordinates

Activities

- Read the material provided below, and as you read:
  a. Define acceleration of a moving particle.
  b. Calculate tangential and normal components of acceleration of a particle moving on a stationary path.
  c. Calculate radial and transverse components of acceleration of a particle moving on a stationary path.
  d. Calculate components of acceleration of a particle moving on a path that is subjected to rotational motion.
  e. Calculate components of acceleration of a particle moving on a path which is subjected to rotational and translational motions.
- Check your answers to the competency items with your instructor.

Reading Material

Part I: Definition of acceleration

Let us examine the motion of a point in a fixed coordinate system XYZ. Point P is traveling along the continuous path "s" shown in Fig. 1a. Positions P_1, P_2, etc., of point P can be described using position vectors R_1, R_2, etc. Note that the magnitude and direction of position vector R are both functions of time. Let us assume that point P moves with a velocity V_1 and V_2 as it passes through positions P_1 and P_2.

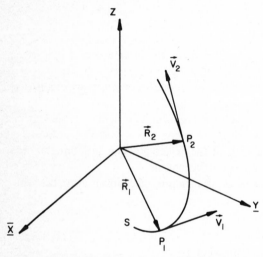

Fig. 1a Motion of a particle P moving along a path S in XYZ frame of reference.

Velocity polygon $O_v P_1 P_2$ is drawn in Fig. 1b. From the veloctiy polygon we can write

$$\Delta V = V_2 - V_1$$

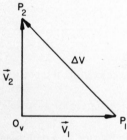

Fig. 1b Velocity polygon to describe motion of particle P.

If the change in the velocity ΔV takes place in Δt interval of time, then the acceleration is defined as,

$$A = \lim_{\Delta t \to 0} \frac{\Delta V}{\Delta t} = \dot{V} = \ddot{R} \qquad (1)$$

Angular acceleration α of the particle moving on path S is defined in a similar manner.

$$\alpha = \dot{\omega} = \lim_{\Delta t \to 0} \frac{\Delta \omega}{\Delta t} = \frac{d\omega}{dt} \qquad (2)$$

Example 1.

Particle P is moving on path $y = \frac{1}{2}at^2$. Determine the acceleration of P.

Solution.

$$V = \frac{dy}{dt} = at$$

$$A = \frac{dv}{dt} = \frac{d^2y}{dt^2} = a.$$

The acceleration of a particle moving on a parabolic path is constant.

Example 2.

Particle P is moving on path $y = a\cos\omega t$. Determine the acceleration of P (ω = constant).

Solution.

$$V = \frac{dy}{dt} = -a\omega\,\sin\omega t$$

$$A = \frac{dv}{dt} = \frac{d^2y}{dt^2} = -a\frac{d\omega}{dt}\sin\omega t - a\omega^2\cos\omega t$$

Thus,

$$A = -a\omega^2\cos\omega t = -\omega^2 y$$

when

$$\alpha = \frac{d\omega}{dt} = 0$$

The acceleration of a particle moving on a cosine curve is proportional to its amplitude.

Part II. Tangential and normal components of **A**

In Unit II, we observed that the velocity of point P moving along path "s" is,

$$V = \frac{ds}{dt}\mathbf{u}_t \tag{3}$$

The time derivative of Eq. (1) will give us the acceleration of point P.

$$A = \dot{V} = \frac{d^2s}{dt^2}\mathbf{u}_t + \frac{ds}{dt}\dot{\mathbf{u}}_t \tag{4}$$

The first term defines the tangential acceleration. The second term defines the radial acceleration. However, we must show the corresponding direction of these components. For this reason we must investigate the nature of the second term. Consider that point P_1 is rotating about an instant center C. As shown in Fig. 2, let ρ be the radius of curvature of path S at P_1. In the interval of time Δt which corresponds to rotation of $\Delta\Phi$ of point P_1 about c, unit vector $\mathbf{u}_t$ associated with velocity $\dot{S}$ changes to $\mathbf{u}_t + \Delta\mathbf{u}_t$. The next change within $\Delta\Phi$ rotation of unit vector $\mathbf{u}_t$ is,

$$\Delta(\Delta\mathbf{u}_t) = (\mathbf{u}_t + \Delta\mathbf{u}_t) - \mathbf{u}_t$$

Then, by definition

$$\lim_{\Delta\Phi\to 0} \frac{\Delta\mathbf{u}_t}{\Delta\Phi} = \frac{d\mathbf{u}_t}{d\Phi}$$

Note that as $\Delta\Phi \to 0$, $\Delta\mathbf{u}_t$ becomes normal to $\mathbf{u}_t$. Therefore,

$$\frac{d\mathbf{u}_t}{d\Phi} = \mathbf{u}_n \tag{5}$$

We may rewrite $\dot{\mathbf{u}}_t$ as,

$$\dot{\mathbf{u}}_t = \frac{d\mathbf{u}_t}{dt} = \frac{d\mathbf{u}_t}{d\Phi} \cdot \frac{d\Phi}{dt} = \dot{\Phi}\mathbf{u}_n \tag{6}$$

Since ρ is the radius of curvature and Δs is the distance traveled by point P along the path s,

$$\Delta s = \rho\Delta\Phi$$

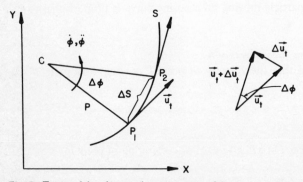

Fig. 2 Tangential and normal components of **A**.

Time derivative of Eq. (7) yields $\dot{s} = \rho\dot{\Phi}$ or $\dot{\Phi} = \dot{s}/\rho$. Equation (4) can be expressed as,

$$A = \underbrace{\ddot{s}\,\mathbf{u}_t}_{\substack{\text{Tangential} \\ \text{component}}} + \underbrace{\frac{\dot{s}^2}{\rho}\,\mathbf{u}_n}_{\substack{\text{Normal} \\ \text{component}}} \tag{7}$$

NOTE: $\ddot{s}$ = the tangential component of acceleration A along the tangent to path "s"

$\dfrac{\dot{s}^2}{\rho}$ = the normal component of acceleration A along the normal to the tangent to path "s"

Example 3.
A particle is moving in a circular path with radius of curvature ρ = 10 in. has a speed of 5 in./sec and is losing speed at the rate of 5 in./sec^2. What is the resultant acceleration of the particle?
Solution.

$$A = -5 \text{ in./sec}^2\,\mathbf{u}_t + \frac{(5 \text{ in./sec})^2}{10 \text{ in.}}\,\mathbf{u}_n$$

$$= -5 \text{ in./sec}^2\,\mathbf{u}_t + 2.5 \text{ in./sec}^2\,\mathbf{u}_n$$

The magnitude of A then is $\sqrt{(-5)^2 + (2.5)^2}$
Choosing an arbitrary orientation for R we can represent the results graphically as shown in Fig. 3.

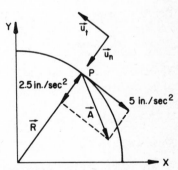

Fig. 3

Part III.   Radial and transverse components of acceleration
The position of particle P is located on path S using position vector $R = r\mathbf{U}_r$, where $\mathbf{U}_r$ is the unit vector associated with R. As shown in Fig. 4, let $\mathbf{U}_\theta$ be the unit vector normal to $\mathbf{U}_r$ such that $\mathbf{U}_Z \times \mathbf{U}_r = \mathbf{U}_\theta$.

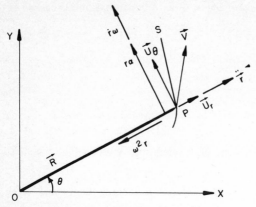

Fig. 4 Radial and transverse components of acceleration of a particle P moving along a path S.

The velocity and acceleration of the particle P are obtained by taking successive time derivatives of position vector $\mathbf{R}$. The radial and transverse components of velocity $\mathbf{V}$ are

$$\dot{\mathbf{R}} = \mathbf{V} = \dot{r}\mathbf{U}_r + r\dot{\mathbf{U}}_r$$

But

$$\dot{\mathbf{U}}_r = \boldsymbol{\omega} \times \mathbf{U}_r$$

where

$$\boldsymbol{\omega} = \frac{d\theta}{dt}\mathbf{U}_z = \omega\mathbf{U}_z$$

Hence, we have

$$\mathbf{V} = \dot{r}\mathbf{U}_r + \boldsymbol{\omega} \times \mathbf{R} \tag{8}$$

The acceleration of particle P is obtained by taking time derivative of Eq. (8). Thus, we have

$$\ddot{\mathbf{R}} = \dot{\mathbf{V}} = \ddot{r}\mathbf{U}_r + \dot{r}\dot{\mathbf{U}}_r + \dot{\boldsymbol{\omega}} \times \mathbf{R} + \boldsymbol{\omega} \times \dot{\mathbf{R}}$$

Since we know that

$$\mathbf{U}_\theta = \mathbf{U}_z \times \mathbf{U}_r$$

$$\dot{\mathbf{U}}_r = \boldsymbol{\omega} \times \mathbf{U}_r$$

$$\dot{\boldsymbol{\omega}} = \frac{d^2\theta}{dt^2}\mathbf{U}_z = \boldsymbol{\alpha}$$

and

$$\dot{\mathbf{R}} = \dot{r}\mathbf{U}_r + \boldsymbol{\omega} \times \mathbf{r}$$

we can obtain the simplified results for

$$\ddot{\mathbf{R}} = (\ddot{r} - \omega^2 r)\mathbf{U}_r + (r\alpha + 2\omega\dot{r})\mathbf{U}_\theta \tag{9}$$

The components $(\ddot{r} - \omega^2 r)$ and $(r\alpha + 2\omega\dot{r})$ are called the *radial* and the *transverse components*. The direction of these components are shown in Fig. 4. The component $-\omega^2 r$ is called the *centripetal acceleration*. The component $2\omega\dot{r}$ is called the *coriolis acceleration*.

Part IV: Components of acceleration of a particle moving on a path that is subjected to rotational motion

Figure 5 shows point P tracing path S which is located in the l-m frame of reference which, in turn, is rotating about the Z axis with angular velocity ω, angular acceleration α. We will obtain vector equation to calculate acceleration of point as observed from the stationary XY frame of reference.

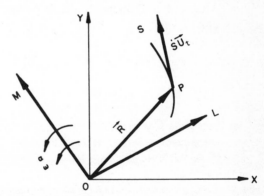

Fig. 5 Acceleration of a particle moving on a path that is subjected to rotational motion.

1. The velocity of particle P within the l-m frame of reference is $d\mathbf{R}/dt$.
2. Since the l-m frame of reference is rotating with angular velocity $\boldsymbol{\omega} = \omega\mathbf{U}_z$, the velocity of particle P in the XY frame of reference is

$$\mathbf{V} = \frac{d\mathbf{R}}{dt} + \boldsymbol{\omega} \times \mathbf{R} \tag{10}$$

3. From Unit II, we note that

$$\frac{d\mathbf{R}}{dt} = \dot{S}\mathbf{U}_t$$

where $\mathbf{U}_t$ is the unit vector tangent to path S.

4. The total acceleration $\mathbf{A}$ of particle P in the XY frame of reference is

$$\mathbf{A} = \left(\ddot{s}\mathbf{U}_t + \frac{\dot{s}^2}{\rho}\mathbf{U}_n \right) + (\dot{\omega}r\mathbf{U}_\theta - \omega^2 r\mathbf{U}_r) + 2\omega\dot{s}\mathbf{U}_n \tag{11}$$

Component $2\omega\dot{s}\mathbf{U}_n$ is the coriolis acceleration.

The components of $\mathbf{A}$ are obtained by using the following steps:

a. Within the $l\text{-}m$ frame of reference, acceleration of particle P is

$$\frac{d}{dt}\frac{d\mathbf{R}}{dt} = \ddot{s}\mathbf{U}_t + \frac{\dot{s}^2}{\rho}\mathbf{U}_n$$

Where $\mathbf{U}_t, \mathbf{U}_n$ are unit tangent and normal vectors to path S, and ρ is the radius of curvature of path S at P.

b. Since path S is rotating along with $l\text{-}m$, the rotational motion will contribute component $\boldsymbol{\omega} \times \dot{s}\mathbf{U}_t{}' = \omega\dot{s}\mathbf{U}n$.

c. Acceleration components due to the time derivative of term $\boldsymbol{\omega} \times \mathbf{R}$ are obtained as

$$\frac{d}{dt}(\boldsymbol{\omega} \times \mathbf{R}) = \dot{\boldsymbol{\omega}} \times \mathbf{R} + \boldsymbol{\omega} \times [\dot{\mathbf{R}}]$$

$$= \dot{\boldsymbol{\omega}} \times \mathbf{R} + \boldsymbol{\omega} \times \dot{s}\mathbf{U}_t{}' = \dot{\omega}r\mathbf{U}\theta + \omega\dot{s}\mathbf{U}n$$

d. Since the $l\text{-}m$ frame of reference is rotating with angular velocity ω, component $\boldsymbol{\omega} \times (\boldsymbol{\omega} \times \mathbf{R})$ will be added to the acceleration components in item c. Note $\boldsymbol{\omega} \times (\boldsymbol{\omega} \times \mathbf{R}) = -\omega^2 r\mathbf{U}r$.

Part V: Components of acceleration of a particle moving on a path which is subjected to rotational and translational motion

Figure 6 shows particle P traveling on a path in an $l\text{-}m\text{-}n$ frame of reference which is rotating with angular velocity ω about axis $\mathbf{U}_z$, and translating along line OO' relative to the fixed frame of reference XYZ.

Let $OO' = \mathbf{H}$ $O'P = \mathbf{G}$ and $\mathbf{R} = \mathbf{G} + \mathbf{H}$.

Let $\mathbf{U}_l$, $\mathbf{U}_m$, and $\mathbf{U}_n$ be the unit vectors associated with axes l, m, n. Then

$$\mathbf{G} = g_l\mathbf{U}_l + g_m\mathbf{U}_m + g_n\mathbf{U}_n$$

where g_l, g_m, g_n are components of $\mathbf{G}$ along the l, m, n axes.

The velocity of particle P as observed in the XYZ frame of reference is

$$\mathbf{V}_P = \mathbf{V}_{O'} + \mathbf{V}_{P/O'} \tag{12}$$

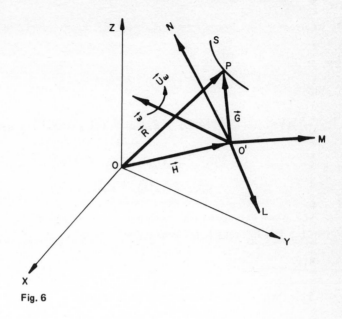

Fig. 6

But

$$V_{P/O'} = \frac{d}{dt}(\mathbf{G})$$

$$= \frac{d}{dt}(g_l\mathbf{U}_l + g_m\mathbf{U}_m + g_n\mathbf{U}_n)$$

$$= \dot{g}_l\mathbf{U}_l + \dot{g}_m\mathbf{U}_m + \dot{g}_n\mathbf{U}_n + g_l\dot{\mathbf{U}}_l + g_m\dot{\mathbf{U}}_m + g_n\dot{\mathbf{U}}_n$$

Since

$$\dot{\mathbf{U}}_l = \boldsymbol{\omega} \times \mathbf{U}_l$$

$$\dot{\mathbf{U}}_m = \boldsymbol{\omega} \times \mathbf{U}_m$$

$$\dot{\mathbf{U}}_n = \boldsymbol{\omega} \times \mathbf{U}_n,$$

$$V_{P/O'} = [\dot{\mathbf{G}}] + \boldsymbol{\omega} \times \mathbf{G}$$

where $[\dot{\mathbf{G}}]$ is the velocity of P relative to the l, m, n, frame of reference. Hence, Eq. (12) becomes

$$\mathbf{V}_P = \mathbf{V}_{O'} + [\dot{\mathbf{G}}] + \boldsymbol{\omega} \times \mathbf{G} \tag{13}$$

The acceleration of particle P in the XYZ frame of reference is

$$\mathbf{A}_P = \mathbf{A}_{O'} + \mathbf{A}_{P/O'} \tag{14}$$

where

$$A_{P/O'} = \frac{d}{dt}[V_{P/O'}] \tag{15}$$

But

$$\frac{d}{dt}[V_{P/O'}] = \frac{d}{dt}[\dot{g}_l U_l + \dot{g}_m U_m + \dot{g}_n U_n + g_l \dot{U}_l + g_m \dot{U}_m + g_n \dot{U}_n]$$

$$= \ddot{g}_l U_l + \ddot{g}_m U_m + \ddot{g}_n U_n + 2(\dot{g}_l \dot{U}_l + \dot{g}_m \dot{U}_m + \dot{g}_l \dot{U}_m)$$

$$+ g_l \ddot{U}_l + g_m \ddot{U}_m + g_n \ddot{U}_n$$

The above expression can be simplified provided we know quantities $\ddot{U}_l$, $\ddot{U}_m$, and $\ddot{U}_n$. For this reason, let us examine

$$\dot{U}_l = \omega \times U_l .$$

Taking the time derivative on both sides we get

$$\ddot{U}_l = \dot{\omega} \times U_l + \omega \times \dot{U}_l$$

$$= \dot{\omega} \times U_l + \omega \times (\omega \times U_l)$$

Similarly

$$\ddot{U}_m = \dot{\omega} \times U_m + \omega \times (\omega \times U_m)$$

$$\ddot{U}_n = \dot{\omega} \times U_n + \omega \times (\omega \times U_n)$$

Hence, Eq. (15) becomes

$$A_{P/O'} = [\ddot{G}] + \omega \times \omega \times G + \dot{\omega} \times G + 2\omega \times [\dot{G}]$$

So the total acceleration of particle P is

$$A = A_{O'} + [\ddot{G}] + \omega \times \omega \times G + \dot{\omega} \times G + 2\omega \times [\dot{G}] \tag{16}$$

In Eq. (16), the term $2\omega \times [\dot{G}]$ is the coriolis accelerations. From this term we observe that the coriolis acceleration will exist when the relative motion of two points is subjected to rotational motion.

Example 4.
As shown in Fig. 7 ring B is sliding along link MA which, in turn, is rotating with angular velocity ω about fixed pivot M. Show the direction of coriolis acceleration for the following cases.

a. Ring B moves toward M, and ω is counter-clockwise.
b. Ring B moves toward M and ω is clockwise.
c. Ring B moves away from M and ω is counter-clockwise.
d. Ring B moves away from M and ω is clock-wise.

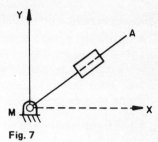

Fig. 7

Solution

Case a: Fig. 8a shows $\mathbf{V}_B$, the velocity of ring B is toward M and ω is counterclockwise. The coriolis acceleration is given by $\mathbf{A}^c = 2\omega \times \mathbf{V}_B$.
Since $\boldsymbol{\omega} = \omega\mathbf{U}_z$, the direction of the coriolis acceleration is normal to MA downward as shown in Fig. 8a.

Case b: Since $\boldsymbol{\omega}$ is clockwise, and $\mathbf{V}_B$ is toward M, $\mathbf{A}^c = 2\boldsymbol{\omega} \times \mathbf{V}_B$ is normal to MA upward as shown in Fig. 8b.

Case c: Since $\boldsymbol{\omega}$ is counterclockwise and $\mathbf{V}_B$ is away from M, $A^c = 2\boldsymbol{\omega} \times \mathbf{V}_B$ is normal to MA upward as shown in Fig. 8c.

Case d: Since $\boldsymbol{\omega}$ is clockwise and $\mathbf{V}_B$ is away from M, $A^c = 2\boldsymbol{\omega} \times \mathbf{V}_B$, is normal to MA downward as shown in Fig. 8d.

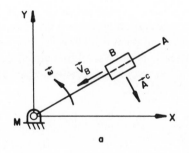

a

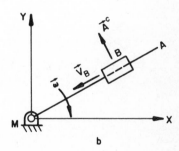

b

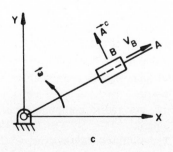

c

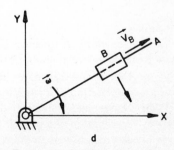

d

Fig. 8 Four cases of coriolis acceleration.

Competency Items

- Particle P is moving on path $y = at^3 + bt$. Calculate the acceleration at time $t = 10$ sec if $a = 1$ in./sec^3 and $b = 5$ in./sec.
- If a particle has constant acceleration given by $A = 5$ in./sec^2 at time $t = 2$ sec, at which time it has velocity $V = 20$ in./sec and position $y = 50$ in., draw the path describing function.

- For a particle moving on a path $y = \frac{1}{2}x^2 + x + 1$ with constant speed $\dot{s} = 5$ units/sec at position $x = 1$, calculate the normal and tangential components of acceleration.

NOTE: $\rho = \dfrac{[1 + (dy/dx)^2]^{3/2}}{d^2y/dx^2}$

- For a particle moving on path $y = \sin x$ with constant speed $\dot{s} = 10$ units/sec, calculate the normal and tangential components of acceleration when $x = \pi/2$ and when $x = \pi$.
- Link MA with slider block B is rotating with angular velocity $\omega = 60$ rad/sec in a counterclockwise direction. Calculate

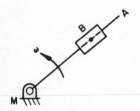

  a. radial and transverse components of the acceleration of point B when it lies on link MA
  b. Relative acceleration between point B on the slider block and point B on link MA
  c. Coriolis acceleration: Sketch the directions of the components of accelerations calculated for the above items.

Objective 2

To calculate, using complex number notation, the acceleration of a particle moving on a path

Activities

- Review your performance on Objective 3 of Unit II.
- Read the material provided below, and as you read:
  a. Derive and calculate the horizontal and vertical components of the acceleration of a particle moving on a path.
  b. Derive and calculate the transverse and radial components of the acceleration of a particle moving on a path.
- Check your answers to the competency items with your instructor.

Reading Material

Radial and transverse components of **A**

The position of particle P is described using position vector $\mathbf{R}$ which is inclined at angle θ with the x axis as shown in Fig. 9. This position vector $\mathbf{R}$ lying in the XY plane can be expressed using the complex number notations as

$$\mathbf{R} = re^{j\theta} \tag{17}$$

The velocity of point P is obtained by taking the time derivative of Eq. (17)

$$\dot{\mathbf{R}} = \dot{r}e^{j\theta} + jr\omega e^{j\theta} \tag{18}$$

where

$$\omega = \frac{d\theta}{dt} = \text{angular velocity}$$

The acceleration of point P is obtained by taking the time derivative of Eq. (2).

$$\mathbf{A} = \ddot{\mathbf{R}} = \ddot{r}e^{j\theta} + j\dot{r}\omega e^{j\theta} + j\dot{r}\omega e^{j\theta} + jr\dot{\omega}e^{j\theta} + j^2r\omega^2 e^{j\theta} \tag{19}$$

Angular acceleration, α, is defined as the time derivative of the angular velocity ω. That is,

$$\alpha = \frac{d\omega}{dt} = \dot{\omega} \tag{20}$$

Since $j^2 = -1$, Eq. (19) can be simplified.

$$\mathbf{A} = \ddot{\mathbf{R}} = \underbrace{(\ddot{r} - \omega^2 r)e^{j\theta}}_{\substack{\text{Radial} \\ \text{component}}} + \underbrace{j(r\alpha + 2\dot{r}\omega)e^{j\theta}}_{\substack{\text{Transverse} \\ \text{component}}} \tag{21}$$

or

$$\mathbf{A} = \mathbf{A}^r + \mathbf{A}^t \tag{22}$$

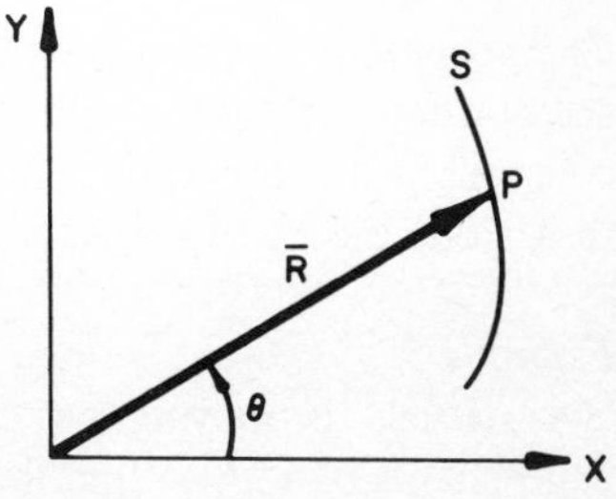

Fig. 9 Complex number notation to represent $\mathbf{R} = re^{j\theta}$.

Where A^r and A^t represent the radial and the transverse components of A. Figure 10 shows all the components of acceleration given in Eq. (21). Note that there are two terms for the radial component of the acceleration A. Component $\ddot{r}$ acts in the direction away from the center of rotation O. Component $\omega^2 r$ is called the *centripetal component* and acts in the direction toward the center of rotation O. Note that there are two terms for the transverse component of acceleration A. Component $r\alpha$, due to the angular acceleration of point P, acts in the direction perpendicular to R. Component $2\dot{r}\omega$, called the *coriolis component*, also acts in the direction perpendicular to R. The coriolis component exists because of two reasons: (a) The position vector R is changing in magnitude with respect to time, and (b) the position vector R is rotating with an angular velocity ω.

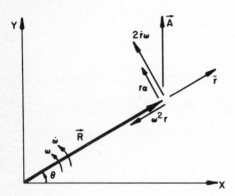

Fig. 10   Radial and transverse components of the acceleration A of the particle P moving on a path t.

Since $e^{j\theta} = \cos\theta + j\sin\theta$, Eq. (21) can be written as

$$A = \ddot{R} = (\ddot{r} - \omega^2 \dot{r})(\cos\theta + j\sin\theta)$$
$$+ j(r\alpha + 2\dot{r}\omega)(\cos\theta + j\sin\theta)$$

Simplifying, the above equation, we get

$$A = [(\ddot{r} - \omega^2 r)\cos\theta - (r\alpha + 2\dot{r}\omega)\sin\theta]$$
$$+ j[(\ddot{r} - \omega^2 r)\sin\theta + (r\alpha + 2\dot{r}\omega)\cos\theta] \tag{23}$$

Component $[(\ddot{r} - \omega^2 r)\cos\theta - (r\alpha + 2\dot{r}\omega)\sin\theta]$ is the x component of acceleration A. Component $[(\ddot{r} - \omega^2 r)\sin\theta + (r\alpha + 2\dot{r}\omega)\cos\theta]$ is the y component of acceleration A. The x and y components of A are the projections of A along the x and y axes.

Example 5.

A particle is moving on path $R = x^2 e^{j\theta}$. Find the radial and transverse components of A. Also find the horizontal and vertical components of A.

Solution.

$$V = \frac{d}{dt}(x^2 e^{j\theta})$$

$$= 2x\dot{x}e^{j\theta} + jx^2\omega e^{j\theta}$$

$$A = \frac{dV}{dt} = 2\dot{x}^2 e^{j\theta} + 2x\ddot{x}e^{j\theta} + j2x\dot{x}\omega e^{j\theta}$$

$$+ j2x\dot{x}\omega e^{j\theta} + jx^2\alpha e^{j\theta} - x^2\omega^2 e^{j\theta}$$

$$A = (2\dot{x}^2 + 2x\ddot{x} - x^2\omega^2)e^{j\theta} + j(4x\dot{x}\omega + x^2\alpha)e^{j\theta}$$

$$A = (2\dot{x}^2 + 2x\ddot{x} - x^2\omega^2)\cos\theta - (4x\dot{x}\omega + x^2\alpha)\sin\theta$$

$$+ j(2\dot{x}^2 + 2x\ddot{x} - x^2\omega^2)\sin\theta + j(4x\dot{x}\omega + x^2\alpha)\cos\theta$$

Then

$$A^r = (2\dot{x}^2 + 2x\ddot{x} - x^2\omega^2)e^{j\theta}$$

$$A^t = j(4x\dot{x}\omega + x^2\alpha)e^{j\theta}$$

$$A_x = [(2\dot{x}^2 + 2x\ddot{x} - x^2\omega^2)\cos\theta - (4x\dot{x}\omega + x^2\alpha)\sin\theta]e^{j0°}$$

$$A_y = [(2\dot{x}^2 + 2x\ddot{x} - x^2\omega^2)\sin\theta + (4x\dot{x}\omega + x^2\alpha)\cos\theta]e^{j90°}$$

Competency Items
- A particle is moving on path $R = (2x + 1)e^{j\theta}$. Write the radial and transverse components of acceleration; also, write the horizontal and vertical components of acceleration.
- Write the components of the acceleration of a particle moving on a circular path. The center of the circle is at the origin.
- Calculate the radial and transverse components of acceleration of a particle moving on path $y = (25 - x^2)^{1/2}$ at position $x = 2.5$ if the particle has a velocity of 10 units/sec and the position vector has an angular acceleration of -2 rad/sec^2.

Objective 3

To describe the acceleration-vector relationship of a rigid body moving in a plane

Activities
- Read the material provided below, and as you read:
  a. Derive the acceleration-vector relationship for a rigid body executing planar motion.
  b. Write rules for solving the acceleration-vector relationship.
  c. Demonstrate the application of the acceleration relationship using the vector polygon.
- Check your answers to the competency items with your instructor.

Reading Material

In Unit II, Objective 4, we have derived that for point B in a rigid body, velocity $\mathbf{V}_B$ is obtained by solving the velocity vector relationship

$$\mathbf{V}_B = \mathbf{V}_A + \mathbf{V}_{B/A}$$

The time rate of change of the velocities of points A and B will yield the acceleration of points A and B. Thus, the time derivative of Eq. (1) gives

$$\frac{d}{dt}(\mathbf{V}_B) = \frac{d}{dt}(\mathbf{V}_A + \mathbf{V}_{B/A})$$

$$= \frac{d}{dt}(\mathbf{V}_A) + \frac{d}{dt}(\mathbf{V}_{B/A})$$

or

$$\dot{\mathbf{V}}_B = \dot{\mathbf{V}}_A + \dot{\mathbf{V}}_{B/A}$$

or

$$\mathbf{A}_B = \mathbf{A}_A + \mathbf{A}_{B/A} \tag{24}$$

From Unit III, Objective 2, Eq. (22), we know that acceleration $\mathbf{A}$ had radial and transverse components. That is,

$$\mathbf{A} = \mathbf{A}^r + \mathbf{A}^t \tag{22}$$

Thus

$$\mathbf{A}_B = \mathbf{A}_B^r + \mathbf{A}_B^t$$

$$\mathbf{A}_A = \mathbf{A}_A^r + \mathbf{A}_A^t$$

$$\mathbf{A}_{B/A} = \mathbf{A}_{B/A}^r + \mathbf{A}_{B/A}^t$$

Using the results of Eq. (22), Eq. (24) can be written as

$$\mathbf{A}_B^r + \mathbf{A}_B^t = \mathbf{A}_A^r + \mathbf{A}_A^t + \mathbf{A}_{B/A}^r + \mathbf{A}_{B/A}^t \tag{25}$$

Equation (25) describes the acceleration-vector relationship for a rigid body motion, and can be solved either analytically using complex number notations, or graphically using the vector polygon technique.

There are four important rules in solving Eq. (25).

1. The direction of radial component $\ddot{r}$ is along the line on which a particle is moving.

2. The magnitude and directions of the centripetal acceleration component are always known. The magnitude of this radial component is $\omega^2 r$ where ω is either known or obtained as a result of the velocity analysis. The direction of

the radial acceleration is toward the center of rotation of the point under consideration.

3. The direction of transverse component $r\alpha$ is perpendicular to position vector **R** of the particle.

4. The magnitude and direction of the coriolis acceleration $\mathbf{A}^c$ are completely known. The magnitude is $2\dot{r}\omega$. The direction of $\mathbf{A}^c$ is obtained by rotating the direction of $\dot{r}$ in the direction of ω.

Example 6.

Given rigid body AB with the components of acceleration shown in the Fig. 11. Construct an acceleration polygon and determine the angular acceleration of AB if $AB = 1$ in.

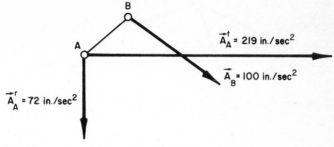

Fig. 11

Solution.

Since the point is moving relative to point B, we can use Eq. (25). Accordingly, we get

$$A_A^r + A_A^t = A_B + A_{A/B}^r + A_{A/B}^t$$

Since we know the magnitudes and directions of A_A^r, A_A^t, and A_B, we can construct a vector polygon to determine the radial and transverse components $A_{A/B}^r$ and $A_{A/B}^t$. The vector polygon is shown in Fig. 12.

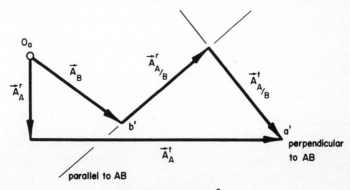

Scale: 1 in. = 50 in./sec^2

Fig. 12

From the polygon we can calculate angular acceleration

$$\alpha = \frac{A^t_{A/B}}{AB} = \frac{96 \text{ in./sec}^2}{1 \text{ in.}} = 96 \text{ rad/sec}^2 \text{ (counterclockwise)}$$

The direction of angular acceleration α of link AB is counterclockwise since it forces point A to rotate about point B in a counterclockwise direction.

Competency Items

- Rigid body AB has acceleration components as shown in the figure. Determine the angular acceleration of AB if $AB = 18$ in.

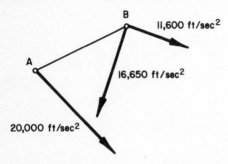

- Can the angular velocity of AB be determined if $A^r_{A/B}$ is known? If your answer is yes, then what is ω_{AB}? If your answer is no, state the reasons.

Performance Test #1—Objectives 1, 2, 3

1. A particle is moving on path $\mathbf{R} = (x + 2)e^{j\theta}$. Write the radial and transverse components of acceleration as well as the horizontal and vertical components of acceleration.
2. Calculate the radial and transverse components of acceleration of a particle moving on path $y = (25 - x^2)^{1/2}$ at position $x = 2.5$ if the position vector has an angular velocity of 2 rad/sec and an angular acceleration of -2 rad/sec^2.
3. For a particle moving on path $y = 2\sin 2x$ with constant speed $\dot{s} = 10$ units/sec, calculate the normal and the tangential components of acceleration when $x = \pi/4$ and when $x = \pi/2$.

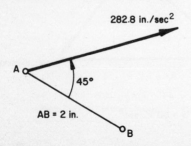

4. If rigid body AB has components of acceleration as shown, calculate the angular acceleration and angular velocity of AB. What can you say about the velocity of B?

Performance Test #2–Objectives 1, 2, 3

1. A particle has an acceleration of $A = 60$ in./sec at time $t = 5$ sec. The velocity of the particle is given by $\dot{y} = Kt^2 - 3$ where K is a constant. Locate position y of the particle at $t = 5$ sec if y at time $t = 0$ sec is 4 in.
2. A particle is moving on path $R = (x - 1)e^{j\theta}$. Calculate the radial and transverse components of acceleration and the horizontal and vertical components of acceleration.
3. A particle is moving on path $y = \frac{1}{2}x^2 + 2x$. The velocity of the particle along the tangent to the path is 10 units/sec and the acceleration along the tangent to the path is 20 units/sec^2. At position $x = 1$ calculate the normal and the tangential components of its acceleration.
4. Link 2 in the figure rotates clockwise as shown. Calculate linear and angular velocity and angular acceleration of the link.

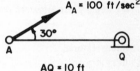

$A_A = 100$ ft/sec^2

30°

A   Q

$AQ = 10$ ft

Performance Test #3–Objectives 1, 2, 3

1. A particle is moving on path $y = \frac{1}{3}ct^4 + bt$. Calculate the acceleration at time $t = 3$ sec if $c = 2$ in./sec^4 and $b = 3$ in./sec.
2. A point is moving on curve $y = \frac{1}{2}x^2 - \frac{3}{2}$. The point is at $x = \sqrt{3}$ in., the velocity of the point (along the tangent to the curve) is $\dot{s} = 20$ in./sec, the angular velocity of the radius vector is negative (clockwise), and the acceleration of the point along the tangent to the curve is $\ddot{s} = 50\sqrt{3}$ in./sec^2. The x component of $\ddot{s}$ is positive. Calculate:

   a. angular acceleration α of the radius vector of the point
   b. $\ddot{r}$ where $R = re^{j\theta}$ locates the point
   c. the acceleration of point $A = A_x + jA_y$

   Proceed using the following steps and sketching *carefully:*

   a. Locate radius vector R. Sketch the curve and the radius vector.
   b. Calculate the components of the acceleration that are normal and tangential to the curve and show them on the sketch.
   c. Calculate acceleration A of the point and show it on the sketch.
   d. Calculate the transverse and radial components of the velocity and show them on a second sketch. Determine angular velocity ω.
   e. Using the acceleration of step (c) calculate the radial and transverse components of the acceleration and show them on a third sketch.
   f. Using the values of the components of step (e) calculate α and $\ddot{r}$.
   g. Show the four components of acceleration on a fourth sketch.

Objective 4

To demonstrate the application of the acceleration-vector polygon technique in order to perform acceleration anslysis of mechanisms

Activities
- Read the material provided below, and as you read demonstrate application of vector polygon technique to perform acceleration analysis of:
  a. a four-link mechanism
  b. a slider-crank mechanism
  c. an inverted slider-crank mechanism
- Check your answers to the competency items with your instructor.

Reading Material
Acceleration analysis of mechanisms usually is a step-by-step procedure beginning with the driving link and progressing from point to point through the mechanism. This solution can be accomplished either using analytical or graphical approach. In this objective, we will demonstrate how the vector polygon technique can be applied to solve the acceleration-vector relationship written for each link of a mechanism.

The reader is asked to repeat each step by constructing the acceleration vector polygon and calculating acceleration components for each of the mechanisms presented in the examples below.

Example 1.

Four-link mechanism $MABQ$ is shown in Fig. 13a. Input crank MA has angular velocity $\omega_2 = 92.4$ rad/sec in a clockwise direction. $\omega_3 = 30$ rad/sec, $\omega_4 = 74.2$ rad/sec, $\omega_2 = 94.2$ rad/sec. Calculate angular accelerations α_3 and α_4. Also calculate the acceleration of points C and D.

Solution.

The following steps will lend to the solution shown in Fig. 13b and c.

1. $A_A = A_M + A_{A/M}$

    Since point M is a stationary point, $A_M = 0$. Therefore,

    $$A_A = A_{A/M} = A^r_{A/M} + A^t_{A/M}$$

    Since the input crank is rotating with a constant angular velocity, the magnitude of $A^t_{A/M}$ is

    $$\alpha_2 \cdot MA = 0$$

    The radial acceleration $A^r_A = \omega_2^2 \cdot AM = (94.2)^2 \frac{4}{12} = 2,960 \text{ ft/sec}^2$

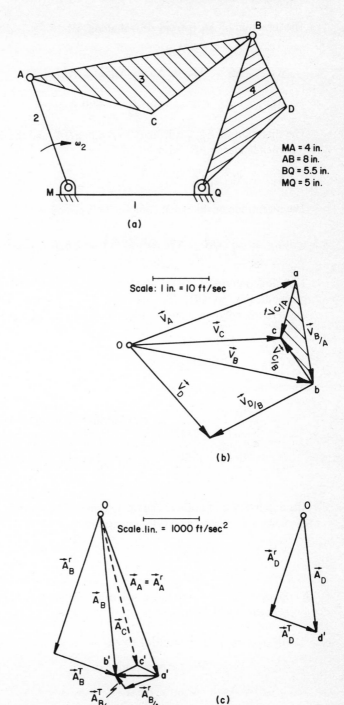

MA = 4 in.
AB = 8 in.
BQ = 5.5 in.
MQ = 5 in.

(a)

Scale: 1 in. = 10 ft/sec

(b)

Scale.lin. = 1000 ft/sec²

(c)

Fig. 13 Acceleration analysis of a four-link mechanism.

The direction of A_A^r is from A to M along AM.

2. $A_B = A_A + A_{B/A}$

$$A_B = A_B^r + A_B^t = A_A^r + A_{B/A}^r + A_{B/A}^t$$

$$A_{B/A}^r = \omega_3^2 \cdot BA = (30)^2 \frac{8}{12} = 600 \text{ ft/sec}^2$$

The direction of $A_{B/A}^r$ is from B to A along BA, and the direction of $A_{B/A}^t$ is perpendicular to AB. Also,

$$A_B = A_B^r + A_B^t$$

The magnitude of the radial component is given by

$$A_B^r = \omega_4^2 \cdot QB = (74.2)^2 \cdot \left(\frac{5.5}{12}\right) = 2,520 \text{ ft/sec}^2$$

The direction of A_B^r is along BQ from B to Q. The direction A_B^t is perpendicular to QB. We are now in a position to construct the acceleration polygon shown in Fig. 13c. From the acceleration polygon

$$A_{B/A}^t = 120 \text{ ft/sec}$$
$$A_B^t = 1,150 \text{ ft/sec}^2$$

We can now calculate α_3 and α_4

$$\alpha_3 = \frac{A_{B/A}^t}{AB} = \frac{120}{8/12} = 180 \text{ rad/sec}^2 \text{ (counterclockwise)}$$

$$\alpha_4 = \frac{A_B^t}{BQ} = \frac{1150}{5.5/12} = 2,510 \text{ rad/sec}^2 \text{ (clockwise)}$$

3. The acceleration of point C can be determined two ways:
    a. Solve graphically two vector equations

$$A_C = A_A + A_{C/A}$$
$$A_C = A_B + A_{C/B}$$

b. Since ω_3 and α_3 are known, we can compute the acceleration of point C in the following manner:

$$A_{C/A}^r = \omega_3^2 \cdot AC, \text{ parallel to } AC$$

$$A_{C/A}^t = \omega_3 \cdot AC, \text{ perpendicular to } AC$$

$$A_C = A_A + A_{C/A}^r + A_{C/A}^t$$

From acceleration polygon $A_C = 2,750 \text{ ft/sec}^2$.

4. The acceleration of point D can be determined by directly computing the tangential and radial acceleration components.

$$A_D = A_D^r + A_D^t$$

$$A_D^r = (\omega_4)^2 DQ = (74.2)^2 \frac{4}{12} = 1{,}835 \text{ ft/sec}^2$$

$$A_D^t = \alpha_4 \cdot DQ = 2{,}510 \times \frac{4}{12} = 837 \text{ ft/sec}^2$$

$$A_D = 2{,}000 \text{ ft/sec}^2$$

5. The acceleration polygon is shown in Fig. 13c.

Example 8.

Input crank MA of the slider-crank mechanism shown in Fig. 13a is driven at constant speed 600 rpm. From the velocity analyses it is found that $\omega_2 = 62.8$ rad/sec, $\omega_3 = 26.16$ rad/sec. Calculate angular accelerations α_3 and A_B.

Solution.

The following steps lead to construction of the acceleration vector polygon shown in Fig. 14c.

1. $A_A = A_M + A_{A/M}$

Since point M is a stationary point, $A_M = 0$, and, therefore,

$$A_A = A_{A/M} = A_{A/M}^r + A_{A/M}^t$$

Since the input crank is rotating at a constant angular velocity $A_{A/M}^t = 0$, and, therefore, the magnitude of A_A is obtained from

$$A_A = \omega_2^2 \cdot MA = 657.6 \text{ ft/sec}^2$$

The direction of A_A is the direction of A_A^r. The direction of A_A^r is along AM from A to M.

2. $A_B = A_A + A_{B/A}$

or

$$A_B = A_A^r + A_{B/A}^r + A_{B/A}^t$$

Since we know

a. $A_{B/A}^r = \omega_3^2 \cdot AB = \dfrac{(V_{B/A})^2}{BA} = \dfrac{(6.54)^2}{3/12} = 171 \text{ ft/sec}^2$

b. the direction $A_{B/A}$ is along BA from B to A

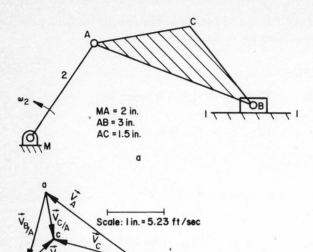

MA = 2 in.
AB = 3 in.
AC = 1.5 in.

a

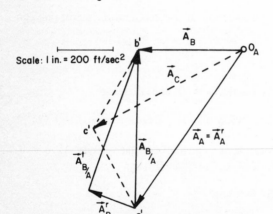

Scale: 1 in. = 5.23 ft/sec

b

Scale: 1 in. = 200 ft/sec^2

c

Fig. 14 Acceleration analysis of a slicer-crank mechanism.

   c. $A_{B/A}^t = \alpha_3 \cdot AB$

   d. the direction of $A_{B/A}^t$ is perpendicular to BA, and

   e. the direction of A_B is along line $l\text{-}l$, we are in a position to construct the acceleration polygon, shown in Fig. 14c. From the polygon,

$$A_{B/A}^t = 496 \text{ ft/sec}^2$$

$$A_B = 394 \text{ ft/sec}^2 \text{ parallel to } l\text{-}l$$

$$\alpha_3 = \frac{A_{B/A}^t}{BA} = \frac{496}{3/12} = 1{,}985 \text{ rad/sec}^2 \text{(counterclockwise)}$$

3. The acceleration of point C is determined using the following vector equations:

$$A_C = A_B + A_{C/B}$$
$$A_C = A_A + A_{B/A}$$

From the vector polygon

$$A_C = 530 \text{ ft/sec}^2$$

Example 9.
Input crank of the inverted slider-crank mechanism shown in Fig. 15a is driven at 600 rpm, counterclockwise. Calculate angular acceleration α_3 of output link QA.
Solution.
   The acceleration analysis of the inverted slider-crank mechanism requires careful understanding of the coriolis acceleration. Point A is a coincident point. It lies on link MA and it also lies on link QA. Point A on link QA is denoted as point A_3. Point A on link MA is denoted as point A_2. Note that point A_2 is traveling along link QA and is rotating about fixed center M. This situation is similar to that in which an existence of coriolis acceleration is noted. The acceleration of point A_2, therefore, has three components, the radial component, the transverse component, and the coriolis component. Hence,

$$A_{A_2} = A_{A_3} + A_{A_2/A_3} + 2\omega_3 \times V_{A_2/A_3}$$

From velocity analysis

$$\omega_2 = 62.8 \text{ rad/sec}$$
$$V_A = 10.46 \text{ fps}$$
$$\omega_3 = 12.55 \text{ rad/sec}$$
$$V_{A_2/A_3} = 9.93 \text{ fps}$$

Rewriting the vector equation, we get

$$A_{A_2}^r + A_{A_2}^t = A_{A_3}^r + A_{A_3}^t + A_{A_2/A_3}^r + A_{A_2/A_3}^t + 2\omega_3 \times V_{A_2/A_3}$$

$$A_{A_2} = A_{A_2}^r = \omega_2^2 \cdot MA_2 = 657 \text{ ft/sec}^2, \text{ parallel to } AM$$

1. Since point A_2 is moving along $A_3 Q$,

$$A_{A_2/A_3}^r = 0$$

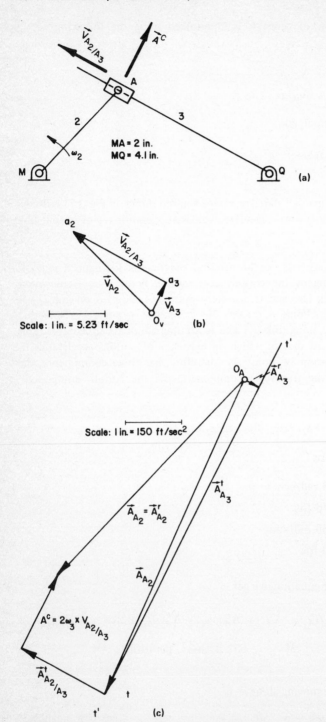

Fig. 15 Acceleration analysis of an inverted slider-crank mechanism.

2. The direction of $A^t_{A_2/A_3}$ is parallel to QA.
3. The direction of the coriolis acceleration $\mathbf{A}^c$ obtained by rotating the direction of V_{A_2/A_3} through 90° about a_3 in the direction of ω_3

$$2V_{A_2/A_3} \cdot \omega_3 = 2 \times 9.93 \times 12.55 = 249 \text{ ft/sec}^2$$

4. $$A^r_{A_3} = (\omega_3)^2 \cdot AQ = (12.55)^2 \cdot \tfrac{3}{12} = 39.4 \text{ ft/sec}^2$$

$$A^r_{A_2} = (\omega_2)^2 \cdot AM = (62.8)^2 \cdot \tfrac{2}{12} = 657 \text{ ft/sec}^2$$

$$A^t_{A_2} = 0, (\alpha_2 = 0)$$

$$A^r_{A_2/A_3} = 0$$

We are now in a position to construct the acceleration polygon shown in Fig. 15c. From the polygon, the magnitude of

$$A_{A_3} = 857 \text{ ft/sec}^2$$

In order to obtain α_3, resolve acceleration A_3 along QA and perpendicular to QA. The component perpendicular to QA yields $A^t_{A_3}$. From the polygon, Fig. 15c,

$$A^t_{A_3} = 857 \text{ ft/sec}^2$$

Calculate

$$\alpha_3 = \frac{A^t_{A_3}}{AQ} = 3{,}428 \text{ rad/sec}^2 \text{ (counterclockwise)}$$

Competency Items
- For the figure shown below:
  a. Construct an acceleration polygon to determine the angular accelerations of links AB and BQ. Assume that link MA is moving clockwise with a constant angular velocity of 50 rad/sec.

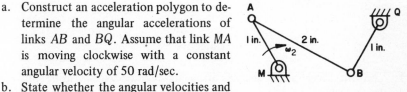

  b. State whether the angular velocities and accelerations remain the same in magnitude but opposite in direction; if MA is moving counterclockwise with a constant angular velocity of 50 rad/sec.
  c. Construct an acceleration polygon to calculate the angular accelerations of links AB and BQ, assuming that MA is moving counterclockwise with an angular velocity of 50 rad/sec but with a clockwise angular acceleration of 10 rad/sec^2.

- For the figure shown below:

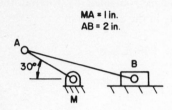

  a. Construct an acceleration polygon to determine the acceleration of point B assuming that link AM is moving clockwise with a constant angular velocity of 50 rad/sec.

  b. Calculate the acceleration of point B if the $30°$ angle is changed to a $60°$ angle and then to a $150°$ angle.

  c. Calculate the angular velocity and angular acceleration of link AM if the slider crank is in the position shown and point B has a constant velocity for 20 in./sec to the right. Compare this to the case of B moving to the left at the rate of 20 in./sec.

- Define the coriolis acceleration and state in what direction it acts.

- State whether the coriolis component of acceleration is present in all inverted slider-crank mechanisms. Write reasons why it is not present in the four-link mechanisms or the slider-crank mechanisms.

- The inverted slider-crank mechanism in the figure shown below has point C fixed to coupler link AB. If link MA has an angular velocity of 10 rad/sec in a counterclockwise direction and an angular acceleration of 2 rad/sec^2 in a clockwise direction, calculate the absolute acceleration of point C and the angular acceleration of link BQ.

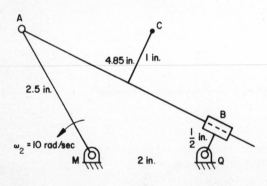

Objective 5

To perform acceleration analysis of a six-link mechanism using the vector polygon technique

Activities

- Read material provided below, and as you read:
  a. Calculate the acceleration of the output link.
  b. Demonstrate the application of the vector polygon technique.
- Check your answers to the competency items with your instructor.

Reading Material

There are 42 different types of six-link mechanisms with revolute and prism pairs. These six-link mechanisms are shown in Figs. 17–22 of Unit II. It is suggested that the reader take the different steps required in order to perform velocity analysis of these mechanisms. In Objective 4, Unit III, you have studied the vector polygon technique to perform acceleration analysis of four-link, slider-crank, and inverted slider-crank mechanisms. Since the six-link mechanisms of Figs. 17–22 are obtained by adding dyads for these basic mechanisms, it is only necessary for the reader to understand the necessary steps in conducting acceleration analysis of any one of these six-link mechanisms. Once the vector polygon technique is thoroughly understood, the reader will be able to carry out the acceleration analysis of the other types of six-link mechanisms. The reader should repeat all the steps described in the example that follows below.

Example 10.

The six-link mechanism shown in Fig. 16a is driven by crank 2 at 900 rpm clockwise. Find the accelerations of points A, B, C, and D. Calculate the angular acceleration of links 3, 4, 5, and 6. The results of the velocity analysis are shown using the velocity polygon in Fig. 16b.

Solution.

1. From the velocity analysis, we know

$$\omega_2 = 94.2 \text{ rad/sec,} \quad V_A = 15.7 \text{ fps,} \quad \omega_3 = 62.8 \text{ rad/sec}$$

$$\omega_4 = 93.6 \text{ rad/sec,} \quad \omega_5 = 17.9 \text{ rad/sec,} \quad \omega_6 = 103.55 \text{ rad/sec}$$

2. We write the vector relationship to obtain acceleration of point B,

$$A_B = A_B^r + A_B^t = A_A^r + A_A^t + A_{B/A}^r + A_{B/A}^t$$

and calculate

$$A_A^r = \frac{(V_A)^2}{AM} = \frac{(15.7)^2}{2/12} = 1,479 \text{ ft/sec}^2$$

$$A_A^t = 0$$

$$A_B^r = (\omega_4)^2 \cdot BQ = (93.6)^2 \cdot \frac{1.75}{12} = 1,278 \text{ ft/sec}^2$$

$$A_{B/A}^r = (\omega_3)^2 \cdot BA = (62.8)^2 \cdot \frac{1.88}{1.2} = 618 \text{ ft/sec}^2$$

3. Draw the vector polygon as shown in Fig. 16c to obtain A_B and its components A_B^r and A_B^t. From Fig. 16c,

$$A_B = 2.6 \times 500 = 1,300 \text{ ft/sec}^2$$

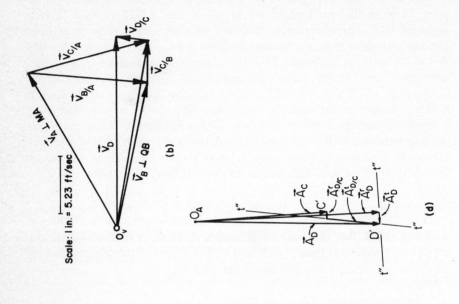

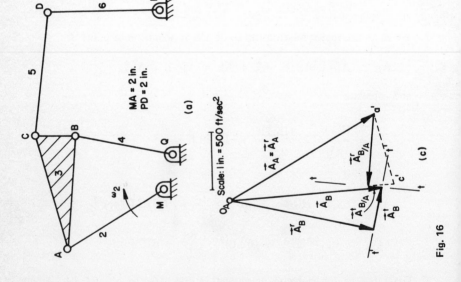

Scale: 1 in. = 5.23 ft/sec

(b)

(d)

MA = 2 in.
PD = 2 in.

(a)

Scale: 1 in. = 500 ft/sec²

(c)

Fig. 16

$$A_{B/A}^t = 0.08 \times 500 = 40 \text{ ft/sec}^2$$

$$\alpha_3 = \frac{a_{B/A}^t}{BA} = \frac{40}{1.88/12} = 255.6 \text{ rad/sec}^2 \quad \text{(clockwise)}$$

$$A_B^t = 0.6 \times 500 = 300 \text{ ft/sec}^2$$

$$\alpha_4 = \frac{a_B^t}{BQ} = \frac{300}{1.75/12} = 2058 \text{ rad/sec}^2 \quad \text{(clockwise)}$$

4. Obtain acceleration of point C using two vector relationships and constructing a vector polygon shown on Fig. 16c.

$$\mathbf{A}_C = \mathbf{A}_B + \mathbf{A}_{C/B}$$
$$\mathbf{A}_C = \mathbf{A}_A + \mathbf{A}_{C/A}.$$

5. Obtain acceleration of point D using the vector relationship

$$\mathbf{A}_D = \mathbf{A}_C + \mathbf{A}_{D/C}$$

or

$$\mathbf{A}_D^r + \mathbf{A}_D^t = \mathbf{A}_C^r + \mathbf{A}_C^t + \mathbf{A}_{D/C}^r + \mathbf{A}_{D/C}^t$$

Calculate from the vector polygon shown in Fig. 16d.

$$a_D^r = (\omega_6)^2 \cdot DP = (103.55)^2 \times \frac{2}{12}$$

$$= 1{,}790 \text{ ft/sec}^2 /\!/DP$$

$$a_{D/C}^r = (\omega_5)^2 \cdot DC = (17.9)^2 \times \frac{2.1}{12}$$

$$= 56.1 \text{ ft/sec}^2 /\!/DC$$

$$a_D = 3.56 \times 500 = 1{,}780 \text{ ft/sec}^2$$

$$a_{D/C}^t = 1.05 \times 500 = 525 \text{ ft/sec}^2$$

$$\alpha_5 = \frac{a_{D/C}^t}{DC} = \frac{525}{2.1/12} = 3{,}000 \text{ rad/sec}^2 \quad \text{(clockwise)}$$

$$a_D^t = 0.23 \times 500 = 115 \text{ ft/sec}^2$$

$$\alpha_6 = \frac{a_D^t}{DP} = \frac{115}{2/12} = 690 \text{ rad/sec}^2 \quad \text{(counterclockwise)}$$

Competency Item
- Assume some parameters for any of the 42 different types of six-link mechanisms shown in Figs. 17-22 of Unit II. Perform an acceleration analysis on as many of these as necessary to become proficien:.

Performance Test #1–Objectives 4, 5

1. In the figure below link *MA* is moving clockwise with an angular velocity of 20 rad/sec and it is accelerating in a counterclockwise sense at the rate of 15 rad/sec^2. Use the polygon technique to determine A_A, A_B, A_C, α_3, and α_4.

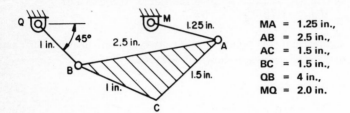

MA	= 1.25 in.,
AB	= 2.5 in.,
AC	= 1.5 in.,
BC	= 1.5 in.,
QB	= 4 in.,
MQ	= 2.0 in.

2. In the figure below the slider is moving to the left at 5 in./sec and is accelerating to the left at 50 in./sec^2. Use the polygon technqiue to determine A_A, A_B, α_2, and α_3.

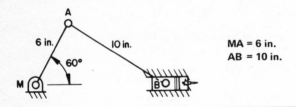

MA = 6 in.
AB = 10 in.

Performance Test #2–Objectives 4, 5

1. An inverted slider-crank mechanism is shown. Input crank A_0A has a uniform angular velocity of 100 rpm. Determine angular acceleration α_4 of link 4. Use the polygon method and label the polygon completely.

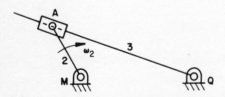

2. Perform an acceleration analysis for the mechanism shown. Use the polygon method. Find angular accelerations α_3 and α_4 and the magnitude of the acceleration of point c.

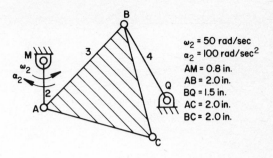

Performance Test #3—Objectives 4, 5

1. The slider in the figure is moving to the right with a velocity of 6 in./sec and is accelerating to the left at 25 in./sec^2. Find the angular acceleration of link 4 and the total acceleration of point *B* using the polygon technique. Show *neatly* the complete polygons, label

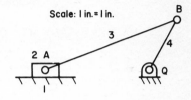

vectors and points, and show all equations used in your solution.

2. An inverted slider-crank mechanism is shown with its velocity and acceleration polygons. Write the vector equations for both velocity and acceleration. Complete the polygons by placing arrows on the vectors and labeling the vectors and the points. Your equations must correspond to your vector labeling.

   Calculate the angular acceleration of link 3 and the relative acceleration between points *A* of link 2 and link 3.

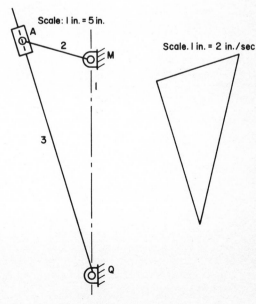

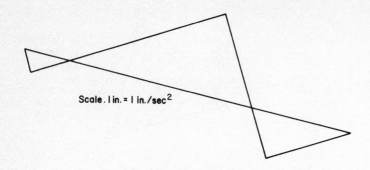

Scale. I in. = I in./sec^2

Supplementary References

1(2.3–3.A.1); **3**(5.1–5.9); **5**(7.1–7.4, 7.7); **7**(5.1–5.5); **8**(2.2–2.3, 6.1–6.5).

Unit IV. Displacement, Velocity, and Acceleration Analysis of Mechanisms Using Analytical Methods

Objectives

1. To perform displacement, velocity, and acceleration analysis of a four-link mechanism
2. To perform displacement, velocity, and acceleration analysis of a slider-crank mechanism
3. To perform displacement, velocity, and acceleration analysis of an inverted slider-crank mechanism

Objective 1

To perform displacement, velocity, and acceleration analysis of a four-link mechanism

Activities

- Read the material provided below, and as you read:
  a. Calculate angles θ_3 and θ_4 for given values of θ_2.
  b. Calculate velocities ω_3 and ω_4 for given values of θ_2.
  c. Calculate accelerations α_3 and α_4 for given values of θ_2.
  d. Demonstrate the application of the complex numbers in obtaining analytical expressions for displacements, velocities, and accelerations of coupler and output links of a four-link mechanism.
- Check your answers to the competency items with your instructor.

Reading Material

Introduction

The graphical approach examined in the previous units is a stepwise procedure. The linkage is placed in a desired position to determine the velocities and accelerations of the coupler and output links. The construction of velocity and acceleration polygons must be repeated if it is desired to determine the velocities and accelerations corresponding to some other position of the linkage. The analytical technique, which will be considered here, has some definite advantages over the graphical technique. The analytical expressions giving displacements, velocities, and accelerations can be derived in terms of the general parameters of a linkage, and calculations can be performed stepwise either using a desk calculator, a slide rule, or a digital computer.

The general procedure in the three objectives of Unit IV involves four steps:

• Write the vector displacement relationship to describe the mechanism.
• Express the vector displacement relationship in terms of complex numbers, and obtain the displacement relationships.
• Take the time derivative of displacement relationship to obtain velocity relationships.
• Take the time derivative of velocity relationship to obtain acceleration relationships.

Displacement analysis of a four-link mechanism

The continuous motion of the follower link must be studied using either a graphical approach or a mathematical approach. In Fig. 1, let links QM, MA, AB, and BQ be represented by vectors $\mathbf{D}$, $\mathbf{A}$, $\mathbf{B}$, and $\mathbf{C}$. Then since the four-link mechanism represents a vector polygon, the following equation holds true:

$$\mathbf{D} + \mathbf{A} + \mathbf{B} - \mathbf{C} = 0 \tag{1}$$

If $\mathbf{U}_D$, $\mathbf{U}_A$, $\mathbf{U}_B$, and $\mathbf{U}_C$ represent the unit vectors, and d, a, b, and c represent the magnitude of vectors $\mathbf{D}$, $\mathbf{A}$, $\mathbf{B}$, and $\mathbf{C}$, then

$$\mathbf{D} = \mathbf{U}_D d \tag{2}$$

$$\mathbf{A} = \mathbf{U}_A a \tag{3}$$

$$\mathbf{B} = \mathbf{U}_B b \tag{4}$$

$$\mathbf{C} = \mathbf{U}_C c \tag{5}$$

These vectors can also be represented using complex notations. Thus, with j as a complex operator ($j^2 = -1$), we have

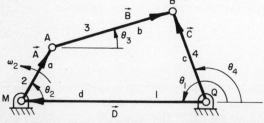

Fig. 1   Vector representation for a four-link mechanism.

$$\mathbf{D} = de^{j\theta_1} \tag{6}$$

$$\mathbf{A} = ae^{j\theta_2} \tag{7}$$

$$\mathbf{B} = be^{j\theta_3} \tag{8}$$

and

$$\mathbf{C} = ce^{j\theta_4} \tag{9}$$

where angles θ_1, θ_2, θ_3, and θ_4 describe the relative positions of vectors $\mathbf{D}$, $\mathbf{A}$, $\mathbf{B}$, and $\mathbf{C}$. Upon substituting Eqs. (6)–(9) in Eq. (1), we get

$$de^{j\theta_1} + ae^{j\theta_2} + be^{j\theta_3} - ce^{j\theta_4} = 0 \tag{10}$$

Using the identity

$$e^{j\theta_i} = \cos\theta_i + j\sin\theta_i \tag{11}$$

Eq. (10) becomes

$$(d\cos\theta_1 + a\cos\theta_2 + b\cos\theta_3 - c\cos\theta_4)$$
$$+ j(d\sin\theta_1 + a\sin\theta_2 + b\sin\theta_3 - c\sin\theta_4) = 0 \tag{12}$$

Equation (12) contains a real part and a complex part. Separating the real and the complex parts, we get two equations.

$$d\cos\theta_1 + a\cos\theta_2 + b\cos\theta_3 - c\cos\theta_4 = 0 \tag{13}$$

$$d\sin\theta_1 + a\sin\theta_2 + b\sin\theta_3 - c\sin\theta_4 = 0 \tag{14}$$

Note, however, that in Fig. 1, $\theta_1 = 180°$. Therefore, the above equations can be simplified to yield

$$-d + a\cos\theta_2 + b\cos\theta_3 - c\cos\theta_4 = 0 \tag{15}$$

$$a\sin\theta_2 + b\sin\theta_3 - c\sin\theta_4 = 0 \tag{16}$$

If a relationship between input link rotation θ_2 and output link rotation θ_4 is desired, then angle θ_3 must be eliminated from Eqs. (15) and (16). Transferring the terms containing θ_3 on the right-hand side of the equations, and squaring both sides, we get

$$(-b\cos\theta_3)^2 = (-d + a\cos\theta_2 - c\cos\theta_4)^2 \tag{17}$$

$$(-b\sin\theta_3)^2 = (a\sin\theta_2 - c\sin\theta_4)^2 \tag{18}$$

Adding Eqs. (17) and (18), and simplifying, the resultant equation becomes

$$K_1 \cos \theta_4 - K_2 \cos \theta_2 + K_3 = \cos (\theta_2 - \theta_4) \tag{19}$$

where

$$K_1 = \frac{d}{a}$$

$$K_2 = \frac{d}{c} \tag{20}$$

$$K_3 = \frac{a^2 - b^2 + c^2 + d^2}{2ac}$$

Equation (19) is called Freudenstein's equation and can be used if it is desired to synthesize a four-link mechanism for a specified three positions of the input-link and three positions of the output link. Note, however, that Eq. (19) is not in convenient form for the computation of the output link position for a specified position of the input link. For this reason, it is necessary to simplify Eq. (19). Equation (19) can be written as

$$K_1 \cos \theta_4 - K_2 \cos \theta_2 + K_3 = \cos \theta_2 \cos \theta_4 + \sin \theta_4 \sin \theta_2 \tag{21}$$

Substituting the trigonometric identities for

$$\sin \theta_4 = \frac{2 \tan (\theta_4/2)}{1 + \tan^2 (\theta_4/2)} \tag{22}$$

and

$$\cos \theta_4 = \frac{1 - \tan^2 (\theta_4/2)}{1 + \tan^2 (\theta_4/2)} \tag{23}$$

in Eq. (21) and simplifying the resultant equation, we get

$$A \tan^2 \frac{\theta_4}{2} + B \tan \frac{\theta_4}{2} + C = 0 \tag{24}$$

Where

$$A = \cos \theta_2 + K_3 - K_1 - K_2 \cos \theta_2$$

$$B = -2 \sin \theta_2 \tag{25}$$

$$C = K_1 + K_3 - (1 + K_2) \cos \theta_2$$

Equation (24) is a quadratic in $\tan (\theta_4/2)$. The two roots of the quadratic equation are

$$\left[\tan \frac{\theta_4}{2}\right]_{1,2} = \frac{-B \pm \sqrt{B^2 - 4AC}}{2A} \tag{26}$$

and, therefore,

$$\left[\theta_4\right]_{1,2} = 2 \tan^{-1}\left(\frac{-B \pm \sqrt{B^2 - 4AC}}{2A}\right) \tag{27}$$

If the link proportions and the position of the input link are known, then Eq. (27) provides the corresponding position of the output link. Thus, using Eq. (27), it is possible to perform a displacement analysis of a four-link mechanism.

If a relationship between input-link rotation θ_2 and the coupler-link position angle θ_3 is desired, angle θ_4 must be eliminated from the Eqs. (15) and (16). Transferring the terms containing θ_4 on the right-hand side of the equations, and squaring both sides, we get

$$(-d + a \cos \theta_2 + b \cos \theta_3)^2 = (c \cos \theta_4)^2 \tag{28}$$

$$(a \sin \theta_2 + b \sin \theta_3)^2 = (c \sin \theta_4)^2 \tag{29}$$

Upon adding Eqs. (28) and (29) and simplifying, the resultant equation becomes

$$K_4 \cos \theta_2 + K_1 \cos \theta_3 + K_5 = \cos (\theta_2 - \theta_3) \tag{30}$$

Where

$$K_4 = \frac{d}{b}$$

and

$$K_5 = \frac{c^2 - d^2 - a^2 - b^2}{2ab} \tag{31}$$

When Eq. (30) is rearranged we will get

$$D \tan^2\left(\frac{\theta_3}{2}\right) + E \tan\left(\frac{\theta_3}{2}\right) + F = 0 \tag{32}$$

where

$$D = K_4 \cos \theta_2 + \cos \theta_2 + K_5 - K_1$$

$$E = -2 \sin \theta_2$$

and

$$F = K_4 \cos \theta_2 - \cos \theta_2 + K_5 + K_1$$

Equation (32) yields two values of mi which are given by

$$\left[\theta_3\right]_{1,2} = 2 \tan^{-1}\left(\frac{-E \pm \sqrt{E^2 - 4DF}}{2D}\right) \tag{33}$$

Thus, Eqs. (27) and (33) will permit us to calculate θ_4 and θ_3 provided the link length a, b, c, d, and θ_2, the position of the input link MA, are known.

Velocity analysis of a four-link mechanism

In Fig. 1, a four-link mechanism is shown. The input crank MA is moving with angular velocity ω_2 in a counterclockwise direction. Our objective is to develop explicit mathematical relationships giving V_A, V_B, α_3, α_4, A_A, and A_B of the coupler and the output links. The relative positions of the four links QM, MA, AB, and QB are described using vectors **D**, **A**, **B**, and **C** where

$$
\begin{aligned}
\mathbf{D} &= de^{j\theta_1} \\
\mathbf{A} &= de^{j\theta_2} \\
\mathbf{B} &= be^{j\theta_3} \\
\mathbf{C} &= ce^{j\theta_4}
\end{aligned}
\tag{34}
$$

and θ_1, θ_2, θ_3, θ_4 are the angles measuring the relative positions of the links as shown in Fig. 1.

For vector polygon $QMAB$, the vector equation can be written as:

$$\mathbf{D} + \mathbf{A} + \mathbf{B} = \mathbf{C} \tag{35}$$

That is,

$$de^{j\theta_1} + ae^{j\theta_2} + be^{j\theta_3} = ce^{j\theta_4} \tag{36}$$

Angles θ_1, θ_2, θ_3, and θ_4 are measured in counterclockwise direction. In order to obtain velocities of each of the moving links, we must take the time derivative of Eq. (36). Since a, b, c, d, and θ_1 ($\theta_1 = 180°$) do not vary with time, the time derivative of Eq. (36) yields

$$ja\frac{d\theta_2}{dt}e^{j\theta_2} + jb\frac{d\theta_3}{dt}e^{j\theta_3} = jc\frac{d\theta_4}{dt}e^{j\theta_4} \tag{37}$$

By definition

$$\omega_2 = \frac{d\theta_2}{dt} \qquad \omega_3 = \frac{d\theta_3}{dt} \qquad \text{and} \qquad \omega_4 = \frac{d\theta_4}{dt}$$

Therefore, Eq. (37) can be written as

$$ja\omega_2 e^{j\theta_2} + jb\omega_3 e^{j\theta_3} = jc\omega_4 e^{j\theta_4} \tag{38}$$

The above relationship defines both the magnitude and direction of the velocities of each link. For example, the magnitude of the velocity of point A is $V_A = a\omega_2$ and its direction is given by $je^{j\theta_2}$, that is, the direction of $\mathbf{V}_A$ is perpendicular to MA. Thus,

$$\mathbf{V}_A = (a\omega_2)(je^{j\theta_2})$$

Similarly

$$\mathbf{V}_{B/A} = (b\omega_3)(je^{j\theta_3})$$

and

$$\mathbf{V}_B = (c\omega_4)(je^{j\theta_4})$$

Hence, Eq. (38) can be expressed vectorally as

$$\mathbf{V}_B = \mathbf{V}_A + \mathbf{V}_{B/A} \tag{39}$$

Separating the real and the complex parts, Eq. (38) yields,

$$c\omega_4 \sin\theta_4 = a\omega_2 \sin\theta_2 + b\omega_3 \sin\theta_3 \tag{40}$$

$$c\omega_4 \cos\theta_4 = a\omega_2 \cos\theta_2 + b\omega_3 \cos\theta_3 \tag{41}$$

Simultaneous solution of Eqs. (40) and (41) yields

$$\omega_4 = \frac{a\omega_2}{c} \frac{\sin(\theta_2 - \theta_3)}{\sin(\theta_4 - \theta_3)} \tag{42}$$

and

$$\omega_3 = \frac{a\omega_2}{b} \frac{\sin(\theta_4 - \theta_2)}{\sin(\theta_3 - \theta_4)} \tag{43}$$

The angular displacements θ_3 and θ_4 of the coupler and the output links may be determined using Eqs. (27) and (33).

Acceleration analysis

The explicit mathematical relationships for the calculation of accelerations of coupler and output links of a four-link mechanism are obtained by taking the time derivative of Eq. (38); this yields

$$ja\alpha_2 e^{j\theta_2} + j^2 a\omega_2^2 e^{j\theta_2} + j^2 b\omega_3^2 e^{j\theta_3} + jb\alpha_3 e^{j\theta_3}$$
$$= j^2 c\omega_4^2 e^{j\theta_4} + jc\alpha_4 e^{j\theta_4} \quad (44)$$

where

$$\alpha_2 = \frac{d\omega_2}{dt}$$

$$\alpha_3 = \frac{d\omega_3}{dt}$$

$$\alpha_4 = \frac{d\omega_4}{dt}$$

We observe that Eq. (44) describe the vector relationship

$$\mathbf{A}_B = \mathbf{A}_A + \mathbf{A}_{B/A}$$

where

$$\mathbf{A}_A = \mathbf{A}_A^r + \mathbf{A}_A^t = j^2 a\omega_2^2 e^{j\theta_2} + ja\alpha_2 e^{j\theta_2} \tag{45}$$

$$\mathbf{A}_{B/A} = \mathbf{A}_{B/A}^r + \mathbf{A}_{B/A}^t = j^2 b\omega_3^2 e^{j\theta_3} + jb\alpha_3 e^{j\theta_3} \tag{46}$$

and

$$\mathbf{A}_B = \mathbf{A}_B^r + \mathbf{A}_B^t = j^2 c\omega_4^2 e^{j\theta_4} + jc\alpha_4 e^{j\theta_4} \tag{47}$$

Substituting

$$e^{j\theta} = \cos\theta + j\sin\theta$$

in Eq. (44), and separating its real and complex parts, we get

$$c\alpha_4 \sin\theta_4 - b\alpha_3 \sin\theta_3$$
$$= a\alpha_2 \sin\theta_2 + a\omega_2^2 \cos\theta_2 + b\omega_3^2 \cos\theta_3 - c\omega_4^2 \cos\theta_4$$

$$c\alpha_4 \cos\theta_4 - b\alpha_3 \cos\theta_3$$
$$= a\alpha_2 \cos\theta_2 - a\omega_2^2 \sin\theta_2 - b\omega_3^2 \sin\theta_3 + c\omega_4^2 \sin\theta_4$$

Simultaneous solution of the above two equations in two unknowns α_3 and α_4 yields

$$\alpha_3 = \frac{CD - AF}{AE - BD} \tag{48}$$

$$\alpha_4 = \frac{CE - BF}{AE - BD} \tag{49}$$

where,

$$A = c \sin\theta_4$$

$$B = b \sin\theta_3$$

$$C = a\alpha_2 \sin\theta_2 + a\omega_2^2 \cos\theta_2 + b\omega_3^2 \cos\theta_3 - c\omega_4^2 \cos\theta_4$$

$$D = c \cos\theta_4$$

$$E = b \cos\theta_3$$

$$F = a\alpha_2 \cos\theta_2 - a\omega_2^2 \sin\theta_2 - b\omega_3^2 \sin\theta_3 + c\omega_4^2 \sin\theta_4$$

Equations (48) and (49) permit us to calculate the angular accelerations α_3 and α_4. The abosolute acceleration of B is computed using Eq. (47).

Competency Items

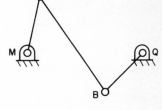

- Define vectors representing the links of the mechanism shown in the figure below. Write the vector equation representing the mechanism.
- If the mechanism has the following dimensions: $MA = 1$ in., $AB = 2$ in., $MQ = 2$ in., calculate the angular accelerations of links AB and BQ when the input link MA rotates with a constant angular velocity of 20 rad/sec in a clockwise sense. Use the complex number method.
- Derive the relationships to calculate θ_6, ω_6, and θ_6 of the six-link mechanism shown below.

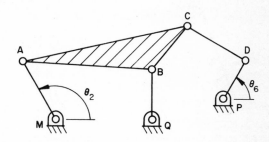

Objective 2

To perform displacement, velocity, and acceleration analysis of a slider-crank mechanism

Activities

- Read the material provided below, and as you read:

a. Calculate angles θ_3 and output displacement s of the slider block for given values of θ_2.
b. Calculate velocities ω_3 of the coupler link and $\dot{s}$ of the slider block for given values of θ_2.
c. Calculate accelerations α_3 of the coupler link and $\ddot{s}$ of the slider block for given values of θ_2.
d. Demonstrate the application of the complex numbers in obtaining analytical expressions for displacements, velocities, and accelerations of the coupler link and the output slider block.

• Check your answers to the competency items with your instructor.

Reading Material

Displacement analysis of a slider-crank mechanism

Let MA, BA, MX, and XB in Fig. 2 represent vectors $\mathbf{A}$, $\mathbf{B}$, $\mathbf{S}$, and $\mathbf{E}$, where

$$
\begin{aligned}
\mathbf{A} &= a\mathbf{U}_A = ae^{j\theta_2} \\
\mathbf{B} &= b\mathbf{U}_B = be^{j\theta_3} \\
\mathbf{S} &= s\mathbf{U}_S = se^{j\theta_1} \\
\mathbf{E} &= e\mathbf{U}_E = ee^{j\theta_4}
\end{aligned}
\tag{50}
$$

$\mathbf{U}_A$, $\mathbf{U}_B$, $\mathbf{U}_S$, and $\mathbf{U}_E$ are the unit vectors associated with $\mathbf{A}$, $\mathbf{B}$, $\mathbf{S}$, and $\mathbf{E}$. Also, $\theta_1 = 180°$ and $\theta_4 = 90°$ in Fig. 2.

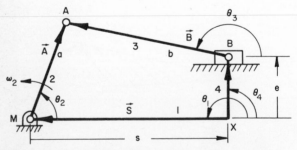

Fig. 2 Vector representation for a slider-crank mechanism.

Writing the vector equation for the vector polygon $MABX$, we get

$$
\mathbf{S} + \mathbf{A} - \mathbf{B} = \mathbf{E}
\tag{51}
$$

Rewriting Eq. (51) using the complex notations, we get

$$
ae^{j\theta_2} + se^{j\theta_1} - ee^{j\theta_4} - be^{j\theta_3} = 0
\tag{52}
$$

Writing the real part and the complex part of the Eq. (52) we get

$$a \cos \theta_2 + s \cos \theta_1 - e \cos \theta_4 - b \cos \theta_3 = 0 \tag{53}$$

$$a \sin \theta_2 + s \sin \theta_1 - e \sin \theta_4 - b \sin \theta_3 = 0 \tag{54}$$

Substituting $\theta_1 = 180°$ and $\theta_4 = 90°$, Eqs. (53) and (54) become

$$a \cos \theta_2 - s - b \cos \theta_3 = 0 \tag{55}$$

$$a \sin \theta_2 - e - b \sin \theta_3 = 0 \tag{56}$$

If a relationship between θ_2 and s is desired, then the angle θ_3 must be eliminated. Transferring the terms containing θ_3 on the right-hand side, and squaring both sides of Eq. (55) and (56), we get

$$(a \cos \theta_2 - s)^2 = (b \cos \theta_3)^2 \tag{57}$$

$$(a \sin \theta_2 - e)^2 = (b \sin \theta_3)^2 \tag{58}$$

Adding the above two equations and simplifying the resultant equation, we get

$$s^2 - 2 as \cos \theta_2 + a^2 + e^2 - b^2 - 2ae \sin \theta_2 = 0$$

or

$$Ls^2 + Ms + N = 0 \tag{59}$$

Where

$$L = 1$$
$$M = -2a \cos \theta_2 \tag{60}$$
$$N = a^2 + e^2 - b^2 - 2ae \sin \theta_2$$

The above quadratic equation has two values of s for a given value of θ_2. These are:

$$s_1 = \frac{-M + \sqrt{M^2 - 4LN}}{2L} \tag{61}$$

$$s_2 = \frac{-M - \sqrt{M^2 - 4LN}}{2L} \tag{62}$$

If the parameters a, b, e, and θ_2 of a slider-crank mechanism are known, then Eqs. (61) and (62) permit one to compute the output displacement of the slider block B.

Velocity analysis of a slider-crank mechanism

The input crank MA of the slider-crank mechanism shown in Fig. 2 is moving with an angular velocity ω_2 in a counterclockwise direction. Let MA, BA, MX, and XB in Fig. 2 represent vectors $\mathbf{A}$, $\mathbf{B}$, $\mathbf{S}$, and $\mathbf{E}$ where

$$\mathbf{A} = a\mathbf{U}_A = ae^{j\theta_2}$$
$$\mathbf{B} = b\mathbf{U}_B = be^{j\theta_3}$$
$$\mathbf{S} = s\mathbf{U}_S = se^{j\theta_1}$$

and

$$\mathbf{E} = e\mathbf{U}_E = ee^{j\theta_4}$$

writing the vector equation for the vector polygon $MABX$, we get

$$\mathbf{A} + \mathbf{S} = \mathbf{E} + \mathbf{B} \tag{63}$$

Rewriting Eq. (63) using the complex notations, we get

$$ae^{j\theta_2} + se^{j\theta_1} - ee^{j\theta_4} - be^{j\theta_3} = 0 \tag{64}$$

In order to obtain the angular velocity ω_3 of the coupler link AB and the linear velocity $\dot{s}$ of the slider block B, we must take the time derivation of Eq. (64). Since a, b, e, θ_1, and θ_4 are constants, the time derivative of Eq. (64) yields,

$$ja\omega_2 e^{j\theta_2} + \dot{s}e^{j\theta_1} = jb\omega_3 e^{j\theta_3} \tag{65}$$

Separating the real and the complex part of Eq. (65) we get,

$$a\omega_2 \sin\theta_2 - \dot{s}\cos\theta_1 = b\omega_3 \sin\theta_3 \tag{66}$$

$$a\omega_2 \cos\theta_2 + \dot{s}\sin\theta_1 = b\omega_3 \cos\theta_3 \tag{67}$$

Since θ_1 is always $180°$, Eq. (67) gives

$$\omega_3 = \frac{a\omega_2}{b}\frac{\cos\theta_2}{\cos\theta_3} \tag{68}$$

Substituting Eq. (68) in Eq. (66), we get

$$\dot{s} = a\omega_2 \frac{\sin(\theta_3 - \theta_2)}{\cos\theta_3} \tag{69}$$

The angular displacement θ_3 of the coupler link AB may be computed using Eq. (56). That is,

$$\theta_3 = \sin^{-1}\left(\frac{e - a\sin\theta_2}{-b}\right) \tag{70}$$

Equations (68) and (69) can be used to calculate the velocities ω_3 and $\dot{s}$ provided a, b, e, and θ_2 are known.

Acceleration analysis of a slider-crank mechanism

In order to obtain explicit relationship for α_3 and $\ddot{s}$, we must take time derivative of Eq. (65) which describes the vector velocity relationship of a slider-crank mechanism. Thus, we have

$$ja\alpha_2 e^{j\theta_2} + j^2 a\omega_2^2 e^{j\theta_2} + \ddot{s}e^{j\theta_1} = jb\alpha_3 e^{j\theta_3} + j^2 b\omega_3^2 e^{j\theta_3} \tag{71}$$

Substituting $e^{j\theta} = \cos\theta + j\sin\theta$ into Eq. (71) and separating it into its real and complex parts, we get

$$a\alpha_2 \sin\theta_2 + a\omega_2^2 \cos\theta_2 - \ddot{s} \cos\theta_1 = b\alpha_3 \sin\theta_3 + b\omega_3^2 \cos\theta_3 \tag{72}$$

$$a\alpha_2 \cos\theta_2 - a\omega_2^2 \sin\theta_2 = b\alpha_3 \cos\theta_3 - b\omega_3^2 \sin\theta_3 \tag{73}$$

From Eq. (73) we can calculate α_3.

$$\alpha_3 = \frac{a\alpha_2 \cos\theta_2 - a\omega_2^2 \sin\theta_2 + b\omega_3^2 \sin\theta_3}{b\cos\theta_3} \tag{74}$$

From Eq. (72) we can calculate $\ddot{s}$.

$$\ddot{s} = \frac{a\alpha_2 \sin\theta_2 + a\omega_2^2 \cos\theta_2 - b\alpha_3 \sin\theta_3 - b\omega_3^2 \cos\theta_3}{\cos\theta_1} \tag{75}$$

Computation of α_3 gives the magnitude of the angular acceleration of the coupler link and computation of $\ddot{s}$ gives the linear acceleration of the output sliding block of a slider-crank mechanism. The directions of these accelerations are obtained from Eq. (71).

Competency Items

- Define the vectors representing the links of the slider-crank mechanism shown in the figure below. Write the vector equation representing the mechanism.

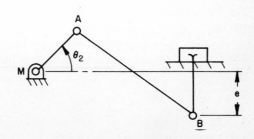

- Perform an acceleration analysis of the mechanism above if $MA = 1$ in., $AB = 2.5$ in., $e = .75$ in., $\theta_2 = 45°$, and MA has an angular velocity of 5 rad/sec; with an angular acceleration of 10 rad/sec^2 (both in a counterclockwise sense).
- Derive the relationships to calculate θ_6, ω_6, and α_6 of the six-link mechanism shown below.

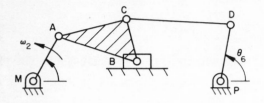

Objective 3

To perform displacement, velocity, and acceleration analysis of an inverted slider-crank mechanism

Activities

- Read the material provided below, and as you read:
  a. Calculate angle θ_4 and stroke b for given values of θ_2.
  b. Calculate velocities ω_4 and $\dot{b}$ for given values of θ_2.
  c. Calculate accelerations α_4 and $\ddot{b}$ for given values of θ_2.
  d. Calculate the coriolis acceleration.
  e. Demonstrate the application of the complex numbers in obtaining analytical expressions for displacements, velocities, and accelerations of the coupler and output links of an inverted slider-crank mechanism.
- Check your answers to the competency items with your instructor.

Reading Material

Displacement analysis of an inverted slider-crank mechanism

An inverted slider-crank mechanism is shown in Fig. 3. Links MA, BA, QB, and QM are represented by vectors $\mathbf{A}$, $\mathbf{B}$, $\mathbf{E}$, and $\mathbf{D}$. Where

$$\mathbf{A} = a e^{j\theta_2}$$
$$\mathbf{B} = b e^{j\theta_3}$$
$$\mathbf{E} = e e^{j\theta_4}$$

and

$$\mathbf{D} = d e^{j\theta_1}$$

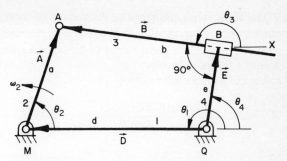

Fig. 3 Vector representation for an inverted slider-crank mechanism

Note that $\theta_1 = 180°$. Since BX and MQ are parallel, $\angle QBX = 180° - \theta_4$. Hence at point B $\theta_3 + (180° - \theta_4) + 90° = 360°$ or $\theta_3 = 90° + \theta_4$

The vector polygon $MABQ$ can be described using Eq. (76).

$$\mathbf{D} + \mathbf{A} - \mathbf{B} - \mathbf{E} = 0 \tag{76}$$

or

$$de^{j\theta_1} + ae^{j\theta_2} - be^{j\theta_3} - ee^{j\theta_4} = 0 \tag{77}$$

Writing the real and the complex part of Eq. (77), we get

$$d \cos\theta_1 + a \cos\theta_2 - b \cos\theta_3 - e \cos\theta_4 = 0 \tag{78}$$

$$d \sin\theta_1 + a \sin\theta_2 - b \sin\theta_3 - e \sin\theta_4 = 0 \tag{79}$$

Substituting the values for θ_1 and θ_3 and simplifying the above equations, we get

$$a \cos\theta_2 - d - e \cos\theta_4 + b \sin\theta_4 = 0 \tag{80}$$

$$a \sin\theta_2 - e \sin\theta_4 - b \cos\theta_4 = 0 \tag{81}$$

The variable coupler link b must be eliminated, if the relationship between θ_2 and θ_4 is desired. From Eq. (81), we get

$$b = \frac{a \sin\theta_2 - e \sin\theta_4}{\cos\theta_4} \tag{82}$$

Substituting Eq. (82) into Eq. (80), and simplifying the resultant equation, we get

$$(a \cos\theta_2 - d) \cos\theta_4 + a \sin\theta_2 \sin\theta_4 - e = 0 \tag{83}$$

The above relationship can be utilized for the synthesis of inverted slider-crank mechanisms. However, Eq. (83) is not in a convenient form to compute output displacement θ_4 for given values of the input displacement θ_2. It is necessary to simplify this relationship.

Let us substitute for

$$\sin\theta = \frac{2\tan(\theta/2)}{1 + \tan^2(\theta/2)}$$

and

$$\cos\theta = \frac{1 - \tan^2(\theta/2)}{1 + \tan^2(\theta/2)}$$

in Eq. (83) and simplify to get

$$K_1 \tan^2 \frac{\theta_4}{2} + K_2 \tan \frac{\theta_4}{2} + K_3 = 0 \tag{84}$$

Where

$$K_1 = d - e - a\cos\theta_2$$

$$K_2 = 2a\sin\theta_2 \tag{85}$$

and $K_3 = a\cos\theta_2 - d - e$

The two roots of θ_4 in Eq. (84) are

$$\left[\theta_4\right]_{1,2} = 2\tan^{-1}\left(\frac{-K_2 \pm \sqrt{K_2^2 - 4K_1K_3}}{2K_1}\right) \tag{86}$$

Equation (86) shows that corresponding to a given value of θ_2, θ_4 has two values. That is, corresponding to one position of input link MA, output link QB can occupy two positions.

Velocity analysis of inverted slider-crank mechanism

Input crank MA of the inverted slider-crank mechanism shown in Fig. 3 is rotating with an angular velocity ω_2 in a counterclockwise direction. Let links MA, AB, BQ, and QM be represented by vectors $\mathbf{A}$, $\mathbf{B}$, $\mathbf{E}$, and $\mathbf{D}$ where,

$$\mathbf{A} = ae^{j\theta_2}$$

$$\mathbf{B} = be^{j\theta_3}$$

$$\mathbf{E} = ee^{j\theta_4}$$

$$\mathbf{D} = de^{j\theta_1}$$

Note that $\theta_1 = 180°$ and $\theta_3 = 90° + \theta_4$. The vector polygon $MABQ$ can be described using the vector polygon equation.

$$D + A - B - E = 0 \tag{87}$$

or

$$de^{j\theta_1} + ae^{j\theta_2} = be^{j\theta_3} + ee^{j\theta_4} \tag{88}$$

Angular velocity ω_4 and relative linear velocity $\dot{b}$ of slider block B can be obtained by taking the time derivative of Eq. (88). Since d, a, e, and θ_1 are constants, the time derivative of Eq. (88) yields,

$$ja\omega_2 e^{j\theta_2} = \dot{b}e^{j\theta_3} + jb\omega_3 e^{j\theta_3} + je\omega_4 e^{j\theta_4} \tag{89}$$

Separating the real and the complex part and substituting $\theta_3 = 90° + \theta_4$ and $\omega_3 = \omega_4$, we get

$$a\omega_2 \sin\theta_2 = \dot{b} \sin\theta_4 + \omega_4(e \sin\theta_4 + b \cos\theta_4) \tag{90}$$

$$a\omega_2 \cos\theta_2 = \dot{b} \cos\theta_4 + \omega_4(e \cos\theta_4 - b \sin\theta_4) \tag{91}$$

Simultaneous solutions of Eqs. (90) and (91) yields,

$$\omega_4 = \frac{a\omega_2}{b} \sin(\theta_2 - \theta_4) \tag{92}$$

and

$$\dot{b} = \frac{a\omega_2}{b} [b \cos(\theta_4 - \theta_2) - e \sin(\theta_4 - \theta_2)] \tag{93}$$

The linear displacement b and the angular displacement θ_4 can be computed using Eqs. (82) and (86). The angular velocity ω_4, and the relative linear velocity $\dot{b}$ of the slider can be computed using Eqs. (92) and (93).

Acceleration analysis of an inverted slider-crank mechanism

The time derivative of Eq. (89) leading to the determination of the accelerations α_4 and $\ddot{b}$ yields

$$ja\alpha_2 e^{j\theta_2} + j^2 a\omega_2^2 e^{j\theta_2} = \ddot{b}e^{j\theta_3} + j\dot{b}\omega_3 e^{j\theta_3} + j\dot{b}\omega_3 e^{j\theta_3} + jb\alpha_3 e^{j\theta_3}$$
$$+ j^2 b\omega_3^2 e^{j\theta_3} + j^2 e\alpha_4 e^{j\theta_4} + je\omega_4^2 e^{j\theta_4} \tag{94}$$

Substituting for

$$e^{j\theta} = \cos\theta + j \sin\theta, \quad \theta_3 = \theta_4 + 90°,$$

$$\omega_3 = \omega_4 \text{ and } \alpha_3 = \alpha_4,$$

in Eq. (94) and separating its real and complex parts, we get

$$\sin\theta_4\ddot{b} + (b\cos\theta_4 + e\sin\theta_4)\alpha_4 = a\omega_2^2\cos\theta_2 + a\alpha_2\sin\theta_2$$
$$- 2\dot{b}\omega_4\cos\theta_4 + b\omega_4^2\sin\theta_4 - e\omega_4^2\cos\theta_4 \tag{95}$$

$$\cos\theta_4\ddot{b} + (-b\sin\theta_4 + c\cos\theta_4)\alpha_4 = -a\omega_2^2\sin\theta_2 + a\alpha_2\cos\theta_2$$
$$+ 2\dot{b}\omega_4\sin\theta_4 + b\omega_4^2\cos\theta_4 + e\omega_4^2\sin\theta_4 \tag{96}$$

The simultaneous solution of Eqs. (95) and (96) will give

$$\alpha_4 = \frac{AF - CD}{AE - BD} \tag{97}$$

and

$$\ddot{b} = \frac{CE - BF}{AE - BD} \tag{98}$$

where

$A = \sin\theta_4$

$B = b\cos\theta_4 + e\sin\theta_4$

$C = a\omega_2^2\cos\theta_2 + a\alpha_2\sin\theta_2 - 2\dot{b}\omega_4\cos\theta_4$
$\quad + b\omega_4^2\sin\theta_4 - e\omega_4^2\cos\theta_4$

$D = \cos\theta_4$

$E = -b\sin\theta_4 + e\cos\theta_4$

$F = -a\omega_2^2\sin\theta_2 + a\alpha_2\cos\theta_2 + 2\dot{b}\omega_4\sin\theta_4$
$\quad + b\omega_4^2\cos\theta_4 + e\omega_4^2\sin\theta_4$

Equations (97) and (98) will permit us to calculate the angular acceleration α_4 of the output link and the relative acceleration $\ddot{b}$ of the slider. All the other quantities such as linkage parameters, displacement angles and velocities are known from the previous computations of displacements and velocities. The student should analyze Eqs. (95) and (96) and identify the components of the coriolis acceleration.

Competency Items

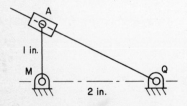

- Define the vectors representing the links of the inverted slider-crank mechanism shown in the figure below. Write the vector equation representing the mechanism.
- Derive an explicit mathematical relationship between the position of input link MA and the position of link AQ.

- Derive the relationships to calculate θ_6, ω_6, and α_6 of the six-link mechanism shown below.

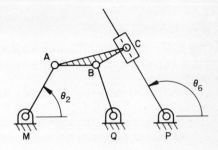

Performance Test #1

Using the vectors shown in the figure to represent a slider-crank mechanism, write the vector equation representing these vectors and derive an expression for the displacement s of the slider in terms of a, b, e, and θ_2. Also derive an expression for $\dot{s}$ in terms of a, b, e, θ_2, and ω_2 (assume ω_2 is constant).

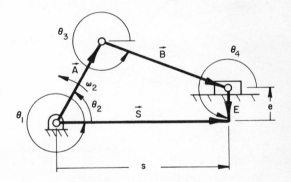

Performance Test #2

On the figure given below, of the four-link mechanism show the angles θ_1, θ_2, θ_3, and θ_4 which correspond to the vectors R_1, R_2, R_3, and R_4 respectively, where $R_i = r_i e^{j\theta_i}$; $i = 1, 2, 3, 4$.

If $r_1 = 0.6$ in., $r_2 = 1.2$ in., $r_3 = 1.8$ in., $r_4 = 1.6$ in. and if vector R_2 is rotating clockwise at 40 rad/sec, derive the relationships for output angle θ_4 in terms of θ_2 and output velocity ω_4 in terms of θ_2, θ_3, and θ_4.

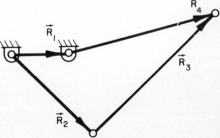

Performance Test #3

The basic mechanism of power loaders used on tractors is shown below in (a); (b) is the equivalent kinematic schematic. The mechanism is an inverted slider crank.

On (b), show angles θ_1, θ_2, and θ_3 that correspond to vectors R_1, R_2, and R_3, where $R_i = r_i e^{j\theta_i}$; $i = 1, 2, 3$. Vector R_1 is vertical.

In this application, the known quantities are $r_1, r_2, r_3, \theta_1, \dot{r}_3$, and $\ddot{r}_3$. Angles θ_2 and θ_3 can be calculated, but you are *not* required to derive the expressions.

1. Derive an expression for the angular velocity ω_2 in terms of $r_1, r_2, r_3, \dot{r}_3, \theta_2$ and θ_3.
2. State how you find ω_3, but do *not* write the expression.
3. Derive an expression for the angular acceleration α_2 in terms of r_1, r_2, r_3, $\dot{r}_3, \theta_2, \theta_3, \omega_2$, and ω_3.

You should use intermediate parameters such as K_1, K_2 etc., to simplify your expressions, and your results may be left in terms of these parameters.

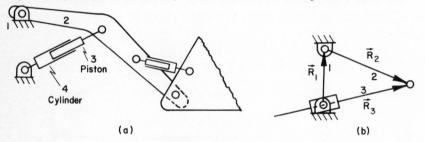

(a)　　　　　　　　　　　　　　　　(b)

Supplementary References

3(6.8); 5(9.3-9.4); 7(4.16, 5.8 [parts])

Unit V. Fundamentals of Uniform Rotary Transmission

Objectives

1. To identify the different ways of synthesizing uniform rotary transmission and to develop the criteria of rotary transmission by direct rolling and sliding contact
2. To design uniform rotary transmission using involute and cycloidal curves
3. To calculate speed ratios for external and internal rolling cylinders and cones

Objective 1

To identify the different ways of synthesizing uniform rotary transmission and to develop the criteria of rotary transmission by direct rolling and sliding contact

Activities

- Read the material provided below, and as you read:
  a. Describe different ways of obtaining uniform rotary motion.
  b. State the advantages and disadvantages of all the alternate design possibilities described in item a.
  c. Define the motion by rolling and sliding contact.
  d. State the criteria that define rolling and sliding contact.
  e. Calculate velocity ratio for uniform rotary transmission by rolling and sliding contact.
- Check your answers to the competency items with your instructor.

Reading Material

Part I: Approaches to design uniform rotary transmission

The uniform rotary motions are normally obtained through mechanical devices that receive and transmit uniform rotary motion. The input and the output members of this device are moving with constant angular velocities. There are three different design situations in which uniform rotary input-output motions are required:

1. Axes of rotations of the input and output members are coaxial.
2. Axes of rotations of the input and output members are parallel or intersecting.
3. Axes of rotations of the input and output members are skew to one another.

These design situations are usually solved using one or a combination of the following mechanical devices:

1. belt-pulley drive
2. chain-sprocket drive
3. hydraulic actuating system
4. gear drive

Belts and chains are more commonly used because of their simplicity in design and low cost in obtaining uniform rotary output motion. The belt-pulley or chain-sprocket drives are selected under the following design requirements:

1. The center distances between the input and output shafts are long.
2. Simpler, low-cost transmission is required.
3. Multiple shafts must be coupled.
4. The center distances of input and output shafts are variable.
5. The coupling of the input and output shaft can endure soft shock.
6. The lubricating facilities are inadequate.

Belt-pulley drives have built-in slippage that works as an advantage, especially when large torque variations must be absorbed. However, this slippage in the belt-pulley drives prevent them from further consideration when the motions of the input and output shafts must be synchronized. The chain-sprocket drives are preferred when synchronization and transmission of large torque are required. However, the chain-sprocket drives require lubrication and, in general, provide small amounts of torque fluctuation which in turn contributes to excessive noise.

Hydraulically actuated systems are often considered in transmitting uniform motion, although these systems prove of practical value when nonuniform motion must be produced.

Gears and gear trains are widely used in producing uniform rotary motion for all the three different types of design situations described above. The widespread use of the gears has forced the gear manufacturers to adopt universal standard notations. This standardization has relieved the designers from the extra burden of choice, and provided opportunities for their wider use.

Part II: Development of criteria of rotary transmission by direct rolling and sliding contact

Two rigid bodies B_1 and B_2 are capable of rotating about fixed centers M and Q. These two bodies have a common point P which is the point of contact. Point P_1 lies in body B_1 and point P_2 lies in body B_2. The rotary motion from body B_1 to B_2 is transmitted by direct contact. The relative motion at the point of contact is either a sliding motion or a rolling motion.

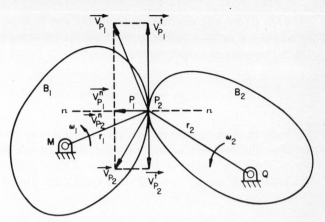

Fig. 1 Rotary transmission under direct contact of two bodies B_1 and B_2.

Let bodies B_1 and B_2 as shown in Fig. 1 rotate with angular velocities ω_1 and ω_2 about fixed centers M and Q. Then, the velocity at the point of contact P_1 lying in body B_1 and P_2 lying in body B_2 are

$$V_{P_1} = r_1\omega_1$$

$$V_{P_2} = r_2\omega_2$$

where $MP_1 = r_1$ and $QP_2 = r_2$. The direction and the sense of V_{P_1} and V_{P_2} are as shown in Fig. 1. Let us resolve $\mathbf{V}_{P_1}$ and $\mathbf{V}_{P_2}$ along the tangent and the normal at point of contact. $\mathbf{V}_{P_1}^n$ and $\mathbf{V}_{P_2}^n$ are normal components of $\mathbf{V}_{P_1}$ and $\mathbf{V}_{P_2}$. $\mathbf{V}_{P_1}^t$ and $\mathbf{V}_{P_2}^t$ are tangential components of $\mathbf{V}_{P_1}$ and $\mathbf{V}_{P_2}$. Since P_1 and P_2 are coincident points,

$$\mathbf{V}_{P_1}^n = \mathbf{V}_{P_2}^n$$

But, since $\mathbf{V}_{P_1}^t \neq \mathbf{V}_{P_2}^t$, relative motion exists between P_1 and P_2. The relative velocity at the point of contact is

$$\mathbf{V}_{P_2/P_1} = \mathbf{V}_{P_2}^t - \mathbf{V}_{P_1}^t$$

Because of the relative velocity, each point on body B_1 will come in contact with more than one point on body B_2. This motion constitutes motion due to sliding contact.

The two bodies transmitting motion by direct contact execute the motion by rolling contact when relative velocity $\mathbf{V}_{P_2/P_1}$ is zero. Since the point of contact has equal velocity in both bodies, Kennedy's theorem of instant center for three bodies moving relative to one another states that the point of contact P, which now becomes an instant center, must lie on center line MQ since M and Q are instant centers of bodies B_1 and B_2.

If body B_1 were to transmit uniform motion to body B_2, then the speed ratio defined as the ratio of output to input angular velocities must remain constant. That is, for uniform output motion

$$SR = \text{speed ratio} = \frac{\omega_2}{\omega_1} = \text{constant}$$

If body B_1 were to transmit by direct sliding contact uniform motion to body B_2, then the common normal to these bodies at the point of contact must intersect the line joining centers M and Q at fixed point F as shown in Fig. 3. The ratio of MF and QF, is equal to the speed ratio. That is

$$SR = \frac{\omega_2}{\omega_1} = \frac{MF}{QF} = \text{constant}$$

For the rolling contact, since the point of contact P lies on MQ and the points P and F are coincident points as shown in Fig. 2,

$$V_{P_1} = V_{P_2}$$

or

$$MF \cdot \omega_1 = QF \cdot \omega_2$$

The speed ratio is given by

$$SR = \frac{\omega_2}{\omega_1} = \frac{MF}{QF}$$

From the above relationship, we can say that for obtaining constant speed ratio in rotary transmission due to rolling contact, the radii of the rolling bodies must be inversely proportional to the angular velocities of the rolling bodies. The reader can easily verify the truth of the above statement by considering the cylindrical bodies.

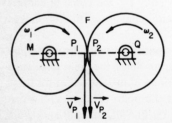

Fig. 2 Rotary transmission due to rolling contact.

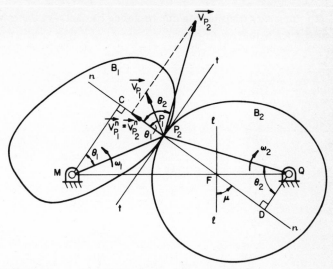

Fig. 3 Rotary transmission due to sliding contact.

Let us now examine Fig. 3, the case of the motion transmission due to sliding contact.

The two rotating bodies make contact at common point P. Draw tangent t-t and normal n-n to the tangent t-t. Draw MC and QD normal to line n-n. Let angle $CMP_1 = \theta_1$ and angle $P_2QD = \theta_2$. The velocity of point P_1 at the point of contact is given by

$$V_{P_1} = MP_1 \cdot \omega_1$$

The velocity of point P_2 at the point of contact is given by

$$V_{P_2} = QP_2 \cdot \omega_2$$

These velocity vectors V_{P_1} and V_{P_2} are as shown in Fig. 3. Since body B_2 is moving due to sliding contact at P, the components of V_{P_1} and V_{P_2} along normal n-n must be equal. That is

$$V_{P_1}^n = V_{P_2}^n$$

Since the direction V_{P_1} is along the line perpendicular to MP_1, and since angle $CP_1M = 90° - \theta_1$, V_{P_1} is inclined at angle θ_1 with normal n-n. Similarly, V_{P_2} is inclined at an angle θ_2 with normal n-n. Therefore,

$$V_{P_1}^n = V_{P_1} \cos\theta_1 = V_{P_2} \cos\theta_2 = V_{P_2}^n$$

Let us calculate the speed ratio and examine the condition that will provide uniform rotary transmission due to sliding contact. Let the common normal n-n intersect MQ in F.

$$\omega_1 = \frac{V_{P_1}}{MP_1}$$

$$\omega_2 = \frac{V_{P_2}}{QP_2}$$

$$SR = \frac{\omega_2}{\omega_1} = \frac{V_{P_2}}{V_{P_1}} \cdot \frac{MP_1}{QP_2}$$

but

$$MP_1 = \frac{MC}{\cos\theta_1} \quad \text{and} \quad QP_2 = \frac{QD}{\cos\theta_2}$$

Therefore,

$$SR = \frac{\omega_2}{\omega_1} = \frac{V_{P_2} \cdot MC \cdot \cos\theta_2}{V_{P_1} \cdot QD \cdot \cos\theta_1}$$

but

$$V_{P_1}^n = V_{P_1}\cos\theta_1 = V_{P_2}^n = V_{P_2}\cos\theta_2$$

Therefore,

$$SR = \frac{\omega_2}{\omega_1} = \frac{MC}{QD}$$

Since triangles MCF and QDF are similar,

$$\frac{MC}{QD} = \frac{MF}{QF} = \frac{CF}{FD}$$

Hence,

$$SR = \frac{\omega_2}{\omega_1} = \frac{MF}{QF} \tag{1}$$

If the speed ratio were to remain constant during the rotary transmission due to sliding, then the ratio MF/QF must remain constant. Point F is called the *pitch point* and angle μ between line l-l, which is perpendicular to MQ, and line n-n is called the *pressure angle*.

Since MQ is of fixed length, it can be divided only one way so that the ratio MF/FQ becomes constant. Thus, pitch point F is uniquely located on MQ.

Summarizing briefly, we can state that if we were to obtain uniform rotary transmission due to sliding contact, then the common normal at the point of contact between two sliding bodies must intersect center line MQ at fixed point F which is called the pitch point. Pressure angle μ need not remain of fixed value during the motion transmission. For rolling contact, the pitch point becomes the instant center, and has equal velocity in each of the two moving bodies.

Competency Items

- Name four different ways of transmitting uniform rotary motion between shafts with parallel axes.
- State several possible disadvantages involved with chain-sprocket drives.
- State if the slippage of belt and pulley drives is considered an advantage or a disadvantage.
- Identify three different types of design situations in which gears or gear trains might be utilized to transmit uniform rotary motion.
- Define rolling contact and state its two properties.
- Define sliding contact and state its two properties.
- Define pressure angle when two bodies are transmitting motion via sliding contact and write criteria for good quality of motion transmission.

Objective 2

To design uniform rotary transmission using involute and cycloidal curves

Activities

- Read the material provided below, and as you read:
  - a. Lay out the two mated involute curves and the two mated cycloidal curves.
  - b. Define conditions that will make the mated involute curves and the mated cycloidal curves generate uniform rotary transmission.
  - c. State advantages and disadvantages of mated involute curves and mated cycloidal curves.
- Check your answers to the competency items with your instructor.

Reading Material

Part I: Involute curves

Figure 4 shows two rolling cylinders having mated involute profiles. The involute surface on the cylinder is obtained by unwinding from the surface of the cylinder strip A_4D so that A_1D_1, A_2D_2, A_3D_3, and A_4D_4 are tangents to the circumference of the cylinder 1. The locus of point D is the involute curve. The involute curve on cylinder 2 is obtained in a similar manner.

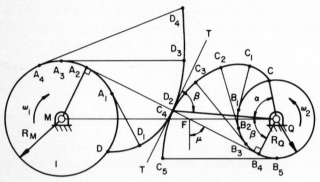

Fig. 4   Two rolling cylinders with mated involute profiles.

The involute surfaces on the two cylinders are mated curves. These curves are obtained so that the amount of unwinding on cylinder 1 equals the amount of winding on cylinder 2. Because of the mated involute surfaces, as cylinder 1 moves in a clockwise rotation, it transmits uniform rotary motion by sliding contact to cylinder 2 which, in turn, moves in a counterclockwise direction. The common points of contact between the two cylinders are D_2 and C_4. D_2 lies on the involute of cylinder 1, and C_4 lies in the involute of cylinder 2. Let

$$\angle C_4 QC = \alpha$$

and

$$\angle C_4 QB_4 = \beta$$

Then, the length of the circumference

$$CB_4 = \rho = R_Q(\alpha + \beta)$$

or

$$\frac{\rho}{R_Q} = \alpha + \beta$$

$$CB_4 = B_4 C_4 \qquad \frac{B_4 C_4}{B_4 Q} = \tan\beta \quad \text{and} \quad B_4 Q = R_Q$$

Therefore,

$$\frac{\rho}{R_Q} = \tan\beta = \alpha + \beta$$

or

$$\alpha = (\tan\beta) - \beta \qquad\qquad\qquad (2)$$

Equation (2) describes the involute curve.

There are two properties of the mated involute surfaces.

1. The common normal at the point of contact of two involute surfaces intersects the center line in fixed point F. This fixed point divides center distance MQ so that

$$\frac{R_M}{R_Q} = \frac{MF}{QF} = \frac{\omega_2}{\omega_1}$$

Since the location of pitch point F is fixed, the speed ratio between the cylinders is a constant.

2. The two mated involute surfaces will continue to transmit uniform motion even when center distance MQ is altered.

Part II: Cycloidal curves

Like an involute curve, a cycloidal curve is also capable of providing uniform rotary transmission for two bodies in direct contact. A cycloidal curve is a curve traced by a point on a circle which is rolling on a fixed circle. Figure 5 shows an

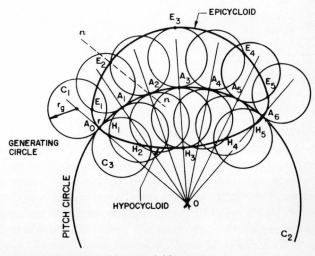

Fig. 5 Epicycloid and hypocycloid curves.

epicycloid curve and a hypocycloid curve obtained by rolling a generating circle C_1 on pitch circle C_2. The epicycloid curve is obtained by following the path of trace point E on generating circle C_1 which is rolling on the outside of pitch circle C_2. The hypocycloid curve is obtained by following the path of the trace point H on generating circle C_3 which is rolling on the inside of pitch circle C_2.

Since arc A_1A_0 on the pitch circle is equal to arc A_1E_1 on the epicycloid curve and equal to arc A_1H_1 on the hypocycloid curve, the two curves and the path of points $A_0, A_1, \ldots, A_6$, are functions of radius r_g of the generating circle. Also, the normal to the epicycloid and hypocycloid at points E_i and H_i passes through points of contact A_i on pitch circle C_2.

Two rigid bodies B_1 and B_2, as shown in Fig. 6, will transmit uniform rotary motion by direct contact if the epicycloid on B_2 is mated with its conjugate hypocycloid on B_1. The generating circle generating epicycloid curve on B_1 and hypocycloid curve on B_2 is the same. Since the common normal to the two mated curves passes through pitch point F which lies on center line MQ; the two bodies will transmit uniform rotary motion as long as center distance MQ does not vary in length, that is, the pitch circles are tangent to one another.

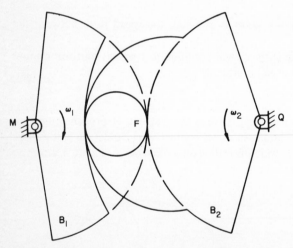

Fig. 6 Rolling cylinders with mated epicycloid and hypocycloid curves.

Part III: Comparison of involute and cycloidal curves

In practice, involute curves have proved to be a better choice than cycloidal curves. The velocity ratio does not change even when the distance changes between the centers of rotations of two rigid bodies with mated involute curves. The velocity ratio, however, is quite sensitive to change when two bodies carry mated cycloidal curves. Because cycloidal profile is relatively easy to fabricate early design of gears had teeth with cycloidal profile. Modern practice virtually excludes the use of cycloidal curves.

We observe from Figs. 4 and 5, that two rigid bodies that have either mated cycloidal or involute curve are not capable of executing full rotation.

Competency Items
- Draw an involute curve on the base circle with radius $r_b = 1$ in. for the region AB.

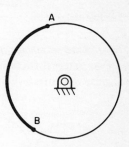

- Draw a pair of mated involute curves on two base circles with centers M and Q.

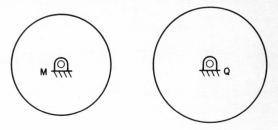

- Draw an epicycloid and its conjugate hypocycloid for the two pitch circles with centers M and Q.

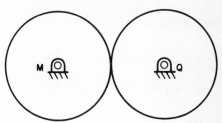

Objective 3

To calculate speed ratios for external and internal rolling cylinders and cones

Activities
- Read the material provided below, and as you read:
  a. Calculate the speed ratio for external rolling cylinders.
  b. Calculate the speed ratio for internal rolling cylinders.

c. Calculate the speed ratio for external rolling cones.

d. Calculate the speed ratio for internal rolling cones.

• Check your answers to the competency items with your instructor.

Reading Material

Part I: External rolling cylinders

Two external friction wheels W_1 and W_2 with radii r_1 and r_2 are rotating with angular velocities ω_1 and ω_2 as shown in Fig. 7.

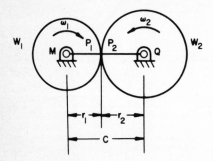

Fig. 7 External friction wheels.

Since the point of contact P lies on center line MQ, the transmission of motion from friction wheel W_1 to friction wheel W_2 is due to pure rolling contact. Since

$$V_{P_1} = V_{P_2}$$

then

$$r_1\omega_1 = r_2\omega_2$$

Hence, the speed ratio is given by

$$SR = \frac{\omega_2}{\omega_1} = \frac{r_1}{r_2}$$

Therefore, the speed ratio in two external friction wheels is inversely proportional to their radii.

Center distance C is given by $C = r_1 + r_2$.

If the speed ratio, the radius of driving wheel r_1, or center distance C are given, then the radius of the driven wheel can be determined. Since

$$\frac{\omega_2}{\omega_1} = \frac{r_1}{r_2}$$

then

$$\frac{\omega_2}{\omega_1} + 1 = \frac{r_1}{r_2} + 1$$

or

$$\left(\frac{\omega_2}{\omega_1} + 1\right) = \left(\frac{r_1 + r_2}{r_2}\right)$$

or

$$r_2 = \frac{r_1 + r_2}{1 + (\omega_2/\omega_1)}$$

or

$$r_2 = \frac{C}{1 + SR}$$

Note that in external friction wheel drive the two wheels are rotating in opposite directions.

Part II: *Internal rolling cylinders*

Two internal friction wheels W_1 and W_2 with radii r_1 and r_2 are rotating with the angular velocities ω_1 and ω_2 as shown in Fig. 8. Center distance C is given by $C = r_2 - r_1$.

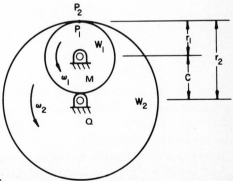

Fig. 8 Internal friction wheels.

Since point of contact P lies on center line MQ,

$$V_{P_1} = V_{P_2}$$

Therefore,

$$r_1\omega_1 = r_2\omega_2$$

The speed ratio is given by

$$SR = \frac{\omega_2}{\omega_1} = \frac{r_1}{r_2}$$

Note that in internal friction wheel drive, the two wheels are rotating in the same direction.

If the speed ratio, the radius of driving wheel r_1, or center distance C are given, then the radius of the driven wheel can be determined for the internal friction wheel drive. Since

$$\frac{\omega_2}{\omega_1} = \frac{r_1}{r_2}$$

then

$$1 - \frac{\omega_2}{\omega_1} = 1 - \frac{r_1}{r_2}$$

or

$$\left(1 - \frac{\omega_2}{\omega_1}\right) = \frac{r_2 - r_1}{r_2}$$

or

$$r_2 = \frac{C}{1 - SR}$$

Note that in the friction drive wheels, it is possible for the two wheels to permit slippage when an overload exists.

Part III: External rolling cones

Two external friction drive rolling cones moving with angular velocities ω_1 and ω_2 are shown in Fig. 9. The axes of the cones intersect at point 0.

The cones are used to transmit motion between the shafts whose axes intersect at an angle. The two cones have a line of contact rather than a point of contact. This line of contact intersects the axes of the cone at point 0. Because of this, the motion is transmitted by rolling contact. The speed ratio is given by

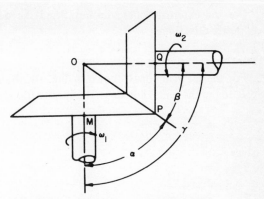

Fig. 9 External friction drive rolling cones.

$$SR = \frac{\omega_2}{\omega_1} = \frac{MP}{QP}$$

where MP and QP are the radii of the cones at the large ends.
   Note

$$\sin\alpha = \frac{MP}{OP}$$

$$\sin\beta = \frac{QP}{OP}$$

where α and β are the angles intercepted by center lines OM and OQ with line of contact OP as shown in Fig. 9. Hence, the speed ratio can be written as

$$SR = \frac{\omega_2}{\omega_1} = \frac{\sin\alpha}{\sin\beta}$$

If $\alpha + \beta = \gamma$, then $\alpha = \gamma - \beta$, and

$$SR = \frac{\omega_2}{\omega_1} = \frac{\sin(\gamma - \beta)}{\sin\beta} = \frac{\sin\gamma\,\cos\beta - \sin\beta\,\cos\gamma}{\sin\beta}$$

or
   Dividing the numerator and the denominator by $\cos\beta$, and solving for $\tan\beta$, we get

$$\tan\beta = \frac{\sin\gamma}{(\omega_2/\omega_1) + \cos\gamma}$$

For the external cones, the input and the output shafts are rotating in the opposite directions.

Part IV: Internal rolling cones

For the internal cones, shown in Fig. 10, the input and the output shafts are rotating in the same direction. The speed ratio is given by

$$SR = \frac{\omega_2}{\omega_1} = \frac{MP}{QP}$$

Since

$$\sin\alpha = \frac{MP}{OP}$$

$$\sin\beta = \frac{QP}{OP}$$

the speed ratio becomes

$$SR = \frac{\omega_2}{\omega_1} = \frac{\sin\alpha}{\sin\beta}$$

Since

$$\beta - \alpha = \gamma$$

then

$$\alpha = \beta - \gamma$$

Hence

$$SR = \frac{\omega_2}{\omega_1} = \frac{\sin(\beta - \gamma)}{\sin\beta}$$

Solving for β, we get

$$\tan\beta = \frac{-\sin\gamma}{(\omega_2/\omega_1) - \cos\gamma}$$

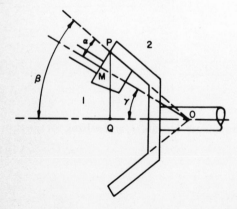

Fig. 10 Internal friction drive rolling cones.

Competency Items

- A friction drive utilizing external cylinders with axes parallel and 10 in. apart is to have a speed ratio of ¼. Calculate the sizes of the driving and driven cylinders.
- Derive a relationship to express the radius of the driver of two external cylinders in terms of the speed ratio and the center distance.
- State if an internal cylinder drive can have a speed ratio of 1 or a speed ratio of 2. Show how you might arrange an internal cylinder drive to achieve these speed ratios.
- If a set of external rolling cones is to have a speed ratio of 3.5 and the angle between the shafts is 60°, calculate the values of β and α.
- State if a set of internal rolling cones can have angle $\beta = 90°$ and $\gamma = 90°$. Sketch these arrangements if $\gamma = 90°$. State if the cones would still be internal.
- Calculate the speed ratio of a set of internal rolling cones with $\alpha = 10°$ and $\gamma = 30°$.

Performance Test #1

1. Derive a relationship to express the radius of the driver of two external cylinders in terms of the speed ratio and the center distance.
2. Referring to Fig. 3 of Unit V, calculate ratio V_{P_2}/V_{P_1}. Measure any lengths directly from the figure.

Performance Test #2

1. Calculate angular velocity ω_3 of body 3 in the position shown if body 2 is rotating clockwise at 10 rad/sec. State whether the bodies are moving due to sliding contact or rolling contact. State reasons for your answer.

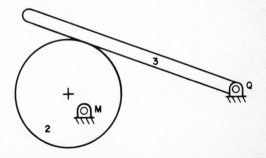

2. The center lines of two shafts intersect at an angle of 60° as shown. The shafts must rotate in the directions shown and the speed ratio of

$\omega_B/\omega_A = 0.5$ must be obtained. Design a pair of rolling cones that will do the job; i.e., calculate half-cone angles α_A and α_B. Show the cones (drawn to scale) on the figure.

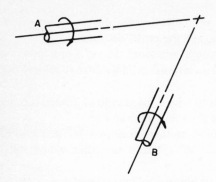

Performance Test #3

1. Calculate the angular velocity of body 2 if body 3 is rotating counter-clockwise at 60 rpm. Write your answer in rpm. State whether the bodies are moving due to rolling or sliding contact. State if body 2 transmits uniform motion to body 3. Write your reasons.
2. Derive a relationship to express the radius of the driver to two internal rolling cylinders in terms of the speed ratio and the center distance.
3. If a set of external rolling cones is to have a speed ratio of 1.75 and the angle between the shafts is $120°$, calculate the values of β and α.

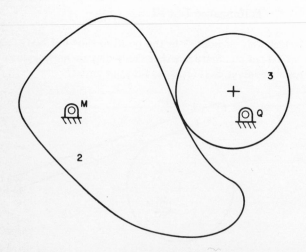

Supplementary References

1 (4.1–4.3, 5.4); **3** (2.16–2.17, 3.5, 9.7); **5** (2.10–2.15, 11.1–11.4, 12.2, 12.4–12.6); **7** (5.6, 8.2–8.3); **8** (5.3, 10.1–10.3)

Unit VI. Gear Tooth Technology

Objectives

1. To define the terminology of spur, helical, bevel, and worm gears
2. To derive relationships to calculate tooth thickness, tooth profile, and contact ratio
3. To calculate the effects of diametral pitch, pressure angle, and speed ratio on interference
4. To name AGMA (American Gear Manufacturers' Association) standards for spur, helical, bevel, and worm gears

Objective 1

To define the terminology of spur, helical, bevel, and worm gears

Activities

- Read the material provided below, and as you read:
  a. Draw a pair of spur gears in action.
  b. Draw a pair of helical gears in action.
  c. Draw worm gears.
  d. Identify the geometry of these gears.
- Check your answers to the competency items with your instructor.

Reading Material

The rolling cylinders are designed when large torque transmission is not of critical importance. However, since the slippage would occur while transmitting large torques, the rolling cylinders are provided with teeth to prevent this slippage. The gear teeth are placed outward along the radius of the rolling cylinders. The profile on the gear teeth is either involute curve or cycloidal curve in order to transmit uniform rotary motion which could take place either due to sliding or rolling contact.

The gear tooth terminology is illustrated in Fig. 1. Draw a sketch of Fig. 1 and show:

Base circle—the circle from which the involute tooth profile begins.

Pitch circle—the circle that corresponds to the circle of equivalent rolling cylinders.

Root circle—the circle with minimum radius of the gear tooth.

Addendum radius—the largest radius of the gear.

Pitch point—the point where the two pitch circles are tangent to one another.

Addendum—measures radially outward the length of the tooth from the pitch point.

Dedendum—measures radially inward the length of the tooth from the pitch circle to the root circle.

Clearance—measures along the center line the distance between addendum radius and root circle radius of the mated gear.

Working depth—the length of the tooth. This length is the sum of addendum and dedendum minus the clearance.

Tooth face—the portion of the tooth that has involute or cycloidal profile.

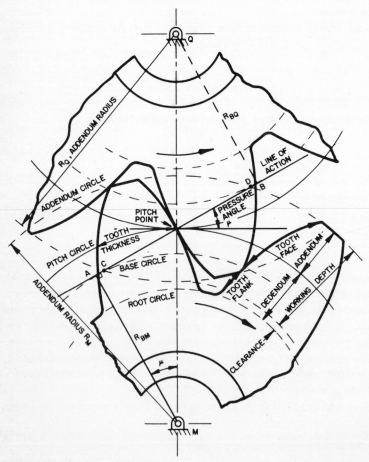

Fig. 1 Gear tooth geometry.

Tooth thickness—measured along the arc of the pitch circle and equal to the length of the arc between the two pitch points.

Tooth flank—the portion of the tooth that extends below the base circle.

Line of action—tangent to the base circle. It passes through the pitch point and is normal to the involute tooth profile at the point of contact.

Pressure angle (μ)—the angle subtended by the line of action and the line normal to the center line of the two mated gears.

Circular pitch (P_c)—the length of the arc along the pitch circle measuring one tooth and one blank space. If D denotes the diameter of the pitch circle, and N denotes the number of teeth of a gear, then the circular pitch P_c is calculated in inches from

$$P_c = \frac{\pi D}{N}$$

Base pitch (P_B)—the length of the arc along the base circle measuring one tooth and one blank space. If R_B denotes the base circle radius and R_p denotes the pitch circle radius, then from Fig. 1,

$$R_B = R_p \cos\mu$$

Hence

$$P_B = \frac{2\pi R_B}{N} = \frac{2\pi R_p \cos\mu}{N}$$

or

$$P_B = P_c \cos\mu$$

Diametral pitch (P_D)—an arbitrary number which relates the diameter of the pitch circle to the number of teeth by the relationship

$$P_D = \frac{N}{D}$$

Note that the gears are standardized by their diametral pitch values.

Angle of action—the angle of rotation of the gear for one tooth to be in contact with the tooth of the mating gear. See Fig. 2.

Path of contact—the distance measured along the line of action and intercepted by the addendum circle of the mating gear. See Fig. 2.

Pitch angle (θ_p)—the angle subtended by the circular pitch at the center of the gear. This is also the angle subtended by the base pitch.

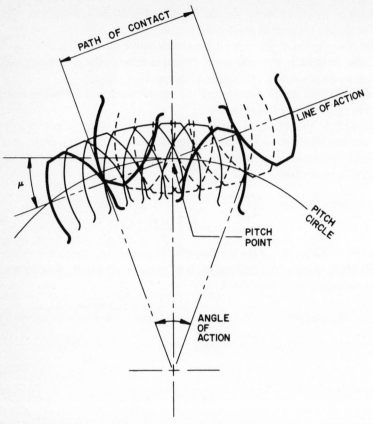

Fig. 2 Action of the mated gears.

Interference—the condition for interference between the mating teeth exists when the common normal at their point of contact does not pass through the pitch point. See Fig. 3. That is, the tooth profile at the point of contact is not an involute profile. Because of the noninvolute profile, the contacting teeth on a gear will have different velocity. This situation will lock two gears. To prevent locking, the noninvolute flank is undercut.

Undercut—the root section of the tooth is reduced to prevent locking due to interference. This reduction in the thickness at the root section is described as undercut.

Contact ratio (G_C)—gear contact ratio provides information pertaining to the number of teeth that are in contact with the teeth of the mating gear. It is calculated as a ratio to the angle of action to the pitch angle. If the contact is unity, then one tooth is in contact with its mating gear tooth. If the contact ratio is 2, then two teeth are in contact with mating gear teeth. The contact ratio is given by

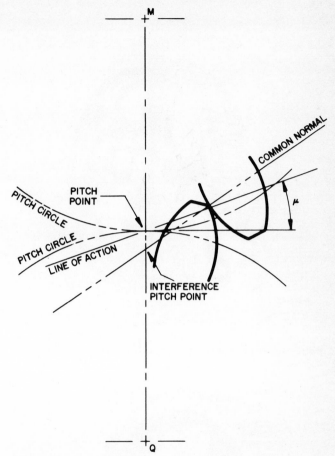

Fig. 3 Interference contact due to noninvolute flank of the mating tooth.

$$G_C = \frac{\text{angle of action}}{\text{pitch angle}}$$

There are different types of gears. Some of the commonly designed gears are spur gears, helical gears, worm gears, hypoid gears, and noncircular gears.

Figures 4 and 5 show a pair of external and compound spur gears. A spur gear has gear teeth with their axes parallel to the axis of rotation. Figures 6 and 7 show a pair of helical gears. A helical gear has gear teeth with their axes inclined at an angle with the axis of rotation.

The angle between a tooth of a helical gear and its axis of rotation is called a *helix angle*. The "hand" of helix of a helical gear is called *right-hand helix* or *left-hand helix* depending upon the direction in which the helix angle is measured. Figure 8 shows the two cases of right-hand and left-hand helix.

Fig. 4 A pair of external spur gears. (*By permission of Boston Gear.*)

Fig. 5 A pair of compound spur gears. (*By permission of Boston Gear.*)

Fig. 6 A pair of helical gears with parallel axes of rotation. (*By permission of Boston Gear.*)

Fig. 7 A pair of helical gears with skew axes. (*By permission of Boston Gear.*)

For helical gears, we define, normal circular pitch, normal diametral pitch, axial pitch, and total contact ratio:

Normal circular pitch (P_{CN}) is the arc length between one tooth and one space in a plane normal to the face of the tooth. If P_C is the circular pitch measured in the plane of rotation, then from Fig. 8a we note that

$$P_{CN} = P_C \cos\psi$$

Normal diametral pitch (P_{DN}) is defined so that

$$P_{CN} P_{DN} = \pi$$

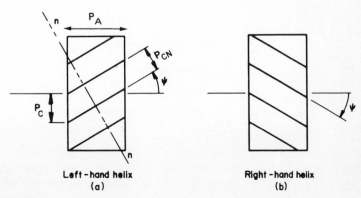

Left-hand helix
(a)

Right-hand helix
(b)

Fig. 8

That is,

$$P_{DN} = \frac{\pi}{P_{CN}}$$

or

$$P_{DN} = \frac{\pi}{P_C \cos\psi}$$

Axial pitch (P_A) measures width of a helical gear. The measurement of width corresponds to advance of one circular pitch. From Fig. 8a, we note that

$$P_A = \frac{P_C}{\tan\psi}$$

Total contact ratio in a helical gear is an addition of two factors: contact ratio in the plane of rotation, and contact ratio due to circular pitch. Hence, the total contact ratio is $G_C + P_C$.

Figures 9–11 show three types of bevel gears. Figure 9 shows a pair of bevel gears with straight teeth. Figure 10 shows a bevel gear with spiral teeth. Figure 11 shows a pair of hypoid bevel gears. A bevel gear with straight teeth is obtained from the frustum of a cone. A face gear is a special case of a bevel gear with straight teeth when the angle of the pitch cone is 90°. A pair of hypoid bevel gears, shown in Fig. 11, is obtained from a frustum of a hyperboloid. The helix angles of tooth correspond to the angles which generate hyperboloid.

Fig. 9 A pair of bevel gears with straight teeth. (*By permission of Gleason Works.*)

Fig. 10 A pair of bevel gears with spiral teeth. (*By permission of Gleason Works.*)

Fig. 11 A pair of hypoid bevel gears. (*By permission of Gleason Works.*)

Figure 12 illustrates the terminology of a bevel gear.

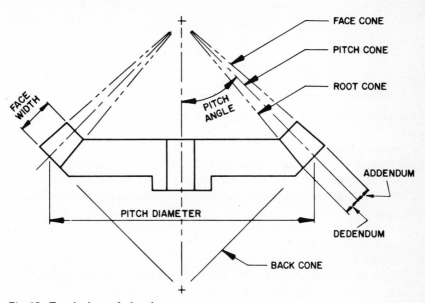

FACE CONE

PITCH CONE

ROOT CONE

FACE WIDTH

PITCH ANGLE

ADDENDUM

PITCH DIAMETER

DEDENDUM

BACK CONE

Fig. 12 Terminology of a bevel gear.

A worm is a helical gear with approximately $90°$ helix angle. A worm may be driven by a helical gear. The combination of worm and helical gear is called a worm gear. See Fig. 13.

Figure 14 shows a rack which is a gear having an infinite pitch radius. The involute surface of the teeth of a rack is a plane to which the line of action is normal.

Spur gears are employed when rotary motion must be transmitted between two parallel shafts. A pair of helical gears can be designed to transmit motion between parallel or nonparallel shafts. A worm and a helical gear can be used to

transmit motion between two skew shafts. Bevel gears can be used when the axes of the shafts are intersecting. A rack and a pinion can be used when it is required to convert rotary motion into translatory motion.

Fig. 13 A worm gear. (*By permission of Boston Gear.*)

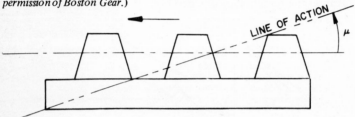

Fig. 14 A rack.

Competency Items

- Draw a pair of spur gears in action and show base circle, pitch circle, root circle, pitch point, line of action, path of contact, and pressure angle.
- Define diametral pitch, circular pitch, and contact ratio for spur and helical gears.
- Draw a bevel gear and show pitch cone, face cone, back cone, pitch angle, and pitch radius.
- Name four types of bevel gears.
- List the advantages and disadvantages of spur gears and helical gears.
- List design situations using worm gears and bevel gears.
- Define interference and undercutting.

Objective 2

To derive relationships to calculate tooth thickness, tooth profile, and contact ratio

Activities

- Read the material provided below, and as you read:
  a. Derive a relationship to calculate tooth thickness.

b. Derive a relationship to calculate contact ratio and minimum number of teeth to avoid undercut.

• Check your answers to the competency items with your instructor.

Reading Material

Part I: Derivation to calculate tooth thickness

Examine Fig. 15 and let:

t_{ad} be the tooth thickness measured on a pitch circle where the pressure angle is μ

t_{bc} be the tooth thickness measured on a circle where the pressure angle is μ'

R_{PM} and R_{BM} be the radii of the pitch circle and the base circle

R be the radius of the circle that passes through point c, e, b

Then,

$$\alpha = \text{angle } eMb = \frac{t_{bc}}{2R}$$

$$\delta = \text{angle } fMa = \frac{t_{ad}}{2R_{PM}}$$

$$\beta = \tan \mu' - \mu' = \text{Inv. } \mu'$$

$$\sigma = \tan \mu - \mu = \text{Inv. } \mu$$

From Fig. 15 we observe that $\alpha + \beta = \delta + \sigma$

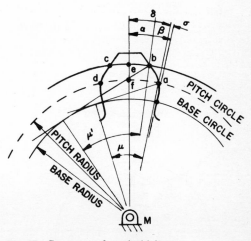

Fig. 15 Geometry of tooth thickness.

That is,

$$\frac{t_{bc}}{2R} + \text{Inv.}\,\mu' = \frac{t_{ad}}{2R_{PM}} + \text{Inv.}\,\mu$$

Hence, tooth thickness

$$t_{bc} = 2R\,\frac{t_{ad}}{2R_{PM}} + (\text{Inv.}\,\mu) - (\text{Inv.}\,\mu')$$

The above relationship can be used to calculate tooth thickness corresponding to any radius $R > R_{BM}$.

Part II: Derivation to calculate contact ratio

By definition, the contact ratio is defined as a ratio of the angle of action to the pitch angle. Thus,

$$\text{Contact ratio} = CR = \frac{\text{angle of action}}{\text{pitch angle}}$$

In Fig. 16, the two mated gears are shown in action. R_{BM}, R_{BQ}, and R_M, R_Q are their base radii and addendum radii. The gears have the common point of contact P. Line AB which is tangent to the base circles of the two mated gears is the line of contact. The tooth of gear 1 initiates contact with its corresponding tooth of gear 2 at point D and terminates the contact at point C. The angle between line t-t which is tangent to the pitch circles, and line of action AB is pressure angle μ.

The contact ratio between the two mated spur gears can also be computed from

$$CR = \frac{\text{path of contact}}{\text{base pitch}} = \frac{\text{length}\,CD}{P_C\,\cos\mu}$$

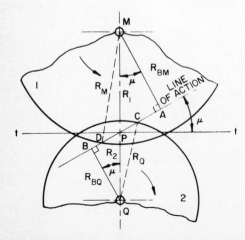

Fig. 16 Pinion and gear in action.

But

$$CD = CP + PD = (CB - PB) + (AD - PA)$$

where

$$CB^2 = R_Q^2 - R_{BQ}^2$$

$$AD^2 = R_M^2 - R_{BM}^2$$

$$PB = QP \sin\mu = R_{PQ} \sin\mu$$

$$PA = MP \sin\mu = R_{PM} \sin\mu$$

Therefore,

$$CD = (R_Q^2 - R_{BQ}^2)^{1/2} + (R_M^2 - R_{BM}^2)^{1/2} - (R_{PQ} + R_{PM}) \sin\mu$$

and

$$CR = \frac{(R_Q^2 - R_{BQ}^2)^{1/2} + (R_M^2 - R_{BM}^2)^{1/2} - (R_{PQ} + R_{PM}) \sin\mu}{P_C \cos\mu}$$

In general, for gears with standard full depth teeth,

$$R_M = R_{PM} + \frac{1}{P_D}$$

$$R_Q = R_{PQ} + \frac{1}{P_D}$$

where R_{PM} and R_{PQ} are the pitch radii of the gears.

In order to achieve smooth and uniform rotary motion, more than one pair of teeth must be in contact. A reasonably good and acceptable design value for contact ratio is 1.4:1.

Competency Items
- State how the contact ratio of spur gears is affected by the pressure angles.
- State in words the contact ratio for a pair of mated spur gears.
- Derive a relationship to calculate contact ratio for bevel gears.
- Derive a relationship to calculate contact ratio for helical gears.
- Derive a relationship to calculate tooth thickness of a bevel gear.

Objective 3

To calculate the effects of diametral pitch, pressure angle, and speed ratio on interference

Activities
- Read the material provided below, and as you read:
  - a. Layout the graphical solution for interference.
  - b. Calculate the effect of diametral pitch on interference.
  - c. Calculate the effect of pressure angle on interference.
  - d. Calculate the effect of speed ratio on interference.
- Check your answers to the competency items with your instructor.

Reading Material

Part I: Conditions that prevent interference

Interference condition will exist when the involute surface of a tooth comes in contact with the noninvolute flank of the other tooth. A repeated occurrence of this condition results in the undercut of the tooth flank. Conditions that will prevent interference are examined more easily by using a graphical construction. The student should repeat the following steps. As shown in Fig. 17,

Let R_{BM} and R_{BQ} be the base radii of the two gears rotating about centers M and Q. Draw the base circles and find the line of action by drawing the tangent to the base circles. AB is the line of action.

Find pitch point P by finding the point of intersection of center line MQ and line of action AB.

Let R_M and R_Q be the radii of the addendum circles. Draw the addendum circles intersecting the line of action in points D and C.

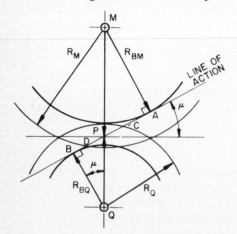

Fig. 17 Graphical determination of interference conditions.

Note that length AB represents the maximum possible length of contact the two gear teeth can have. However, the two gears are in contact for length CD on the line of action. This is due to addendum circles. To avoid the interference, the two mated teeth must always have involute contact. This condition is guaranteed as long as the points C and D lie within length AB.

Since the location of points C and D are directly controlled by the addendum radii, interference condition is greatly controlled by the diametral pitch.

Part II: The effect of diametral pitch on interference

In Fig. 18, the pitch radii and the pressure angle μ are assumed to be of fixed values. Since addendum radius R_M is of a large value, the addendum circle intersects the line of action at point D which lies outside interval AB. In order to prevent this interference condition, the addendum radius must be made smaller. That is, the diametral pitch must be increased in its value. This increase in the diametral pitch increases the number of teeth on each gear since the pitch radii are held constant.

Part III: The effect of pressure angle on interference

The pressure angle is controlled by the inclination of line of action with the tangent to the pitch circles. For constant pitch radii and diametral pitch, the only way to eliminate interference conditions is to make the base radii smaller. The smaller base radii will increase the pressure angle and eliminate the interference condition as shown in Fig. 19. Note that by changing μ to μ', the interference condition is eliminated.

Part IV: The effect of speed ratio on interference

The speed ratio is inversely proportional to the pitch radii of the gears. The interference condition can be eliminated by increasing the length of the pitch radius of the driven gear if it is required to have the same pressure angle and diametral pitch. Figure 20 shows that by increasing the pitch radius of gear 2, the interference condition is eliminated. However, the speed ratio has decreased with the increase in the pitch radius.

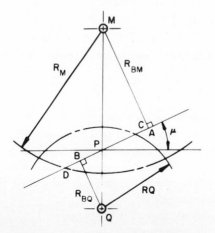

Fig. 18 Interference due to small diametral pitch.

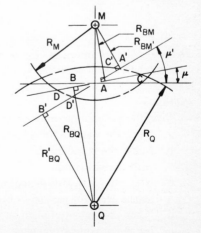

Fig. 19 Interference due to smaller pressure angle.

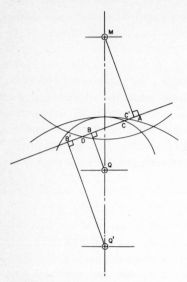

Fig. 20 Interference due to larger speed ratio.

Competency Items
- Define interference.
- Draw sketches and show how interference is affected by:
  a. diametral pitch
  b. pressure angle
  c. speed ratio

Objective 4

To name AGMA (American Gear Manufacturers' Association) standards for spur, helical, bevel, and worm gears

Activities
- Read the material provided below, and as you read:
  a. Name the standard diametral pitch and pressure angles for spur gears.
  b. Name the standard diametral pitch and pressure angles for bevel gears with straight teeth.
  c. Name the standard diametral pitch, pressure angles, and helix angle for helical gears for parallel or nonparallel shafts.
  d. Name the diametral pitch and pressure angles for worm gears.
  e. Name the standard involute tooth proportions.
- Check your answers to the competency items with your instructor.

Reading Material

The American Gear Manufacturers' Association has proposed standards for gears with objective that a designer is in a position to interchange gears manufactured by different manufacturers. There are four requirements that gears could be interchanged without affecting the speed ratio of gear trains.

The interchangeable gears must have equal diametral pitch, equal addendum and dedendum, equal tooth thickness, and equal pressure angles.

Table 1 presents more commonly used diametral pitch and pressure angles for spur, helical, bevel, and worm gears. Table 2 presents the standard tooth proportions for different pressure angles.

TABLE 1   Commonly Used Pitch and Pressure Angle

Type of gears	Pitch	Pressure angles	Helix angle
Spur	Diametral pitch $\begin{pmatrix} 1, 1.25, 1.5, 2 \\ 2.5, 3, 4, 5 \\ 6, 8, 10, 16 \\ 20, 24, 32, 48 \\ 64, 94, 128 \end{pmatrix}$	$\left.\begin{matrix} 14\frac{1}{2}° \\ 20° \\ 22° \\ 25° \end{matrix}\right\}$ 20°	0°
Bevel	same as spur gears	$14\frac{1}{2}°$ $17\frac{1}{2}°$ 20°	0°
Helical	same as spur gears	20° (in plane of rotation)	15°
Worm	Axial pitch varies from 0.25 to 1.50 with lead angles up to 30°	$14\frac{1}{2}°, 20°, 25°$ (in normal planes)	

TABLE 2   Tooth Proportion Standards Recommended by AGMA

	20° Full depth	20° Stub	14½° Full depth	14½° Composite
Full depth	$2.157/P_D$	$18/P_D$	$2.157/P_D$	$2.157/P_D$
Filet radius	$0.235/P_D$	$0.3/P_D$	$0.209/P_D$	$0.209/P_D$
Clearance	$0.157/P_D$	$0.2/P_D$	$0.157/P_D$	$0.157/P_D$
Addendum	$1/P_D$	$0.8/P_D$	$1/P_D$	$1/P_D$
Dedendum	$1/P_D$	$0.8/P_D$	$1/P_D$	$1/P_D$

Performance Test # 1

1. Pressure angles for standard gears are either $14.5°$, $20°$, or $25°$. Two gears have standard full depth teeth and a diametral pitch of 3. One has 27 teeth

and the other 18. Calculate the minimum pressure angle of these values that will result in mesh without interference. Show all work.

2. A gear has a pitch diameter of 6 inches and a diametral pitch of 2. It has standard full depth teeth and a pressure angle of $20°$. Calculate the maximum speed ratio that can result if this is the driver gear. Calculate the contact ratio of this pair of gears.

Performance Test #2

1. A pair of spur gears have 16 and 22 teeth, respectively, diametral pitch of 2, 1/2-in. addendum, 9/16-in. dedendum, and $20°$ pressure angle. Determine the following:
   a. pitch circle radii
   b. base circle radii
   c. circular pitch
   d. base pitch
   e. length of path of contact
   f. contact ratio
2. A gear has 30 teeth, diametral pitch of 3, 3/8-in. addendum, and $20°$ pressure angle. Determine:
   a. the smallest pinion that will mesh with the gear without interference
   b. the contact ratio
3. As a pair of gears turn together do the teeth exhibit sliding contact, rolling contact, or a combination of both? Give the reason for your answer.

Performance Test #3

1. Two gears having an angular velocity ratio of 1.4:1 are mounted on shafts whose centers are 6.00 in. apart. If the diametral pitch of the gears is 3, calculate the number of teeth on each gear.
2. A pair of mating gears have pitch circle diameters of 5 in. and 6 in. The diametral pitch is 4, the pressure angle is $20°$, and the gears have standard full depth teeth. Calculate the contact ratio. State whether this ratio is satisfactory and why.
3. A pair of mating gears have pitch circle diameters of 4 in. and 6 in. The pressure angle is $20°$. Calculate the maximum addendum that the gears may have if interference is not to occur.

Supplementary References

1 (Chap. 6, 7.4); 3 (7.1-7.13, 7.16-7.17); 5 (12.1-12.11); 7 (8.1-8.8, 8.10-8.12); 8 (10.3-10.7)

Unit VII. Design of Gear Trains

Objectives

1. To design a simple gear train
2. To design a compound gear train
3. To design a reverted compound train
4. To design a planetary gear train

Objective 1

To design a simple gear train

Activities

- Read the material provided below, and as you read:
  a. Name different types of simple gear trains.
  b. State the reasons why simple gear trains are designed.
  c. Calculate the speed ratio for a simple gear train.
  d. Calculate the number of teeth required for the different gears of the designed simple gear train.
- Check your answers to the competency items with your instructor.

Reading Material

A simple gear train is a series of gears, which are capable of rotating about independent axes and are capable of receiving and transmitting motion from one gear to another. An example of a simple gear train is shown in Fig. 1.

Let $R_1, R_2, R_3, \ldots, R_6$ be the pitch radii of these gears. $A, B, C, D,$ and E are the pitch points of gears 1 and 2, 2 and 3, 3 and 4, 4 and 5, and 5 and 6. Gear 1 drives gear 2 which in turn drives gear 3. Gear 3 drives gear 4 which in turn drives gear 5. Gear 5 drives gear 6. Thus, gear 1 is the driver and gears 2, 3, 4, 5, and 6 are the driven gears. The driver gear is the input gear and all the other gears are the output gears.

The speed ratio (SR) of the simple gear train can be calculated in the following manner.

1. If gears 1 and 2 are moving with angular velocities ω_1 and ω_2 as shown in Fig. 1, then

$$SR_{1,2} = \frac{\omega_2}{\omega_1} = \frac{R_1}{R_2}$$

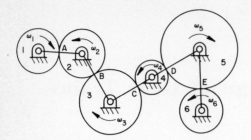

Fig. 1 Simple gear train.

2. If gear 3 is rotating with angular velocity ω_3, as shown in Fig. 1, then

$$SR_{2,3} = \frac{\omega_3}{\omega_2} = \frac{R_2}{R_3}$$

3. Similarly,

$$SR_{3,4} = \frac{\omega_4}{\omega_3} = \frac{R_3}{R_4}$$

$$SR_{4,5} = \frac{\omega_5}{\omega_4} = \frac{R_4}{R_5}$$

and

$$SR_{5,6} = \frac{\omega_6}{\omega_5} = \frac{R_5}{R_6}$$

4. If gear 1 is the input gear and gear 6 is the output gear, then, by definition, the speed ratio is given by

$$SR_{1,6} = \frac{\omega_6}{\omega_1}$$

But ω_6/ω_1 can be written as

$$\frac{\omega_6}{\omega_1} = \frac{\omega_6}{\omega_5} \times \frac{\omega_5}{\omega_4} \times \frac{\omega_4}{\omega_3} \times \frac{\omega_3}{\omega_2} \times \frac{\omega_2}{\omega_1}$$

$$= (SR_{5,6}) \times (SR_{4,5}) \times (SR_{3,4}) \times (SR_{2,3}) \times (SR_{1,2})$$

$$= \frac{R_5}{R_6} \times \frac{R_4}{R_5} \times \frac{R_3}{R_4} \times \frac{R_2}{R_3} \times \frac{R_1}{R_2} = \frac{R_1}{R_6}$$

Hence, for a simple gear train, the speed ratio is

$$SR_{1,6} = \frac{\omega_6}{\omega_1} = \frac{R_1}{R_6}$$

The student should examine the directions in which these six gears are rotating. Note that a pair of mated external gears will always move in the opposite directions. Thus, the odd-numbered gears 1, 3, 5 are moving in a clockwise direction. All the even-numbered gears 2, 4, 6 are rotating in a counterclockwise direction. The speed ratio of a gear system is negative when the input and the output gears are rotating in opposite directions. The speed ratio of a gear system is positive, when the input and output gears are rotating in the same directions.

For the gear system in Fig. 1, if the directions of rotation of input and output gears are taken into consideration, then the speed ratio is

$$SR_{1,6} = -\frac{\omega_6}{\omega_1} = -\frac{R_1}{R_6} \tag{1}$$

In a simple gear train, the size of the intermediate gears 2, 3, 4, and 5 do not contribute to the calculation of the speed ratio of the train. These intermediate gears are called *idler gears*. The idler gears are used when

A longer distance must be spanned between the input and the output gears, and gears with large diameters are not permitted in the gear train design.
Direction of rotation of the output gear must be changed.

The speed ratio of gear systems can be computed also by counting the number of teeth of the individual gears. We know that the circular pitch P_C is defined as

$$P_C = \frac{\pi D}{N} = \frac{\pi}{N/D}$$

and diametral pitch P_D is defined as

$$P_D = \frac{N}{D}$$

Hence,

$$D = \frac{N}{P_D}$$

Now,

$$SR_{1,6} = \frac{\omega_6}{\omega_1} = \frac{R_1}{R_6} = \frac{D_1}{D_6} = \frac{(N/P_D)_1}{(N/P_D)_6}$$

Since the diametral pitch is

$$P_{D_1} = P_{D_2} = P_{D_3} = P_{D_4} = P_{D_5} = P_{D_6}$$

$$SR_{1,6} = \frac{\omega_6}{\omega_1} = \frac{N_1}{N_6}$$

where N_1 and N_6 are the number of teeth on gears 1 and 6. We observe from the above relationship that the speed ratio for the gear train, shown in Fig. 1, is inversely proportional to the ratio of number of teeth on driving and driven gears.

Competency Items
- Define a speed ratio for a simple gear train having four gears with pitch diameters D_1, D_2, D_3, D_4.
- Calculate the speed ratio for a simple gear train with $N_1 = 20, N_2 = 30, N_3 = 60, N_4 = 10$.
- Write the assumption under which the speed ratio of a simple gear train is inversely proportional to the number of teeth on the driver and the driven gear.
- Identify design situations when idler gears are used in a gear train.

Objective 2

To design a compound gear train

Activities
- Read the material provided below, and as you read:
  a. Identify a compound gear train.
  b. Calculate the speed ratio of a compound gear train.
  c. Calculate the number of teeth on the individual gears of a compound gear train.
- Check your answers to the competency items with your instructor.

Reading Material
A compound gear train shown in Fig. 2 is a series of gears connected in such a way that the train has one or more gears rotating about an axis with the same angular velocity.

The compound gear train shown in Fig. 1 has nine gears in series. Let the pitch radii of these gears by $R_1, R_2, \ldots, R_9$. Also, let the angular velocities of these gears be $\omega_1, \omega_2, \ldots, \omega_9$. Note that in this compound gear train, gears 2 and 3, 4

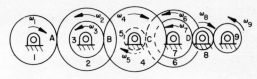

Fig. 2 Compound gear train.

and 5, 6 and 7 are rigidly mounted on the same shaft so that $\omega_2 = \omega_3$, $\omega_4 = \omega_5$, $\omega_6 = \omega_7$. That is, gears 2 and 3 are rotating with the same angular velocity, gears 4 and 5 are rotating with the same angular velocity, etc.

Gears 1, 3, 5, and 7 are the driver gears and gears 2, 4, 6, and 9 are the driven gears. Gear 8 is the idler gear. The input to the compound gear train is via gear 1 and the output is obtained from the motion of gear 9.

The speed ratio for gears 1 and 2 is

$$(SR)_{1,2} = \frac{\omega_2}{\omega_1} = \frac{R_1}{R_2}$$

The speed ratio for gears 3 and 4 is

$$(SR)_{3,4} = \frac{\omega_4}{\omega_3} = \frac{R_3}{R_4}$$

The speed ratio for gears 5 and 6 is

$$(SR)_{5,6} = \frac{\omega_6}{\omega_5} = \frac{R_5}{R_6}$$

The gear train containing gears 7, 8, and 9 is a simple gear train. The speed ratio for this simple gear train is

$$(SR)_{7,9} = \frac{\omega_9}{\omega_7} = \frac{R_7}{R_9}$$

The speed ratio for the compound gear train is

$$(SR)_{1,9} = \frac{\omega_9}{\omega_1} = \frac{\omega_9}{\omega_7} \cdot \frac{\omega_7}{\omega_6} \cdot \frac{\omega_6}{\omega_5} \cdot \frac{\omega_5}{\omega_4} \cdot \frac{\omega_4}{\omega_3} \cdot \frac{\omega_3}{\omega_2} \cdot \frac{\omega_2}{\omega_1}$$

Since gears 2 and 3, 4 and 5, 6 and 7 are rotating with the same angular velocities,

$$\frac{\omega_3}{\omega_2} = 1 \qquad \frac{\omega_5}{\omega_4} = 1 \quad \text{and} \quad \frac{\omega_7}{\omega_6} = 1$$

Therefore, the speed ratio becomes

$$(SR)_{1,9} = \frac{\omega_9}{\omega_1} = \frac{\omega_9}{\omega_7} \cdot \frac{\omega_6}{\omega_5} \cdot \frac{\omega_4}{\omega_3} \cdot \frac{\omega_2}{\omega_1}$$

$$= \left(\frac{R_7}{R_9}\right)\left(\frac{R_5}{R_6}\right)\left(\frac{R_3}{R_4}\right)\left(\frac{R_1}{R_2}\right)$$

$$= \frac{R_7 R_5 R_3 R_1}{R_9 R_6 R_4 R_2} \tag{2}$$

The numerator in the speed ratio is a product of the radii of driver gears 1, 3, 5, and 7. The denominator in the speed ratio is a product of the radii of driven gears 2, 4, 6, and 9. Thus, the speed ratio for a compound gear train can be written as

$$(SR)_{1,9} = \frac{\text{product of the pitch radii of driver gears}}{\text{product of the pitch radii of driven gears}} \tag{3}$$

Since diametral pitch P_D is given by

$$P_D = \frac{N}{D}$$

where N denotes the number of teeth on a gear, and D denotes its diameter. From the above relationship, we have

$$D = \frac{N}{P_D}$$

Hence, the speed ratio for a compound gear can be written in terms of number of teeth and diametral pitch. Thus,

$$(SR)_{1,9} = \frac{(N/P_D)_7 (N/P_D)_5 (N/P_D)_3 (N/P_D)_1}{(N/P_D)_9 (N/P_D)_6 (N/P_D)_4 (N/P_D)_2}$$

If all the gears of the compound gear train have the same pitch value, then the speed ratio becomes

$$(SR)_{1,9} = \frac{\omega_9}{\omega_1} = \frac{N_7 N_5 N_3 N_1}{N_9 N_6 N_4 N_2} \tag{4}$$

where $N_1, N_2, \ldots, N_9$ are the number of teeth on each of the gears. Speed ratio $(SR)_{1,9}$ will remain the same even when

$$\left(P_D\right)_1 = \left(P_D\right)_2$$
$$\left(P_D\right)_3 = \left(P_D\right)_4$$
$$\left(P_D\right)_5 = \left(P_D\right)_6$$

and

$$\left(P_D\right)_7 = \left(P_D\right)_8 = P_{D\,9}$$

This shows that the diametral pitch of gears 2 and 3, 4 and 5, 6 and 7 need not be the same.

Competency Items
- State the difference between a simple gear train and a compound gear train.
- Identify the design situations when all the gears of a compound gear train should have the same diametral pitch value.
- Show that the expression for speed ratio of the gear train in terms of tooth numbers as presented in this objective will remain the same even though four different diametral pitch values are involved:

$$P_{D_1} = P_{D_2}, \ P_{D_3} = P_{D_4}, \ P_{D_5} = P_{D_6}, \ P_{D_7} = P_{D_8} = P_{D_9}$$

- List reasons why the size of gear 8 does not affect the speed ratio of the gear train. Could gear 8 be eliminated from the train without affecting the output of the train at gear 9?
- Calculate the speed ratio of the compound gear train described in the objective if $N_1 = 20, N_2 = 35, N_3 = 15, N_4 = 45, N_5 = 10, N_6 = 40, N_7 = 30, N_9 = 12$.

Objective 3

To design a reverted compound gear train

Activities
- Read the material provided below, and as you read:
  a. Identify a reverted compound gear train.
  b. Calculate the speed ratio of a reverted compound gear train.
  c. Calculate the number of teeth of gears of a reverted compound gear train.
- Check your answers to the competency items with your instructor.

Reading Material

A reverted compound train is a special case of a compound gear train. The input and the output gears in the reverted compound gear train, however, have the same axis of rotation as shown in Fig. 3.

Gears 1 and 3 are the driver gears and gears 2 and 4 are the driven gears. Input gear 1 and output gear 4 are rotating about an axis. Gears 2 and 3 are mounted rigidly on the same shaft, so that $\omega_2 = \omega_3$.

In the previous section, we have demonstrated that the speed ratio for a compound gear train is

$$SR = \frac{\text{product of pitch radii of the driver gears}}{\text{product of pitch radii of the driven gears}}$$

Since the reverted compound gear train is a special case of the general case of the compound gear train examined in the previous objective, the speed ratio for the reverted compound train shown in Fig. 3 is given by

$$(SR)_{1,4} = \frac{\omega_4}{\omega_1} = \frac{R_1 R_3}{R_2 R_4} \qquad (5)$$

If the gears have the same diametral pitch values, then

$$(SR)_{1,4} = \frac{\omega_4}{\omega_1} = \frac{N_1 N_3}{N_2 N_4} \qquad (6)$$

From Fig. 3, the center distance between gears 1 and 2 and gears 3 and 4 are related as

$$R_1 + R_2 = R_3 + R_4 \qquad (7)$$

If these gears have the same diametral pitch, then the above relationship yields

$$N_1 + N_2 = N_3 + N_4 \qquad (8)$$

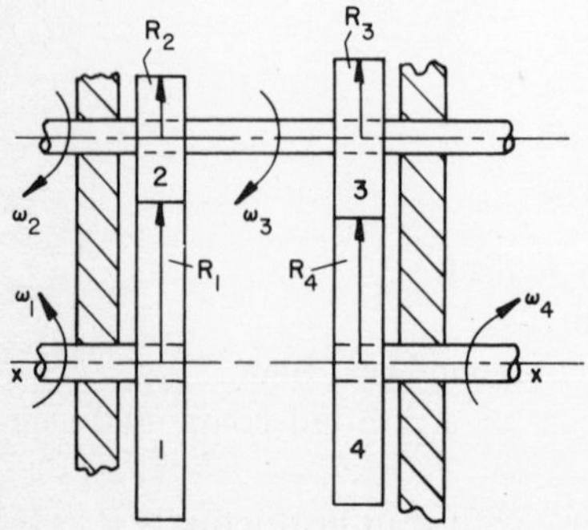

Fig. 3 Reverted compound gear train.

Equation (5) and (7) can be used to determine pitch radii of the reverted compound gear train shown in Fig. 3 if speed ratio is known. Equation (6) and (8) can be used to determine the number of teeth on each gear if the pitch radii of these gears are known.

Competency Items
- Draw a reverted compound gear train and write two design situations where it can be used.
- State which gears of your reverted compound gear train must have equal diametral pitch values.
- Calculate the speed ratio of the gear train shown in Fig. 3, if $N_1 = 150, N_2 = 100, N_3 = 300, N_4 = 500$.
- Calculate R_1, R_2, R_3, and R_4 for the gear train in Fig. 3 for which $N_1 = 150, N_2 = 100, N_3 = 300, N_4 = 500$.

Objective 4

To design a planetary gear train

Activities
- Read the material provided below, and as you read:
  a. Identify different types of planetary gear trains.
  b. Calculate the speed ratio of a planetary gear train under different possible conditions.
  c. Calculate the number of teeth of gears of a planetary gear train.
- Check your answers to the competency items with your instructor.

Reading Material
Different types of planetary gear trains are shown in Fig. 4. These planetary gear trains are often called *epicyclic gear trains* and each gear train has two degrees of freedom. Constrained output motion can be obtained from these trains when they are driven using two external inputs.

A planetary gear train consists of four elements: sun gear, planet arm, planet gears, and ring gear. Two input motions can be either given to sun gear and planet arm or sun gear and ring gear, and output motion in these cases can be obtained either from the ring gear or the planet arm. The input can also be given to planet arm and ring gear and output motion is obtained via sun gear. For the cases described above, the planetary gear will provide different speed ratios. Because of the simplicity in design, the planetary gears are used in designing automotive transmissions.

We will now study the mathematical approach for deriving the equation of motion for a simple planetary gear train. A schematic diagram for the planetary gear train in Fig. 4a is shown in Fig. 5.

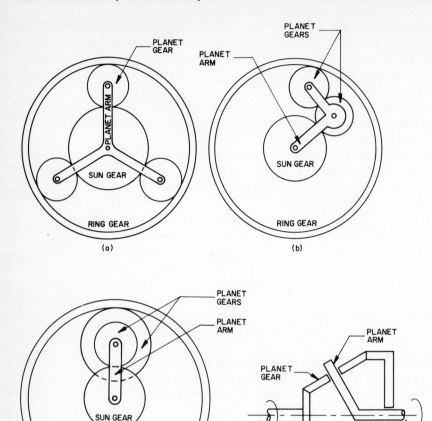

Fig. 4 Examples of planetary gear trains.

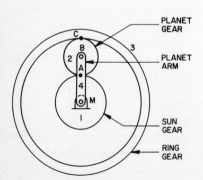

Fig. 5 A planetary gear train.

The sun gear and the ring gear are rotating about the axis passing through M and perpendicular to the plane of the paper. Let the sun gear, the planet gear, the ring gear, and the planet arm be described using numbers 1, 2, 3, and 4. Accordingly, the radii and angular velocities of the different elements of the planetary gear train of Fig. 5 are R_1, R_2, R_3, R_4 and ω_1, ω_2, ω_3, and ω_4.
Velocity of point A on sun gear = $R_1 \cdot \omega_1$.
Velocity of point B on planet arm = $R_4 \cdot \omega_4 = (R_1 + R_2)\omega_4$.
For point A lying on the planet gear

$$\mathbf{V}_A = \mathbf{V}_B + \mathbf{V}_{A/B} \tag{9}$$

For point C lying on the planet gear

$$\mathbf{V}_C = \mathbf{V}_B + \mathbf{V}_{C/B} \tag{10}$$

Note that points A and C are moving in opposite directions. Therefore,

$$\mathbf{V}_{C/B} = -\mathbf{V}_{A/B} \tag{11}$$

but

$$\mathbf{V}_{C/B} = \mathbf{V}_C - \mathbf{V}_B \tag{12}$$

and

$$\mathbf{V}_{A/B} = \mathbf{V}_A - \mathbf{V}_B \tag{13}$$

Hence,

$$(\mathbf{V}_A - \mathbf{V}_B) = -(\mathbf{V}_C - \mathbf{V}_B) \tag{14}$$

but

$$V_A = R_1\omega_1$$
$$V_B = (R_1 + R_2)\omega_4$$
$$V_C = R_3\omega_3$$

Substituting the above quantities in Eq. (14) we get

$$R_1\omega_1 - (R_1 + R_2)\omega_4 = -[R_3\omega_3 - (R_1 + R_2)\omega_4]$$

Simplifying the above relationship, we get

$$R_1\omega_1 + R_3\omega_3 = 2(R_1 + R_2)\omega_4 \tag{15}$$

Equation (15) describes equation of motion of planetary gear train shown in Fig. 5

Let us examine speed ratio of the planetary gear train under special conditions.
Sun gear is fixed: $\omega_1 = 0$. If the input is through the planet arm,

$$SR = \frac{\omega_3}{\omega_4} = \frac{2(N_1 + N_2)}{N_3}$$

also $R_1 + 2R_2 = R_3$ or $N_1 + 2N_2 = N_3$.
Planet arm is fixed: $\omega_4 = 0$. The planetary gear train becomes a reverted gear train. When the sun gear is the input gear, the speed ratio is given by

$$SR = \frac{\omega_3}{\omega_1} = -\frac{R_1}{R_3} = -\frac{N_1}{N_3}$$

The negative sign in the speed ratio suggests that the input and the output gears are moving in opposite directions.
Ring gear is fixed: $\omega_3 = 0$. If the sun gear is the input gear, the speed ratio is given by

$$SR = \frac{\omega_4}{\omega_1} = \frac{R_1}{2(R_1 + R_2)} = \frac{N_1}{2(N_1 + N_2)}$$

A planet-arm gear often carries more than one planet gear. The following criteria are utilized:

For equally spaced planet gears,

The sum of N_1 and N_3 must be an even number.
If η denotes the number of planet gears, then the ratio $(N_1 + N_3)/\eta$ must be an integer value.

Competency Items
- Derive the motion equations for the different types of planetary gear trains shown in Fig. 4b and c.
- What are the differences between arrangements of the different types of planetary gear train shown in Fig. 4a, b, c, and d?
- In Fig. 5, if the sun gear is fixed and $N_1 = 70, N_2 = 50$, calculate N_3.
- In Fig. 5, if the ring gear is fixed and $N_1 = 70, N_2 = 50$, calculate the speed ratio ω_4/ω_1.
- Draw all possible arrangements of a planetary gear train with the four elements of the gear trains shown in Fig. 4.

Performance Test #1

1. Explain how interference is affected by diametral pitch.

2. Derive an expression for the speed ratio ω_6/ω_1 of the compound gear train shown in the figure below, in terms of R_1, R_2, etc.

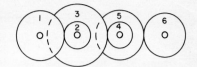

3. Derive expressions for the speed ratios of the compound planetary gear train shown in the figure below. Use ω_1, ω_2, ω_3, and ω_4 to represent the angular velocities of the sun gear, planet arm, planet gears, and ring gear, respectively. Use R_1, R_2, R_{3A}, R_{3B}, and R_4 to represent the radii of these gears (R_{3A} is the larger planet gear radius). Assume that the ring gear is fixed and that the input is through the sun gear. $SR = \omega_2/\omega_1$.

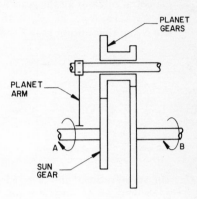

PLANET GEARS

PLANET ARM

SUN GEAR

A   B

Performance Test #2

1. Design a reverted spur gear train of velocity ratio 1:30. The available stock gears are: all tooth numbers from 12 to 24 and all even numbers from 24 to 120. All gears have the same diametral pitch, 7.
   a. Calculate the number of teeth of all the gears.
   b. Locate the centerline distance between the shafts.
2. a. For the planetary gear train shown below, derive a relationship between the variables ω_1, ω_2, ω_4, R_1, R_2, R_{3A}, R_{3B}, and R_4 where ω represents angular velocity and R represents gear radius. The sun gear, planet arm, planet gear, and ring gear are represented by 1, 2, 3, and 4 respectively.
   b. If the sun gear turns at 10 rpm clockwise and the planet arm rotates at 20 rpm counterclockwise, find the angular velocity of the ring gear in terms of the radii of the gears. Note that R_{3A} and R_{3B} correspond to the gears in contact with the sun and ring gears respectively.

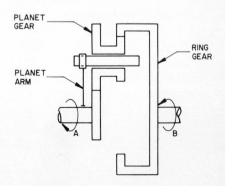

PLANET GEAR

RING GEAR

PLANET ARM

A   B

Performance Test #3

1. In the compound gear train shown below, the input gear 1 rotates at 900 rpm. The distance between the centers of gears 2 and 7 is 33.5 in. The diametral pitch of all the gears is 4, and tooth numbers on the gears are $N_2 = 72$, $N_3 = 18$, $N_4 = 80$, $N_6 = 48$, and $N_7 = 54$. Calculate tooth numbers N_1 and N_5 so that output gear 7 will rotate at 25 rpm.

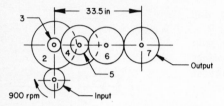

2. The speed ratio of the reverted gear train shown is to be 12. The diametral pitch of gears A and B is 8 and of gears C and D is 10. Calculate the suitable numbers of teeth for the gears. No gear is to have less than 24 teeth.

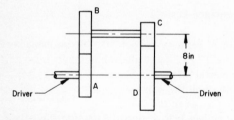

3. In the planetary gear train of Fig. 5, if the sun gear is fixed, the speed ratio of $\omega_3/\omega_4 = 1.5$ and $N_1 = 60$, what are the tooth numbers N_2 and N_3 ? Note that $R_1 + 2R_2 = R_3$.

Supplementary References

1 (5.1–5.3); **3** (11.1–11.9); **5**; **7** (10.1–10.7); **8** (11.1–11.7)

Unit VIII. Cams and Followers

Objectives
1. To identify different types of followers
2. To identify different types of cams
3. To identify the notations of a cam follower system

Objective 1

To identify different types of followers

Activities
- Read the material provided below, and as you read:
  - a. Classify the followers by their motion.
  - b. Classify the followers by their physical shape.
  - c. Classify the followers by their orientation.
- Check your answers to the competency items with your instructor.

Reading Material

A cam is a mechanical member that transmits motion to a follower by direct contact. The driving member is called a *cam* and the driven member is called the *follower*. The cam may rotate, translate, oscillate, or even remain stationary. The follower may, on the other hand, have either rotary or translatory motion.

The followers are classified either according to their shapes, the types of motions displayed by them, and the actual locations of the line of movement.

The knife edge, the roller, and the flat face followers are shown in Figs. 1, 2, and 3. The knife edge follower is quite simple in construction. However, since it produces extreme wear of surface at the point of contact, its use is limited.

The roller follower shown in Fig. 2 is a practical form of the knife edge follower shown in Fig. 1. The roller follower is a cylindrical body free to rotate about the pin joint at P as shown in Fig. 2. The action of the roller at low speed is pure rolling. However, as the speed increases, the action of the roller is a combination of rolling and sliding. The rolling followers normally become a problem case when a cam has a steep rise. In this situation, the roller follower will jam the cam.

The flat face follower shown in Fig. 3 is a further refinement of the roller follower. The flat face follower will not jam when the cam has a steep rise. The

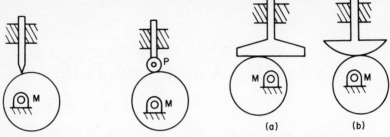

Fig. 1 Knife edge follower.   **Fig. 2** Roller follower.   **Fig. 3** Flat face follower.

type of flat face follower shown in Fig. 3a gives rise to high surface stresses and wear especially due to deflection and misalignment. For this reason, the flat face follower of Fig. 3b is often used.

A follower is said to be a *radial* follower when the follower translates along the axis which passes through the center of rotation of the cam. Two types of offset followers are shown in Figs. 4 and 5.

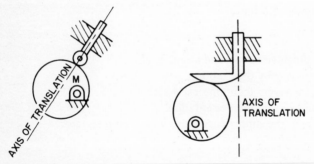

Fig. 4 Offset roller follower.   **Fig. 5** Offset flat face follower.

If a follower is to reproduce exactly the motion transmitted by a cam, then it must remain in the contact with the cam at all speeds at all times. The contact between the cam and its follower is assured by having a preloaded compression spring, a positive drive condition, or by gravity. There are occasions when hydraulic or pneumatic means are considered. The positive drive condition is achieved by letting the roller follower sit in a cam groove or by using conjugate followers as shown in Figs. 6 and 7.

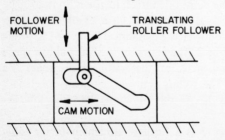

Fig. 6 Positive drive cam with translating roller follower.

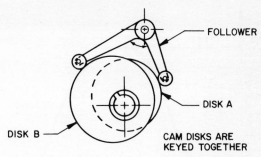

Fig. 7 Positive drive cam with conjugate followers.

Competency Items
- State the advantages or disadvantages that would apply to an offset flat face follower.
- Sketch an offset knife edge follower.
- List three different methods of forcing a cam follower to remain in contact with a cam profile.
- Sketch a positive drive cam mechanism for a translating flat face follower.

Objective 2

To identify different types of cams

Activities
- Read the material provided below, and as you read:
  a. Classify the cams in terms of follower motion.
  b. Classify the cams in terms of their shape.
  c. Classify the cams in terms of the manner of constraint of the follower.
- Check your answers to the competency items with your instructor.

Reading Material

Cam is usually a driving member of the cam and follower assembly. Hence, cams are classified in any one of the three possible ways: according to their follower motion program, according to their shape, and according to the constraints of their followers.

A cam may have any one of the following three important motion programs. These are (1) rise-return (R-R), (2) dwell-rise-return (D-R-R), or (3) dwell-rise-dwell-return (D-R-D-R) motion program.

The rise-return motion program of the follower in terms of angular displacement of the cam is shown in Fig. 8.

The rise-return program makes it possible for a follower to go up and down according to the displacement program shown in Fig. 8. A follower could have either linear or angular displacement. A cycle of cam rotation is completed when

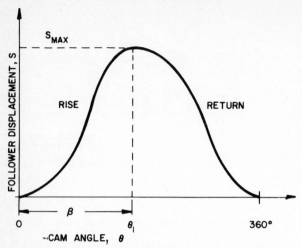

Fig. 8   Rise-return (R-R) displacement program of follower.

the cam rotates $360°$. The cycle begins when $\theta = 0°$, completes the rise segment when $\theta = \theta_1 = \beta$, and completes the return segment when $\theta = 360°$. From Fig. 3, we can write the following boundary conditions for the motion of the follower: at $\theta = 0°$, $S = 0$; at $\theta = \theta_1$, $S = S_{max}$; at $\theta = 360°$, $S = 0$.

The dwell-rise-return motion program of the follower shown in Fig. 9 begins with a dwelling motion of the follower in addition to the rise-return motion program. The dwelling conditions of a follower implies that the follower has no net displacement, has zero velocity, and zero acceleration. The following boundary conditions can be written to describe the follower displacements:

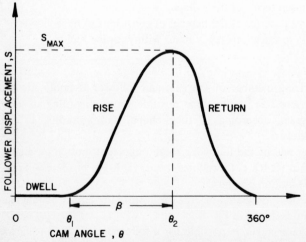

Fig. 9   Dwell-rise-return (D-R-R) displacement program of follower.

for

$$0 \leq \theta \leq \theta_1 \quad S = 0$$

at

$$\theta = \theta_2 \quad S = S_{max}$$

at

$$\theta = 360° \quad S = 0$$

Note that $\beta = \theta_2 - \theta_1$ is the angle that describes the amount of cam rotation that corresponds to the maximum displacement of the follower during its rise segment of the motion.

The dwell-rise-dwell-return motion program of the follower begins with a dwelling motion of the follower; this dwelling motion is followed by a rise and dwelling motion as shown in Fig. 10. The following boundary conditions can be written to describe the follower displacements:

for

$$0 \leq \theta \leq \theta_1 \quad S = 0$$

for

$$\theta_2 \leq \theta \leq \theta_3 \quad S = S_{max}$$

at

$$\theta = 360° \quad S = 0$$

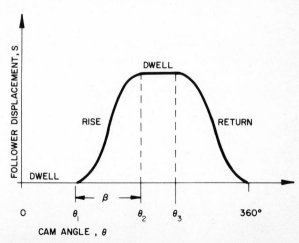

Fig. 10 Dwell-rise-dwell-return (D-R-D-R) motion program of follower.

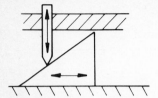

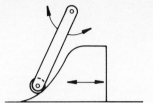

Fig. 11 Wedge cam with translating follower.

Fig. 12 Wedge cam with oscillating follower.

The cams can be also classified by their shapes. The shape of a cam may be a wedge, cylindrical, spiral, conical, spherical, globoidal, radial, conjugate, or three dimensional. The cam may have either rotational or translational motion. A wedge-shape cam having translatory and oscillating types of followers are shown in Figs. 11 and 12.

The wedge-shaped cam is simple in its design. The follower is kept in contact with the cam surface either by spring loading the follower or by providing a positive drive groove.

The radial or the disk cam has followers which move radially from the center of rotation of the cam. The two types of disk cams are shown in Fig. 13a and b; Fig. 13a has translating roller follower; Fig. 13b has oscillating roller follower. The followers are kept in contact with the cams by means of preloaded springs. The disk or plate cams are more popular cams because of their simplicity and compactness.

The conjugate cam has double disk cams which are constantly in contact with one follower. An example of oscillating roller follower conjugate cam is shown in Fig. 14. The conjugate cam is preferred when high speed, high dynamic loads, low noise, low wear, and high degree of control of the follower are design requirements.

The spiral cam, shown in Fig. 15, is a face cam with a spiral groove which controls the motion of either oscillating or translating follower. The application of this type of cam is limited because the cam has to rotate in the reverse direction to reset the follower position.

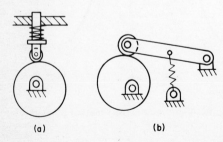

(a)

(b)

Fig. 13 a. Disk cam with translating roller follower. **b.** Disk cam with oscillating roller follower.

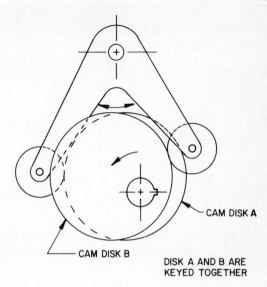

Fig. 14 Conjugate cam.

CAM DISK A

CAM DISK B

DISK A AND B ARE
KEYED TOGETHER

The globoidal or barrel cam is shown in Fig. 16. The motion to the follower is imported by the circumferential contour cut into the surface of rotation of the cam. There are two types of globoidal cams. The type is determined by the surface of the cam. The surface could be either convex or concave as shown in

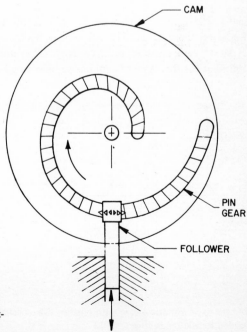

CAM

PIN
GEAR

FOLLOWER

Fig. 15 Spiral cam with translating follower.

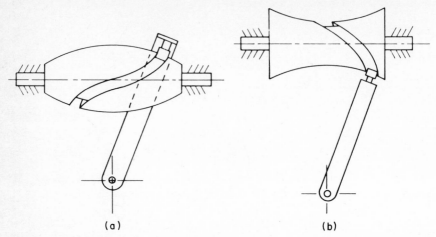

(a)   (b)

Fig. 16 a. Convex globoidal cam with oscillating follower. b. Concave globoidal cam with oscillating follower.

Fig. 16a and b. The globoidal cam is preferred when the angle of oscillation of the follower is large. Because of the groove on the cam surface, the application of the cam is limited to moderate speed.

The barrel cam shown in Fig. 17 is often called a cylindrical or a drum cam. This type of cam has circumferential contour cut in the surface of the cylinder. The cam rotates about the axis of the cylinder. There are two types of barrel cams. The type of the cam is determined by the manner in which the follower is imparted the motion. The cylinder cam with a groove is shown in Fig. 17a. This type of cam assumes positive drive. The end cam is shown in Fig. 17b. The follower for this type of cam has to be spring loaded.

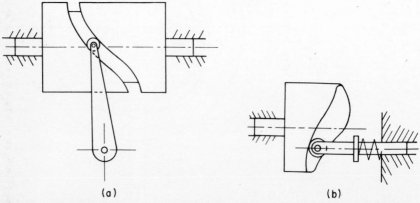

(a)   (b)

Fig. 17 a. Cylindrical cam with positive drive oscillating follower. b. End cam with spring-loaded translating follower.

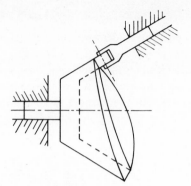

Fig. 18 Conical cam.

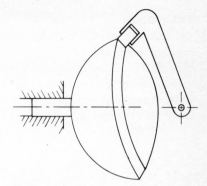

Fig. 19 Spherical cam with oscillating follower.

The conical cam is shown in Fig. 18. The follower of the conical cam translates along the line which generates the cone if revolved about the axis of the cone. The conical cams are expensive to fabricate and, therefore, have limited use.

The spherical cam is shown in Fig. 19. The cam is made from a spherical surface which transmits motion to the follower. The follower oscillates about an axis which is perpendicular to the axis of rotation of the cam. The spherical cams are similar to conical cams in cost and fabrication and have limited use. The advantage of using a spherical cam rahter than a disk cam is that with the spherical cam one can obtain oscillatory motion about an axis which is not parallel to the axis of rotation of the cam.

Competency Items
- Name the two main groups into which followers fall according to their motion characteristics.
- Name six different types of cam shapes.
- Write an application for a cam of each type.

Objective 3

To identify the notations of a cam follower system

Activities
- Read the material provided below, and as your read:
   a. Define all the terminologies of a cam.
   b. Lay out a disk cam profile with different types of followers.
- Check your answers to the competency items with your instructor.

Reading Material

In order to design a cam and a follower, the student should first define the necessary terminologies.

The final contour or shape of a cam is decided by the displacement program of its follower. In the previous objective we have studied the different types of displacement programs a follower may have. For the purpose of constructing a cam profile, examine a dwell-rise-return motion program, shown in Fig. 20.

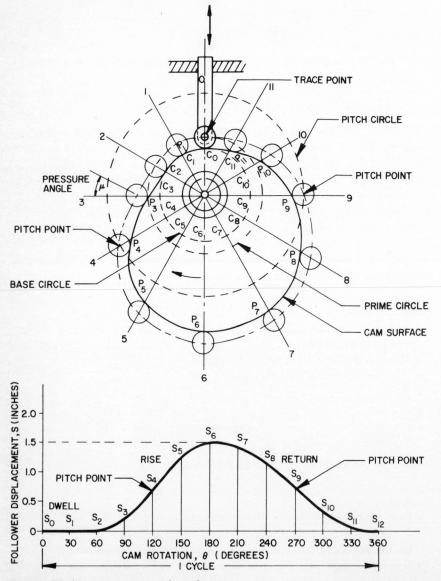

Fig. 20 Displacement program and cam layout.

From the displacement program, we can write the following boundary conditions for the follower motion:

$\theta = 0°$ $S = 0.0$ in.

$\theta = 60°$ $S = 0.0$ in.

$\theta = 180°$ $S = 1.5$ in.

$\theta = 360°$ $S = 0.0$ in.

The cam profile for a roller translating follower is constructed using the following steps:

1. Select a radius for a base circle with $r_b = 0.8$ in. and draw a base circle as shown in Fig. 20.
2. Divide the circumference of the base circle into 12 equal parts since we know from Fig. 20 the displacements of the follower for every 30° of the cam rotation angle. These 12 parts are shown as $C_0, C_1, C_2, \ldots, C_{11}$. These points $C_0, C_1, \ldots, C_{11}$ are located in the direction opposite to that of cam rotation.
3. Select a radius of roller follower $r_f = 0.25$ in. The radius of the roller follower is required to be smaller than the minimum radius of curvature of the cam profile. Hence, a student is required to perform some calculations before a radius of a roller follower is selected.
4. Locate centers $P_0, P_1, \ldots, P_{11}$ of the roller follower by measuring radially from points $C_0, C_1, \ldots, C_{11}$ distances $Y_i = S_i + r_f$ for $i = 0, \ldots, 11$.
5. With $P_0, P_1, \ldots, P_{11}$ as centers and r_f as radius draw circles to represent positions of roller follower.
6. Draw a continuous curve which is tangent to the circles obtained in Step 5. This curve describes the cam profile for a translating roller follower.

In laying out the cam profile, we have selected only 12 positions of the follower. Often, it becomes necessary to have positions of the follower located at every one-degree increment. It is important to note that the basic principle employed in developing the cam surface is the principle of inversion. The cam is held stationary and the follower is rotated in the direction opposite to that of the cam rotation.

The reader should become familiar with the following notations in describing the cam profile characteristics:

Base circle—the smallest circle tangent to the cam profile. See Fig. 20.

Pitch circle—the circle passing through the pitch point and concentric with the base circle.

Pitch curve—the curve traced by the center of the roller follower or the knife-edge follower. Thus, the trace point traces the pitch curve.

Pitch point—describes the position of the point on the pitch curve at which the pressure angle is maximum.

Pressure angle–the angle between a normal to the pitch curve and the direction of follower motion. The pressure angle changes its magnitude at every instant of follower motion. The maximum value of pressure angle plays a significant role in the design of cam; the smaller the maximum pressure angle, the better the design. With too large a pressure angle, it is possible that the follower might jam in its bearing.

Prime circle–the smallest circle tangent to the pitch curve.

Trace point–the point coinciding with the knife edge of the follower or the center of the roller follower.

Competency Items

- Make up a dwell-rise-dwell-return motion program and lay out a cam for a translating roller follower.
- Select different values for a prime circle radius and lay out the corresponding cam profile for an assumed displacement program. Observe critically and state:
  a. how the maximum pressure angle changes with different radii of the prime circle radius
  b. how the size of the cam changes with the maximum values of pressure angles

Performance Test #1

For each of the curves below describing follower displacements, lay out a cam profile.

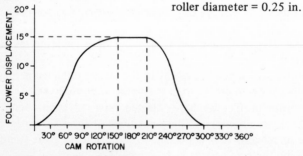

roller diameter = 0.25 in.

a. Swinging roller follower

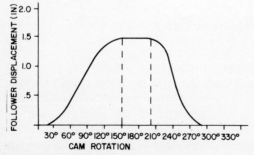

b. Translating flat face follower

Performance Test #2

Using the displacement diagram given, lay out the cam profile for the flat face follower shown. Find the minimum value of angle α so that contact will always exist between the cam and the circular face of the follower.

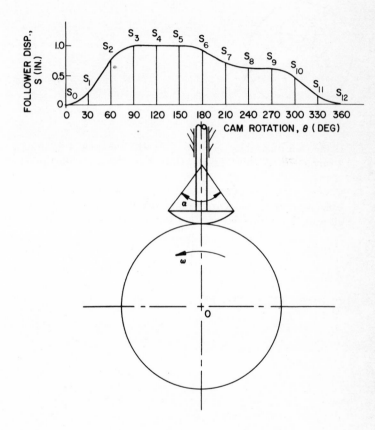

Performance Test #3

1. Sketch an offset knife edge follower.
2. Sketch a positive drive cam mechanism for a translating flat face follower.
3. State the type of cam and follower that is obtained when the half-cone angle β becomes $90°$.

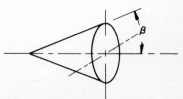

β is the half-cone angle

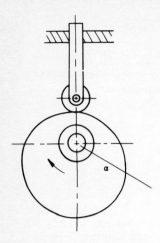

4. On the sketch indicate the pressure angle of the cam-follower system when radial line (a) is vertically upward.

5. If the cam of problem 4 were made with the same motion program but with a larger prime circle radius, state how the maximum pressure angles will change.

6. Choose correct answer within the parenthesis and explain your choice: A cam has several consecutive points where the pressure angle is zero. Describe all situations when this pressure angle becomes zero.

Supplementary References

1 (9.1-9.6); **3** (7.1, 7.9-7.13); **6** (1.1-2.2); **7** 7.1-7.2); **8** (8.1-8.3, 8.5)

Unit IX. Motion Programs

Objectives
1. To identify applications of custom-made motion programs in cam design
2. To synthesize motion programs using the analytical technique

Objective 1

To identify applications of custom-made motion programs

Activities
- Read the material provided below, and as you read:
  - a. Describe the criteria to test a given motion program.
  - b. Name the characteristics of the following motion programs:

    constant velocity motion program
    constant acceleration motion program
    simple harmonic motion program
    modified harmonic motion program
    cycloidal motion program
    cubic curve motion program
    polynomial curves for follower motion program
    trapezoidal motion program

  - c. Demonstrate their application in designing cams.
  - d. Name the advantages and disadvantages of using custom-made motion programs.

- Check your answers to the competency items with your instructor.

Reading Material

A cam is usually fabricated using the displacement program. However, such a displacement program is arrived at by using a series of boundary conditions describing displacement, velocity, acceleration, and jerk at different intervals of cam rotation angle.

The quality of a motion program is measured from the different types of boundary conditions that it satisfies. If a displacement program is available, then the velocity, acceleration, and jerk programs of the follower are obtained by taking time derivative of the displacement program. Thus, if s describes the displacement at any time t, then

$$\text{Velocity} = v = \frac{ds}{dt} \text{ in./sec}$$

$$\text{Acceleration} = a = \frac{d^2 s}{dt^2} \text{ in./sec}^2$$

$$\text{Jerk} = J = \frac{d^3 s}{dt^3} \text{ in./sec}^3$$

In this section, we will study several different motion programs which are synthesized in the part by other kinematicians and recommended to adopt in cam design when special requirements are satisfied. Our purpose is to determine the situations when these programs could be best put to work. A good working knowledge of the existing motion program will save time and labor from synthesizing some unwanted motion programs.

The three most often encountered motion programs for a follower are (1) rise-return (R-R), (2) dwell-rise-return (D-R-R), and (3) dwell-rise-dwell-return (D-R-D-R) motion programs. Before selecting any one of the custom-made programs, one is required to check the boundary conditions describing displacement, velocity, and acceleration requirements of the follower. The jerk is undesirable for a follower motion. However, it is difficult to eliminate the jerk. The most one can accomplish is to keep the jerk at a tolerable value. The existence of large values of jerk prohibit high speed operation for a cam. Let us now examine different custom-made motion programs. The following notations are used:

s = displacement of follower, in.
H = maximum displacement of follower, in.
θ = cam rotation angle
t = time to rotate cam angle θ
β = cam rotation angle for maximum rise displacement H
τ = time for cam to rotate through angle β
v = velocity of follower in./sec
a = acceleration, in./sec^2
J = jerk, in./sec^3

Constant velocity motion program

Let us examine the characteristic of dwell-rise-dwell-return program synthesized using linear relationship between input cam rotation and follower displacement. Equation of straight line passing through the origin is given by

$$s = C\theta \tag{1}$$

where C is the slope of the straight line.

If the follower were to have a maximum displacement of H corresponding to cam rotation angle β, then

$$H = C\beta \tag{2}$$

or

$$C = \frac{H}{\beta}$$

Therefore, the motion program is described by

$$s = H\left(\frac{\theta}{\beta}\right) \tag{3}$$

Successive time derivative of Eq. (3) will give the velocity and the acceleration of the follower.

$$v = \frac{ds}{dt} = \frac{H\omega}{\beta} \tag{4}$$

where ω = angular velocity of cam.
Since, in general, a cam is driven with a constant angular velocity,

$$v = \frac{H\omega}{\beta} = \text{constant}$$

Therefore,

$$a = \frac{dv}{dt} = 0 \tag{5}$$

Figure 1 shows a plot of displacement, velocity, and acceleration of the constant velocity program.

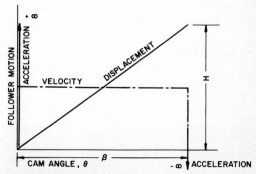

Fig. 1 Displacement, velocity, and acceleration of a constant velocity motion program.

Note that the constant velocity program has a straight-line displacement curve. The displacement of the follower is uniform, has constant velocity, and zero acceleration. The dwelling conditions at the beginning and at the end will create an impractical situation; there will be sudden changes in velocity thus giving infinite acceleration to the follower. This in turn will give rise to infinite forces. Such a situation, however, must be prevented unless the cam is to be driven at a very slow speed. The constant velocity motion program is often used in combination with other types of motion programs.

Constant acceleration motion program

The constant acceleration program yields a second-degree curve for the displacement of the follower. See Fig. 2. The first half of the rise cycle describes positive constant acceleration and the second half describes negative constant acceleration. The displacement relation for the first half may be expressed as

$$s = C\theta^2 \tag{6}$$

We need to replace constant C in terms of parameters H and β. For the first half of the rise programs, the above equation must satisfy the following condition: at

$$\theta = \frac{\beta}{2} \quad s = \frac{H}{2}$$

Substituting the above conditions in Eq. (6), we get

$$C = \frac{2H}{\beta^2}$$

Thus, Eq. (6) becomes

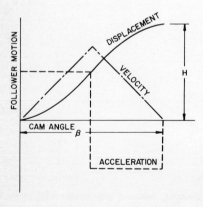

Fig. 2 Displacement, velocity, and acceleration of constant acceleration program.

$$s = 2H\left(\frac{\theta}{\beta}\right)^2 \tag{7}$$

Time derivatives of Eq. (7) will give the velocity and the acceleration of the constant acceleration motion program:

$$v = \frac{ds}{dt} = \frac{4H\omega}{\beta^2} \cdot \theta \tag{8}$$

$$a = \frac{4H\omega^2}{\beta^2} \tag{9}$$

For the second half of the rise portion of the motion program, the displacement relationship can be written as

$$s = C_1\theta^2 + C_2\theta + C_3 \tag{10}$$

Constants C_1, C_2, and C_3 must be determined in a manner that the total displacement program will appear as one smooth curve. Since there are three constants to be evaluated, we must have three independent conditions. These conditions are:

$$\text{at} \quad \theta = \beta \quad s = H$$

$$\text{at} \quad \theta = \beta \quad \frac{ds}{dt} = v = 0$$

$$\text{at} \quad \theta = \frac{\beta}{2} \quad v = \frac{2H\omega}{\beta}$$

The last condition is obtained from the first half of the rise program. Substituting these three conditions in Eq. (10), we get

$$C_1 = -\frac{2H}{\beta^2} \quad C_2 = \frac{4H}{\beta} \quad C_3 = -H$$

and the displacement program for $\beta/2 < \theta < \beta$ becomes

$$s = H\left[1 - 2\left(1 - \frac{\theta}{\beta}\right)^2\right] \tag{11}$$

Successive time derivatives of Eq. (11) will give velocity and acceleration. Thus, for $\beta/2 < \theta < \beta$

$$v = \frac{ds}{dt} = \frac{4H\omega}{\beta}\left(1 - \frac{\theta}{\beta}\right) \tag{12}$$

$$a = \frac{d^2s}{dt^2} = -\frac{4H\omega^2}{\beta^2} \tag{13}$$

Note there are several apparent advantages in selecting the constant acceleration motion program. There are abrupt changes in the acceleration at the beginning and end of the dwells. The maximum acceleration is smaller in comparison with other programs. However, the acceleration has a discontinuity at $\theta = \beta/2$. This discontinuity produces an infinite jerk. Also, since it is not possible to build perfectly rigid members with no backlash or clearance in the system, the constant acceleration program is not expected to provide satisfactory performance. This program is normally recommended for moderate speeds.

Simple harmonic motion program

A simple harmonic motion program is shown in Fig. 3. Note that the displacement program is a cosine function given as

$$s = C(1 - \cos\phi) \tag{14}$$

If the follower were to have a maximum lift or rise of H within cam rotation angle β, then

$$\phi = \frac{\pi\theta}{\beta} \quad \text{and} \quad C = \frac{H}{2}$$

Thus, the displacement motion program is given by

$$s = \frac{H}{2}\left(1 - \cos\frac{\pi\theta}{\beta}\right) \tag{15}$$

Time derivatives of Eq. 15 yield velocity and acceleration for the simple harmonic motion program

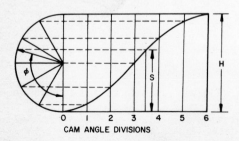

Fig. 3 Construction of a simple harmonic motion program.

$$v = \frac{ds}{dt} = \frac{H\pi\omega}{2\beta} \sin \frac{\pi\theta}{\beta} \qquad (16)$$

$$a = \frac{d^2s}{dt^2} = \frac{H}{2} \left(\frac{\pi\omega}{\beta}\right)^2 \cos \frac{\pi\theta}{\beta} \qquad (17)$$

Note that velocity is a sine function and acceleration is a cosine function. For the purpose of comparison, displacement, velocity, and acceleration of a simple harmonic motion program are plotted in Fig. 4. Because of the cosine function, the displacement program is fairly smooth. However, because of the dwelling conditions and because of the nature of the simple harmonic motion program, there is an abrupt change in acceleration at the beginning and at the end of dwell. These abrupt changes cause infinite jerk, noise, and intolerable vibration in the cam and follower system. If the cam were to rotate at a moderate speed, then the simple harmonic motion program is an acceptable motion program. Because of its simplicity in layout and because it is described by a simple trigonometric function, the simple harmonic motion program has become one of the most popular motion programs.

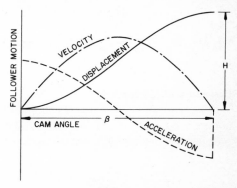

Fig. 4 Displacement, velocity, and acceleration of a simple harmonic motion program.

Modified harmonic motion program

The simple harmonic motion program may be modified by considering the motion program synthesized by taking the difference of two harmonic motion programs. Such a modified simple harmonic function may be expressed by

$$s = \frac{H}{2} \left[\left(1 - \cos \frac{\pi\theta}{\beta}\right) - \frac{1}{4}\left(1 - \cos \frac{2\pi\theta}{\beta}\right) \right] \qquad (18)$$

Time derivative of Eq. (18) will give velocity and acceleration.

$$v = \frac{ds}{dt} = \frac{H\pi\omega}{2\beta} \left(\sin \frac{\pi\theta}{\beta} - \frac{1}{2} \sin \frac{2\pi\theta}{\beta}\right) \qquad (19)$$

$$a = \frac{d^2 s}{dt^2} = \frac{H}{2}\left(\frac{\pi\omega}{\beta}\right)^2 \left(\cos\frac{\pi\theta}{\beta} - \cos\frac{2\pi\theta}{\beta}\right) \tag{20}$$

For the purpose of comparison, Fig. 5 shows the displacement, velocity, and acceleration plots of a modified simple harmonic motion program. Note that the abrupt changes in the acceleration and velocity of the simple harmonic motion program at the beginning of the stroke are eliminated in the modified harmonic motion program. However, these changes do exist only at the beginning of the rise. Hence, the modified simple harmonic motion program is found quite suitable for the dwell-rise-return-dwell motion program.

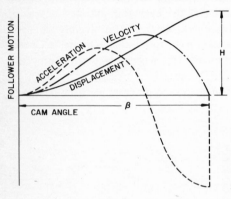

Fig. 5 Displacement, velocity, and acceleration of a modified harmonic motion program.

Cycloidal motion program

The cycloidal motion program is physically visualized when the locus of a point on a circular disc rolling on a straight line. Mathematically the cycloidal program is expressed by

$$s = \frac{H}{\pi}\left(\frac{\pi\theta}{\beta} - \frac{1}{2}\sin\frac{2\pi\theta}{\beta}\right) \tag{21}$$

Successive time derivatives of Eq. (21) will give velocity and acceleration of the follower having the cycloidal motion program.

$$v = \frac{ds}{dt} = \frac{H\omega}{\beta}\left(1 - \cos\frac{2\pi\theta}{\beta}\right) \tag{22}$$

$$a = \frac{d^2 s}{dt^2} = \frac{2H\pi\omega^2}{\beta^2}\sin\frac{2\pi\theta}{\beta} \tag{23}$$

Figure 6 shows the plots of displacement, velocity, and acceleration of a cycloidal motion program.

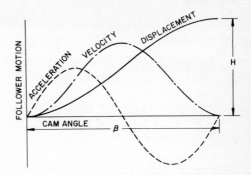

Fig. 6 Displacement, velocity, and acceleration of a cycloidal motion program.

Note there are no abrupt changes in the velocity and the acceleration either at the beginning of the stroke or at the end of the rise portion of the program. The cycloidal motion program is, therefore, the most suitable program for the D-R-D-R motion program. Also, if accuracy can be maintained in machining the cam surface, especially at the beginning and at the end, then, the cycloidal motion program is acceptable for high speed operation.

Cubic curve motion program

The cubic curve displacement motion programs are primarily developed to improve the motion characteristic of the constant acceleration motion program. There are three different types of cubic curve programs.* Despite the improvements over the constant acceleration motion program, these cubic curve motion programs are poor for high speed operation because the jerk is infinite either at $\theta = \beta/2$ or $\theta = 0$, high maximum acceleration, and large velocities. These factors force the cams to be of large size, and make the machining of cams that much more difficult.

The first type of cubic curve has the following displacement, velocity, and acceleration programs:

For the interval $0 < \theta < \beta/2$,

$$s = 4H \left(\frac{\theta}{\beta}\right)^3 \tag{24}$$

$$v = \frac{12H\omega}{\beta} \left(\frac{\theta}{\beta}\right)^2 \tag{25}$$

$$a = \frac{24H\omega^2}{\beta^2} \left(\frac{\theta}{\beta}\right) \tag{26}$$

*Only two of the three programs will be considered here.

$$J = \frac{24H\omega^3}{\beta^3} \tag{27}$$

For the interval $\beta/2 < \theta < \beta$,

$$s = H\left[1 - 4\left(1 - \frac{\theta}{\beta}\right)^3\right] \tag{28}$$

$$v = \frac{12H\omega}{\beta}\left(1 - \frac{\theta}{\beta}\right)^2 \tag{29}$$

$$a = -\frac{24H\omega^2}{\beta^2}\left(1 - \frac{\theta}{\beta}\right) \tag{30}$$

$$J = \frac{24H\omega^3}{\beta^3} \tag{31}$$

The displacement, velocity, and acceleration characteristics of cubic curve, Type I, are shown in Fig. 7.

The cubic curve, Type 2, has continuous acceleration characteristic rather than the abrupt change as in the case of cubic curve, Type 1. However, there is abrupt change in acceleration at the beginning and the end of the rise.

The displacement, velocity, and acceleration relationship of cubic curve, Type 2, are given below:

$$s = H\frac{\theta^2}{\beta^2}\left(3 - \frac{2\theta}{\beta}\right) \tag{32}$$

$$v = \frac{6H\omega\theta}{\beta^2}\left(1 - \frac{\theta}{\beta}\right) \tag{33}$$

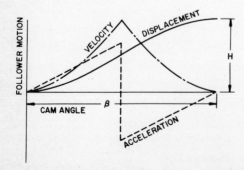

Fig. 7 Cubic curve, Type 1.

$$a = \frac{6H\omega^2}{\beta^2}\left(1 - \frac{2\theta}{\beta}\right) \tag{34}$$

$$J = -\frac{12H\omega^3}{\beta^3} \tag{35}$$

The displacement, velocity, and acceleration characteristics of cubic curve, Type 2, are shown in Fig. 8.

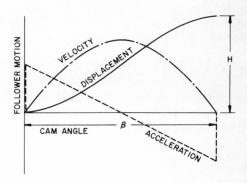

Fig. 8 Cubic curve, Type 2.

Polynomial curves for follower motion program

The custom-made motion programs examined in the previous section are capable of satisfying the general conditions such as H, β, etc. If one needs a motion program which can satisfy some specific conditions in addition to the general conditions, then one is required to look at the polynomial curves which may satisfy the given conditions. The general expression for a polynomial is given by

$$s = D_0 + D_1\theta + D_2\theta^2 + D_3\theta^3 + \cdots + D_n\theta^n \tag{36}$$

where
  s = displacement of follower
  D_i = constants $(i = 0, 1, \ldots n)$
  θ = cam rotation angle to correspond to s

The constants in the above polynomial are evaluated by making use of the boundary conditions involving displacement, velocity, and acceleration of follower motion.

The constant velocity and constant acceleration programs examined earlier are the special cases of the polynomical curve given by Eq. (36).

Let us now consider a problem in which a D-R-D motion program is to be synthesized. Since there is a dwell at the beginning of the rise, we have the following boundary conditions:

at

$$\theta = 0 \quad s = 0$$
$$\dot{s} = 0$$
$$\ddot{s} = 0$$

Since there is a dwell at the end of the rise, we have the following boundary conditions:

at

$$\theta = \beta \quad s = H$$
$$\dot{s} = 0$$
$$\ddot{s} = 0$$

There are six boundary conditions. We can, therefore, evaluate six constants $C_0, C_1, C_2, C_3, C_4,$ and C_5. Hence, the polynomial curve satisfying the above six conditions must be of fifth order. That is,

$$s = D_0 + D_1\theta + D_2\theta^2 + D_3\theta^3 + D_4\theta^4 + D_5\theta^5 \tag{37}$$

The time derivatives of Eq. (37) to obtain velocity and acceleration of the follower are

$$\dot{s} = D_1\omega + 2D_2\omega\theta + 3D_3\omega\theta^2 + 4D_4\omega\theta^3 + 5D_5\omega\theta^4 \tag{38}$$
$$\ddot{s} = 2\omega^2 D_2 + 6D_3\omega^2\theta + 12D_4\omega^2\theta^2 + 20D_5\omega^2\theta^3 \tag{39}$$

The six boundary conditions when substituted in Eqs. (37), (38), (39), will obtain six equations

$$0 = D_0 \tag{40}$$
$$H = D_0 + D_1\beta + D_2\beta^2 + D_3\beta^3 + D_4\beta^4 + D_5\beta^5 \tag{41}$$
$$0 = \omega D_1 \tag{42}$$
$$0 = \omega D_1 + 2\omega D_2\beta + 3\omega D_3\beta^2 + 4\omega D_4\beta^3 + 5\omega D_5\beta^4 \tag{43}$$
$$0 = 2\omega^2 D_2 \tag{44}$$
$$0 = 2\omega^2 D_2 + 6\omega^2 D_3\beta + 12\omega^2 D_4\beta^2 + 20\omega^2 D_5\beta^3 \tag{45}$$

Simultaneous solution of these equations yields

$$D_0 = 0$$
$$D_1 = 0$$
$$D_2 = 0$$

$$D_3 = \frac{10H}{\beta^3}$$

$$D_4 = -\frac{15H}{\beta^4}$$

$$D_5 = \frac{6H}{\beta^5}$$

After substituting the values of the constants, the displacement equation for D-R-D motion program becomes

$$s = \frac{10H}{\beta^3}\theta^3 - \frac{15H}{\beta^4}\theta^4 + \frac{6H}{\beta^5}\theta^5 \qquad (46)$$

The corresponding velocity, acceleration, and jerk expressions are

$$v = \frac{30H\omega}{\beta^3}\theta^2 - \frac{60H\omega}{\beta^4}\theta^3 + \frac{30H\omega}{\beta^5}\theta^4 \qquad (47)$$

$$a = \frac{60H\omega^2}{\beta^3}\theta - \frac{180H\omega^2}{\beta^4}\theta^2 + \frac{120H\omega}{\beta^5}\theta^3 \qquad (48)$$

$$J = \frac{60H\omega^3}{\beta^3} - \frac{360H\omega^3}{\beta^4}\theta + \frac{360H\omega^3}{\beta^5}\theta^2 \qquad (49)$$

The motion program described by the polynomial curve described by Eq. (46) is shown in Fig. 9.

The polynomial curves with power greater than 2 display small displacements at the beginning and end of the follower motion. Extreme care is, therefore, required in cutting cam profile describing polynomial curve.

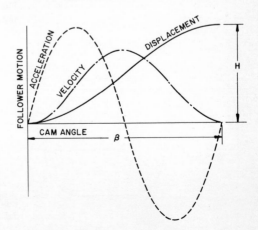

Fig. 9 3-4-5 Polynomial motion curves.

Trapezoidal motion program

There are several different types of trapezoidal motion programs. These are shown in Figs. 10, 11, and 12. The simplest type of trapezoidal motion program is shown in Fig. 10. This trapezoidal acceleration program of the follower is made up of straight lines. The trapezoidal acceleration has finite and tolerable amount of jerk, low value for its maximum acceleration. However, because of the discontinuities at $\theta = \beta/8$, $3\beta/8$, $5\beta/8$, and $7\beta/8$, the trapezoidal acceleration program of Fig. 10 will prevent using it in high speed operation.

The trapezoidal acceleration program of Fig. 10 is modified to get the modified trapezoidal acceleration program shown in Fig. 11. This program

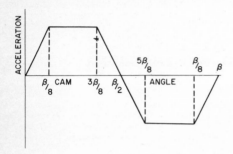

Fig. 10 Trapezoidal motion program.

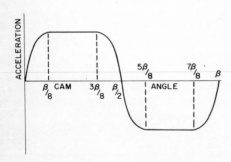

Fig. 11 Modified trapezoidal motion program.

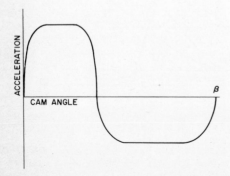

Fig. 12 Modified skew trapezoidal motion program.

consists of cycloidal and constant acceleration program. The mathematical relationships for displacement, velocity, and acceleration are: For $0 \leq \theta \leq \beta/8$,

$$s = 0.09725 \left[4 \frac{\theta}{\beta} - \frac{1}{\pi} \sin\left(4\pi \frac{\theta}{\beta}\right) \right] H \tag{50}$$

$$v = 0.38898 \left[1 - \cos\left(4\pi \frac{\theta}{\beta}\right) \right] \frac{H}{\beta} \tag{51}$$

$$a = 4.88812 \sin\left(4\pi \frac{\theta}{\beta}\right) \frac{H}{\beta^2} \tag{52}$$

For $\beta/8 \leq \theta \leq 3\beta/8$,

$$s = \left[2.44406 \left(\frac{\theta}{\beta}\right)^2 - 0.22203 \frac{\theta}{\beta} + 0.00723 \right] H \tag{53}$$

$$v = \left(4.88812 \frac{\theta}{\beta} - 0.22203 \right) \frac{H}{\beta} \tag{54}$$

$$a = 4.88812 \frac{H}{\beta^2} \tag{55}$$

For $3\beta/8 \leq \theta \leq \beta/2$,

$$s = \left[1.61102 \frac{\theta}{\beta} - 0.03095 \sin\left(4\pi \frac{\theta}{\beta} - \pi\right) - 0.30551 \right] H \tag{56}$$

$$v = \left[1.61105 - 0.38898 \cos\left(4\pi \frac{\theta}{\beta} - \pi\right) \right] \frac{H}{\beta} \tag{57}$$

$$a = 4.88812 \sin\left(4\pi \frac{\theta}{\beta} - \pi\right) \frac{H}{\beta^2} \tag{58}$$

For $\beta/2 \leq \theta \leq 5\beta/8$,

$$s = \left[1.61102 \frac{\theta}{\beta} + 0.93095 \sin\left(4\pi \frac{\theta}{\beta} - 2\pi\right) - 0.30551 \right] H \tag{59}$$

$$v = \left[1.61102 + 0.38898 \cos\left(4\pi \frac{\theta}{\beta} - 2\pi\right) \right] \frac{H}{\beta} \tag{60}$$

$$a = -4.88124 \sin\left(4\pi \frac{\theta}{\beta} - 2\pi\right) \frac{H}{\beta^2} \tag{61}$$

For $5\beta/8 < \theta < 7\beta/8$,

$$s = \left[4.66609 \frac{\theta}{\beta} - 2.44406\left(\frac{\theta}{\beta}\right)^2 - 1.22926 \right] H \tag{62}$$

$$v = \left(4.666091 - 4.88812 \frac{\theta}{\beta} \right) \frac{H}{\beta} \tag{63}$$

$$a = -4.88812 \frac{H}{\beta^2} \tag{64}$$

For $7\beta/8 < \theta < \beta$,

$$s = \left[0.61102 + 0.38898 \frac{\theta}{\beta} + 0.03095 \sin\left(4\pi \frac{\theta}{\beta} - 3\pi\right) \right] H \tag{65}$$

$$v = \left[0.38898 + 0.38898 \cos\left(4\pi \frac{\theta}{\beta} - 3\pi\right) \right] \frac{H}{\beta} \tag{66}$$

$$a = -4.88812 \sin\left(4\pi \frac{\theta}{\beta} - 3\pi\right) \frac{H}{\beta^2} \tag{67}$$

It should be noted that Eqs. (50)–(67) give y, v, and a in in., in./deg., and in./deg^2, respectively, thus v must be multiplied by $180\omega/\pi$ to give in./sec and α must be multiplied by $[(180/\pi)\omega]^2$ to give in./sec^2.

The displacement, velocity, and acceleration of a modified trapezoidal acceleration program are shown in Fig. 13.

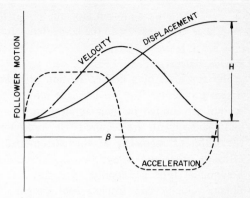

Fig. 13 Modified trapezoidal acceleration program.

The modified trapezoidal motion program presented above is a symmetric curve. Because of the symmetry, the velocity becomes maximum at $\theta = \beta/2$. Quite often, it is required that the follower have a maximum velocity when either $\theta > \beta/2$ or $\theta < \beta/2$. This requirement provides opportunity to further modify the modified trapezoidal motion program. Such a modified program is called a *skew modified trapezoidal program*. The general method of synthesizing such an acceleration motion program is presented in the next objective.

Competency Items
- For given values of H and β, state which of the basic motion programs involves the lowest value of peak acceleration. List the limitations of this motion program.
- For given values of ω, H, and β, derive expressions for maximum velocity and maximum acceleration of the following motion programs:
  a. trapezoidal acceleration
  b. modified trapezoidal acceleration
  c. cycloidal
  d. simple harmonic
- Compare the four motion programs above by listing each in order of maximum velocities, again for maximum acceleration.
- Name the constraint that must be imposed when skewing a motion program for a rise or a return between two dwells.

Objective 2

To synthesize motion programs using the analytical technique

Activities
- Read the material provided below, and as you read:
  a. Synthesize a motion program with predefined acceleration.
  b. Synthesize a motion program to match velocity.

   c. Synthesize a motion program to match displacement.

• Check your answers to the competency items with your instructor.

Reading Material

The synthesis of a desired motion program is quite often a requirement in design problems. The custom-made motion programs examined in the previous section are limited in their application when control of acceleration, velocity, and displacements are required. The motion program could be synthesized using the different segments of the custom-made motion programs. However, an efficient procedure in synthesizing the motion program is to stay with the polynomial type acceleration program and proceed in a stepwise manner to obtain velocity and displacement program. The control of the acceleration program provides the control of force, jerk, velocity, and displacement.

In this section we will examine the procedure of synthesizing a motion program from a selected acceleration program. The following steps are required to synthesize the motion program:

Write an equation for each curve of the acceleration versus time diagram.
Integrate each equation successively to obtain velocity versus time and displacement versus time diagrams.
Use end conditions of each curve to solve for constraints of integration.
Write all equations in terms of maximum acceleration A_{max} and rise time T.

We will now apply this technique to obtain velocity and displacement programs which obey the A - t program shown in Fig. 14.

We will derive the velocity and displacement equation for the three time intervals $0 < t < T/4$, $T/4 < t < T/2$, and $T/2 < t < T$.

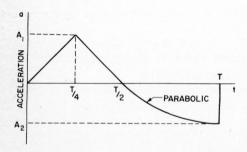

Fig. 14 A-t diagram.

First interval $0 < t < T/4$
From Fig. 15,

$$a = Kt \tag{68}$$

The component K can be evaluated by using the following boundary conditions:

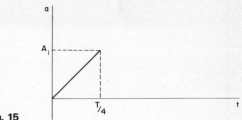

Fig. 15

at

$$t = \frac{T}{4} \qquad a = A_1$$

Hence,

$$K = \frac{4A_1}{T}$$

The equation for the A - t program is

$$a = 4A_1 \frac{t}{T} \tag{69}$$

Integration of Eq. (69) gives the velocity program

$$v = \frac{4A_1}{T} \frac{t^2}{2} + C_1$$

At

$$t = 0 \qquad v = 0$$

Therefore,

$$C_1 = 0$$

Hence,

$$v = \frac{2A_1}{T} t^2 \qquad 0 \leq t \leq \frac{T}{4} \tag{70}$$

Integration of Eq. (70) gives the displacement program

$$s = \frac{2A_1}{T} \frac{t^3}{3} + C_2 \tag{71}$$

At

$$t = 0, \qquad s = 0$$

Therefore,

$$C_2 = 0$$

Hence,

$$s = \frac{2}{3}\frac{A_1}{T}t^3 \qquad 0 < t \le \frac{T}{4} \tag{72}$$

Second interval $T/4 < t < T/2$
From Fig. 16,

$$a = -K\left(t - \frac{T}{2}\right)$$

At

$$t = \frac{T}{4} \quad a = A_1$$

Therefore,

$$K = \frac{4A_1}{T}$$

Hence,

$$a = -\frac{4A_1}{T}\left(t - \frac{T}{2}\right) \tag{73}$$

Integration of Eq. (73) gives the velocity program

$$v = -\frac{4A_1}{T}\frac{(t - T/2)^2}{2} + C_1' = -\frac{2A_1}{T}t^2 + 2A_1 t + C_1 \tag{74}$$

At

$$t = \frac{T}{4}$$

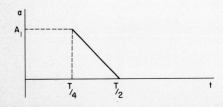

Fig. 16

from Eq. (70),

$$v = \frac{2A_1}{T}\left(\frac{T}{4}\right)^2 = \frac{A_1 T}{8}$$

From Eq. (74)

$$v = -\frac{A_1 T}{8} + 2A_1 \frac{T}{4} + C_1$$

Since the velocity at $t = T/4$ must be equal from either side of the A-t diagram,

$$\frac{A_1 T}{8} = -\frac{A_1 T}{8} + 2A_1 \frac{T}{4} + C_1$$

or

$$C_1 = -\frac{A_1 T}{4}$$

Hence Eq. (74) becomes

$$v = -\frac{2A_1}{T} t^2 + 2A_1 t - \frac{A_1 T}{4} \qquad \frac{T}{4} \le t \le \frac{T}{2} \tag{75}$$

Integration of Eq. (75) gives the displacement program

$$s = -\frac{2A_1}{T}\frac{t^3}{3} + \frac{2A_1 t^2}{2} - \frac{A_1 T}{4}t + C_2 \tag{76}$$

At

$$t = \frac{T}{4}$$

Eq. (72) gives

$$s = \frac{A_1 T^2}{96}$$

and Eq. (76) gives

$$s = -\frac{2A_1}{3T}\frac{T^3}{64} + \frac{A_1 T^2}{16} - \frac{A_1 T}{4}\frac{T}{4} + C_2 \tag{77}$$

Since the displacement of the follower must be the same at $t = T/4$, even though it is computed from two different relationships,

$$C_2 = \frac{A_1 T^2}{48}$$

Hence,

$$s = -\frac{2A_1}{3T} t^3 + A_1 t^2 - \frac{A_1 T}{4} t + \frac{A_1 T^2}{48} \qquad \frac{T}{4} \leq t \leq \frac{T}{2} \qquad (78)$$

Third time interval $T/2 < t < T$

The A-t diagram is shown in Fig. 17. Since the velocity is zero at $t = 0$ and $t = T$, and it becomes a maximum at $t = T/2$, $(a = 0)$, the areas under the acceleration curve from $t = 0$ to $t = T/2$ and from $t = T/2$ to $t = T$ must be equal. That is

$$-\frac{A_1 T}{4} = \frac{2}{3} A_2 \frac{T}{2}$$

or

$$A_2 = -\frac{3}{4} A_1$$

The equation for the acceleration program for the time interval $T/2 < t < T$ is given by

$$a = -\frac{3A_1}{4} + K(t - T)^2 \qquad (79)$$

At

$$t = \frac{T}{2} \qquad a = 0$$

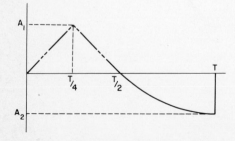

Fig. 17

Hence, from Eq. (79),

$$K = \frac{3A_1}{T^2}$$

and

$$a = -\frac{3A_1}{4} + \frac{3A_1 t^2}{T^2} - \frac{6A_1 t}{T} + 3A_1 \tag{80}$$

Integration of Eq. (80) gives velocity

$$v = \frac{A_1}{T^2} t^3 - \frac{6A_1}{T} \frac{t^2}{2} + 3A_1 t - \frac{3A_1}{4} t + C_1$$

at

$$t = T \qquad v = 0$$

Therefore,

$$O = \frac{A_1}{T^2} T^3 - 3A_1 T + 3A_1 T - \frac{3A_1}{4} T + C_1$$

or

$$C_1 = -A_1 T + \frac{3A_1 T}{4} = -\frac{A_1 T}{4}$$

and

$$v = \frac{A_1}{T^2} t^3 - \frac{3A_1}{T} t^2 + \frac{9}{4} A_1 t - \frac{A_1 T}{4} \tag{81}$$

Integration of Eq. (81) gives displacement

$$s = \frac{A_1}{T^2} \frac{t^4}{4} - \frac{3A_1}{T} \frac{t^3}{3} + \frac{9A_1}{4} \frac{t^3}{2} - \frac{A_1 T}{4} t + C_2 \tag{82}$$

At

$$t = \frac{T}{2}$$

from Eq. (78)

$$y = \frac{A_1 T^2}{16}$$

Hence from Eq. (82)

$$\frac{A_1 T^2}{16} = \frac{A_1}{4T^2}\frac{T^4}{16} - \frac{A_1 T^3}{8T} + \frac{9A_1}{8}\frac{T^2}{4} - \frac{A_1 T}{4}\frac{T}{2} + C_2$$

and

$$s = A_1\left(\frac{t^4}{4T^2} - \frac{t^3}{T} + \frac{9}{8}t^2 - \frac{T}{4}t + \frac{T^2}{64}\right)$$

The maximum displacement can be determined by setting $s = H$ at $t = T$. Thus,

$$H = \frac{9A_1 T^2}{64}$$

or

$$A_1 = \frac{64H}{9T^2}$$

Competency Items

- Replace the second order curve in the $T/2 < t < T$ interval of the example with a program like that in the $0 < t < T/2$ interval, but inverted ($A_2 = -A_1$). Develop the equations to describe displacement, velocity, and acceleration for this motion program.
- Replace the two straight lines on the $0 < t < T/2$ interval of the example with a second order curve like that in the $T/2 < t < T$ interval, but inverted ($A_1 = -A_2$). Develop the equations to describe displacement, velocity, and acceleration for this motion program.
- Show that the area at $x = x_1$ for the polynomial curve $y = kx^n$ is given by $y_1 x_1/(n + 1)$.

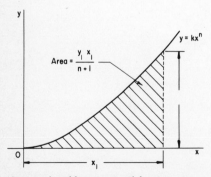

Area enclosed by exponential curve

Performance Test #1

1. Derive expressions for the maximum velocity and maximum acceleration of the 3-4-5 polynomial motion program. Express in terms of H, β and ω.
2. Derive the equations to describe displacement, velocity, and acceleration for a motion program to match the following boundary conditions.

at

$$\theta = 0 \quad s = 0 \quad v = 0$$
$$\theta = \beta \quad s = H \quad v = 0$$

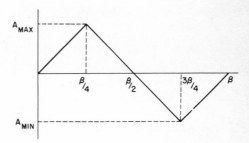

Performance Test #2

1. Derive expressions for the maximum velocity and acceleration of the simple harmonic motion program and the cycloidal motion program. Express in terms of H, β, and ω. Compare the values between the two programs.
2. Derive the equations for the motion program shown for the interval $T/2 < t < T$. The boundary conditions are:

at

$$t = 0 \quad s = 0 \quad v = 0$$
$$t = T \quad s = H \quad v = 0$$

Determine the value of A_1 and A_2 in terms of H and T.

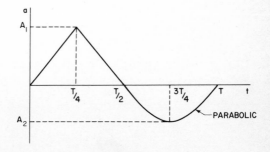

Performance Test #3

1. For the motion program whose displacement is given by

$$s = \frac{H}{2}\left[1 + \sin\left(\frac{\pi}{4} + \frac{\pi\theta}{\beta}\right)\right]$$

   derive relationships for the maximum velocity and acceleration in terms of H, β, and ω. Determine in terms of β at what angle these maximums occur. Compare these maximums with the maximums for simple harmonic motion and cycloidal motion.

2. Derive the equations for the motion program shown. The boundary conditions are:

   at ⁓

$$t = 0 \quad s = 0 \quad v = 0$$
$$t = T \quad s = H \quad v = 0$$

   Determine the value of A_1 in terms of H and T.

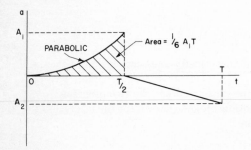

Supplementary References

1 (8.2–8.7); **3** (7.3–7.8); **4** (Ch. 2 and 3); **6** (2.1–2.17, 6.1–6.13); **7** (7.5–7.7); **8** (8.3–8.4)

Unit X. Sizing the Cams

Objectives
1. To derive relationships to calculate coordinates to generate cam profiles for different types of followers
2. To calculate the radius of a roller of translating or oscillating roller follower
3. To calculate the influence of the pressure angle in sizing a cam with a translating roller follower

Objective 1

To derive relationships to calculate coordinates to generate cam profiles for different types of followers

Activities
- Read the material provided below, and as you read:
  a. Derive the mathematical relationship to obtain the pitch profile of a cam having the following types of followers:*

     translating roller follower
     translating offset roller follower
     swinging or oscillating roller follower
     flat-face in-line oscillating follower
     offset flat face oscillating follower
     flat face follower for radial cam

  b. Compare the cam profiles generating a given follower motion.
  c. Select the cam follower system satisfying the requirements of the cam.
- Check your answers to the competency items with your instructor.

Reading Material

Cam profile determination using analytical technique is of special importance when high speed, heavy inertia loads, and accurate positioning of the follower are critical factors. Although a designer can calculate the cutter coordinates to

---

*For further information read R. S. Hanson and T. Churchill, "Theory of Envelopes Provides New Cam Design Equations." *Product Engineering*, August 20, 1962, pp. 45–55.

any degree of accuracy, the actual cutting of the cam is limited because of the ability of the present day machine tools to reproduce the accurate specification of the cam profile.

In this section we will examine the analytical technique based on the theory of envelopes in arriving at the analytical expression for the cutter coordinates of a specific cam profile having a specific type of follower.

Specifically, we will examine the technique for two types of followers:

1. Flat face follower
   a. Swinging in-line follower
   b. Swinging off-set follower
   c. Translating follower
2. Roller follower
   a. Translating follower
   b. Translating offset follower
   c. Swinging follower

Let us examine the theory of envelopes by studying two examples. The envelope can be defined mathematically as a tangent curve to a family of curves.

Example 1.

Determine the envelope of a curve generated by a linearly moving circle.

Solution.

Let the equation of a circle be

$$(x - x_0)^2 + y^2 = 1 \tag{1}$$

The circle has unit radius, with centers at $x = x_0$, $y = 0$. As x_0 varies, a series of circles are obtained.

Equation (1) can be written in the form

$$f(x, y, x_0) = 0 \tag{2}$$

Slope of any member of the family of circles is given by

$$\frac{dy}{dx} = -\frac{\partial f/\partial x}{\partial f/\partial y} \tag{3}$$

or

$$\frac{\partial f}{\partial x} dx + \frac{\partial f}{\partial y} dy = 0 \tag{4}$$

Equation (4) can be written as

$$\frac{\partial f}{\partial x}\frac{dx}{dx_0} + \frac{\partial f}{\partial y}\frac{dy}{dx_0} = 0 \tag{5}$$

Equation (5) holds true for all the members of the family. Since the enveloping curve is a tangent curve to each member of the family curve, Eq. (5) must be satisfied by the enveloping curve.

From calculus, we know that the total derivative of Eq. (2) will yield

$$df = \frac{\partial f}{\partial x}dx + \frac{\partial f}{\partial y}dy + \frac{\partial f}{\partial x_0}dx_0 = 0 \tag{6}$$

or

$$\frac{\partial f}{\partial x}\frac{dx}{dx_0} + \frac{\partial f}{\partial y}\frac{dy}{dx_0} + \frac{\partial f}{\partial x_0} = 0 \tag{7}$$

From Eqs. (5) and (6), we can say that the general equation for the enveloping curve is

$$\frac{\partial f(x, y, x_0)}{\partial x_0} = 0 \tag{8}$$

The envelope is determined either by eliminating the parameter x_0 or by expressing x and y in terms of x_0.

For the given problem of linearly moving circles, the envelope expression is obtained as

$$\frac{\partial f}{\partial x_0}(x, y, x_0) = 2(x - x_0)\left(-\frac{\partial x_0}{\partial x_0}\right) + \frac{\partial(y^2)}{\partial x_0} - 0 = 0 \tag{9}$$
$$= 2(x - x_0)(-1) + 0 - 0 = 0$$

Equation (9) gives

$$x = x_0 \tag{10}$$

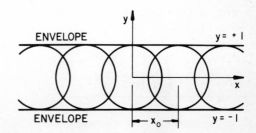

Fig. 1

Substituting $x = x_0$, in Eq. (1), we get $y^2 = 1$ or

$$y = \pm 1 \tag{11}$$

The envelope as shown in Fig. 1 is a pair of straight lines given by $y = +1$ and $y = -1$.

Example 2.

A bullet is fired at any angle β in a vertical plane with a muzzle velocity V_m. Find the envelope that gives the maximum range. Neglect the resistance of air. The equation for the trajectory for any angle β is given by

$$y = x \tan \beta - \frac{gx^2}{2V_m^2} (1 + \tan^2 \beta) \tag{12}$$

where (x, y) defines the position of the bullet in any plane and g is the gravity of earth.

Solution.

Let

$$f(x, y, \beta) = x \tan \beta - \frac{gx^2}{2V_m^2} (1 + \tan^2 \beta) - y = 0 \tag{13}$$

Thus

$$\frac{\partial f(x, y, \beta)}{\partial \beta} = x \sec^2 \beta \left(1 - \frac{gx \tan \beta}{V_m^2} \right) = 0 \tag{14}$$

or

$$\tan \beta = \frac{V_m^2}{gx} \tag{15}$$

Using Eq. (15), we can eliminate β from Eq. (12) to yield the equation for the envelope

$$y = \frac{V_m^2}{2g} - \frac{gx^2}{2V_m^2} \tag{16}$$

The parabolic envelope given by Eq. (16) is shown in Fig. 2.

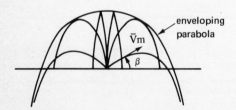

enveloping parabola

$\bar{V}m$

β

Fig. 2 Parabolic envelope.

The theory of envelopes studied in the above two examples will now be applied to obtain the pitch profile of cams with flat face followers. These steps should be followed:

1. Select a convenient coordinate system. There are two systems to choose from: (a) rectangular coordinate system and (b) polar coordinate system.
2. Derive the equation to obtain the envelope having one variable parameter.
3. Take the derivative of the equation derived in Step 2 with respect to the variable parameter. Equate the resultant expression to zero.
4. Use the results of Step 3 to eliminate the variable parameter from the equation obtained in Step 2. The resultant expression is the envelope.
5. Vary the parameter for its entire range to obtain the cam profile.

The flat face swinging follower is shown in Fig. 3. The tangent line at the point of contact of the cam surface passes the center of rotation of the follower. For this reason, this follower is called flat face in-line swinging follower.

The initial position of the follower before the rise cycle begins is measured by angle λ and is given by

$$\lambda = \tan^{-1} \frac{r_b}{r_a} \tag{17}$$

where
r_a = distance between the two pivot points. This distance is measured along the x axis.
r_b = base circle radius

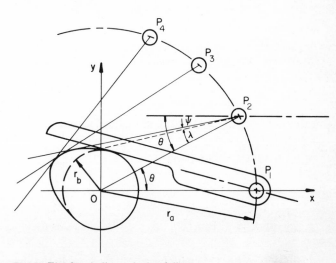

Fig. 3 Flat face in-line swinging follower.

Let the output displacement of the follower be measured by angle ψ in terms of cam rotation angle θ.

The cam profile is usually generated using the inversion technique. In this technique, the cam is assumed stationary and the follower is rotated around the cam in the direction opposite to the one that is required. If we vary θ and ψ and hold λ a constant, we will obtain a function involving x, y, θ, and ψ describing a family of straight lines. Since ψ and θ are related, the function essentially becomes

$$f(x, y, \theta) = 0 \tag{18}$$

The envelope which is the cam profile is obtained by solving simultaneously Eq. (18) and

$$\frac{\partial f(x, y, \theta)}{\partial \theta} = \frac{\partial f}{\partial \theta} = 0 \tag{19}$$

Now, equation of a straight line is

$$y = kx + m \tag{20}$$

Where k is the slope and m is the intercept of the straight line with the y axis. From Fig. 3,

$$k = \tan[180 - (\psi + \lambda - \theta)]$$
$$= -\tan(\psi + \lambda - \theta) \tag{21}$$

Also,

$$x = r_a \cos\theta \tag{22}$$

$$y = r_a \sin\theta \tag{23}$$

Substituting Eqs. (21), (22), (23), in Eq. (20) we get

$$m = r_a[\sin\theta + \cos\theta \tan(\lambda - \theta + \psi)] \tag{24}$$

or

$$f(x, y, \theta) = y + [\tan(\lambda - \theta + \psi)](x - r_a \cos\theta) - r_a \sin\theta = 0 \tag{25}$$

The derivative of Eq. (25) with respect to θ to obtain the envelope will yield

$$\frac{\partial f}{\partial \theta} = \tan(\lambda - \theta + \psi) r_a \sin \theta$$

$$+ (x - r_a \cos \theta)[\sec^2 (\lambda - \theta + \psi)] \left(\frac{d\psi}{d\theta} - 1\right) - r_a \cos \theta = 0 \tag{26}$$

Let $N = \lambda - \theta + \psi$

Simultaneous solution of Eqs. (25) and (26) will give the cam profile coordinates. These are

$$x = r_a \left[\cos \theta + \frac{\cos (\theta + N) \cos N}{d\psi/d\theta - 1}\right] \tag{27}$$

$$y = r_a \left[\sin \theta + \frac{\cos (\theta + N) \sin N}{d\psi/d\theta - 1}\right] \tag{28}$$

For the offset flat face swinging follower, the procedure for obtaining the coordinates of the cam profile is similar. If e is the offset distance (see Fig. 4), then these coordinates are given by

$$x = r_a \left[\cos \theta + \frac{\cos (\theta + N)}{d\psi/d\theta - 1} \cos N\right] + e \sin N \tag{29}$$

$$y = r_a \left[\sin \theta + \frac{\cos (\theta + N)}{d\psi/d\theta - 1} \sin N\right] + e \cos N \tag{30}$$

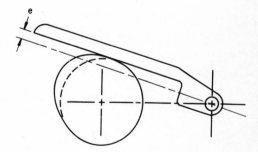

Fig. 4 Offset flat face swinging follower.

Cam profile for flat face translating follower

The flat face in-line translating follower moves radially. See Fig. 5.
The equation of the family of lines generating the envelope is given by

$$y = kx + m \tag{31}$$

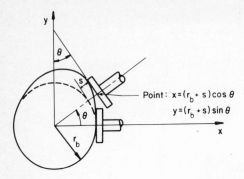

Point: $x = (r_b + s)\cos\theta$
$y = (r_b + s)\sin\theta$

Fig. 5 Flat face in-line translating follower.

where

k = slope = $-\cot\theta$

m = intersection of the line with y coordinate

Let s denote the displacement of the follower. Then

$$x = (r_b + s)\cos\theta$$

$$y = (r_b + s)\sin\theta$$

and

$$m = \frac{r_b + s}{\sin\theta}$$

Equation (31) then becomes

$$y = \frac{r_b + s - x\cos\theta}{\sin\theta} \qquad (32)$$

The family of straight lines generating the envelope is given by

$$f(x, y, \theta) = y\sin\theta + x\cos\theta - (r_b + s) = 0 \qquad (33)$$

The envelope is obtained by solving simultaneously Eq. (33) and

$$\frac{\partial f}{\partial \theta} = y\cos\theta - x\sin\theta - \frac{ds}{d\theta} = 0 \qquad (34)$$

The profile coordinates for cam with translating in-line flat face follower are

$$x = (r_b + s)\cos\theta - \frac{ds}{d\theta}\sin\theta \qquad (35)$$

$$y = (r_b + s)\sin\theta - \frac{ds}{d\theta}\cos\theta \qquad (36)$$

Cam profile coordinates for translating roller follower

The translating radial follower is shown in Fig. 6.
Let

$$M = r_b + r_f + s \qquad (37)$$

where

r_f = radius of roller of the follower
r_b = base circle radius
s = displacement of the follower

The equation of the envelope is given by

$$(x - M \cos \theta)^2 + (y - M \sin \theta)^2 - r_f^2 = 0 \qquad (38)$$

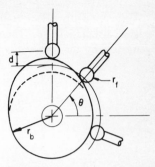

Fig. 6 Translating radial follower.

The profile coordinates are obtained by using the method described above.

$$x = M \cos \theta \pm \cfrac{r_f}{\left\{ 1 + \left[\cfrac{M \sin \theta - (ds/d\theta) \cos \theta}{M \cos \theta + (ds/d\theta) \sin \theta} \right]^2 \right\}^{1/2}} \qquad (39)$$

$$y = \frac{x [M \sin \theta - (ds/d\theta) \cos \theta] + M(ds/d\theta)}{M \cos \theta + (ds/d\theta) \sin \theta} \qquad (40)$$

NOTE: $\dfrac{dM}{d\theta} = \dfrac{ds}{d\theta}$

The plus sign establishes the outer envelope and the negative sign establishes the
inner envelope. Thus, Eqs. (39) and (40) can be used to obtain the cam profile
for positive action cam. See Fig. 7.

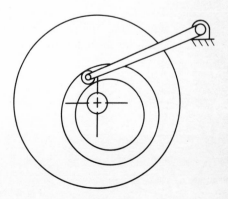

Fig. 7 Positive action cam.

Cam profile coordinates for offset translating roller follower

The cam with offset translating roller follower is shown in Fig. 8. Let the offset distance be e. Then the general equation for the family of curves generated by the follower is

$$[x - e \sin \theta - (k + s) \cos \theta]^2$$
$$+ [y + e \cos \theta - (k + s) \sin \theta]^2 - r_f^2 = 0 \tag{41}$$

where

$$k = [(r_b + r_f)^2 - e^2]^{1/2}$$

The cam profile coordinates for the cam with the offset roller follower are obtained using the technique described earlier.

$$y = \frac{x\left[(k + s) \sin \theta - \left(e + \dfrac{d(s)}{d\theta}\right) \cos \theta\right] + (k + s) \dfrac{d(s)}{d\theta}}{(k + s) \cos \theta + \left(e + \dfrac{d(s)}{d\theta}\right) \sin \theta} \tag{42}$$

$$x = e \sin \theta + (k + s) \cos \theta$$
$$\pm \frac{r_f}{\sqrt{1 + \left[\dfrac{(k + s) \sin \theta - \left(e + \dfrac{d(s)}{d\theta}\right) \cos \theta}{(k + s) \cos \theta + \left(e + \dfrac{d(s)}{d\theta}\right) \sin \theta}\right]^2}} \tag{43}$$

Equations (42) and (43) permit a designer to obtain the cam profile for positive action cam. The plus sign in Eq. (43) is for the outer envelope and the negative sign is for the inner envelope.

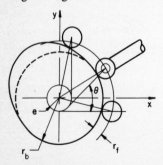

Fig. 8 Offset translating roller follower.

Cam profile coordinates for swinging roller follower

The cam with swinging roller follower is shown in Fig. 9.

The initial position of the follower is defined by the angle λ which is computed from

$$\lambda = \cos^{-1} \frac{r_r + r_a - (r_b + r_f)}{2r_a r_r} \tag{44}$$

The general equation for the family of curves generated by the swinging roller follower is

$$(x - r_a \cos\theta + r_r \cos J)^2 + (y - r_a \sin\theta + r_r \sin J)^2 - r_f^2 = 0 \tag{45}$$

where

$$J = \theta - \lambda - \psi$$

The cam profile coordinates for the cam with swinging roller follower are obtained using the technique described earlier.

$$y = \frac{x[r_a \sin\theta - r_r(1 - d\psi/d\theta)\sin J]}{r_a \cos\theta - r_r(1 - d\psi/d\theta)\cos J} \tag{46}$$

$$x = r_a \cos\theta - r_r \cos J$$

$$\pm \frac{r_f}{\left\{1 + \left[\dfrac{r_a \sin\theta - r_r(1 - d\psi/d\theta)\sin J}{r_a \cos\theta - r_r(1 - d\psi/d\theta)\cos J}\right]^2\right\}^{1/2}} \tag{47}$$

Here again, the plus sign in Eq. (47) gives the outer envelope and the negative sign gives the inner envelope.

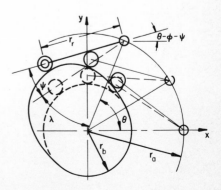

Fig. 9 Cam with swinging roller follower

Competency Items
- A cycloidal motion program is specified for the D-R-D-R motion program with H = 2.0 in., β = 120°. Calculate the cam profile at a 1-degree increment of cam rotation if:
  a. a translating roller follower is used
  b. a translating flat face follower is used
  c. a translating offset roller follower is used
- Compare your cam profiles just obtained with those obtained using the graphical approach.

Objective 2

To calculate the radius of a roller of a translating or an oscillating roller follower

Activities
- Read the material provided below, and as you read:
  a. Calculate the size of the roller.
  b. Identify the conditions that contribute to undercutting the cam.
- Check your answers to the competency items with your instructor.

Reading Material

In this section we will examine the importance of the radius of curvature of the cam profile in selecting the proper size of roller of the translating or oscillating roller follower.

The overall shape of a cam changes significantly when it is made successively smaller and yet satisfying the same set of boundary conditions. See Fig. 10a, b, and c. The shape of the cam, however, in general controls the pressure angle. As the cam becomes smaller, the pressure angle becomes larger. The large value of a pressure angle reduces the efficiency of motion transmission to the follower. In some extreme cases, the follower no longer follows the cam profile.

To prevent undercutting, (the case described in Fig. 10c) the roller of a follower must have its radius smaller than the minimum radius of curvature of

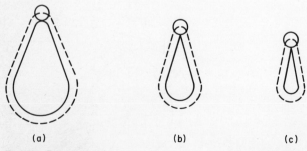

(a)　　　　　(b)　　　　　(c)

Fig. 10 Convex cams.

the pitch curve of the cam. The undercutting phenomenon exists only in the cam having convex surfaces. It is not possible for a follower to undercut its motion when the motion is transmitted by a concave cam. In a concave cam, the contact stresses due to a given force are smaller than in the convex surface. See Fig. 11.

Since the minimum radius of curvature of the pitch curve of a cam determines the size of the roller follower, we are required to learn the technique to calculate the radius of curvature. We will assume that a functional relationship exists between the pitch curve of a cam and cam rotation angle, and the first two derivatives of this function are continuous.

Let us express the pitch curve of the cam using the polar coordinates r and ϕ. Let

$$r = F(\phi) \quad r' = \frac{dr}{d\phi} = F'(\phi) \quad r'' = \frac{d^2r}{d\phi^2} = F''(\phi)$$

Then, radius of curvature ρ of the cam profile at any given angle ϕ is calculated using

$$\rho = \frac{\{F^2(\phi) + [F'(\phi)]^2\}^{3/2}}{F(\phi)^2 + 2[F'(\phi)]^2 - F(\phi)F''(\phi)} \tag{48}$$

The reader should study carefully the following examples that illustrate the importance of minimum ρ in selecting the size of roller follower.

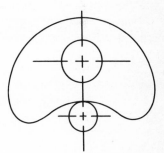

Fig. 11 Concave cam.

Example 3.

Calculate the minimum radius of curvature for a radial cam having a translating roller follower with a cycloidal motion program.

Solution.

A cycloidal motion program of given H and β is

$$f(\theta) = r_0 + H\left(\frac{\theta}{\beta} - \frac{1}{2\pi} \sin 2\pi \frac{\theta}{\beta}\right) \tag{49}$$

$$\frac{df}{d\theta} = f'(\theta) = \frac{H}{\beta}\left(1 - \cos 2\pi \, \frac{\theta}{\beta}\right) \tag{50}$$

$$\frac{d^2f}{d\theta} = f''(\theta) = \frac{2\pi H}{\beta^2}\left(\sin 2\pi \, \frac{\theta}{\beta}\right) \tag{51}$$

where

$$\begin{aligned} r_0 &= r_f + r_b \\ R &= r_0 + y = f(\theta) \end{aligned} \tag{52}$$

$$\frac{dR}{d\theta} = f'(\theta) \tag{53}$$

$$\frac{d^2R}{d\theta^2} = f''(\theta) \tag{54}$$

r_f and r_b denote radius of the roller follower and radius of the base circle.

For the translating roller follower, the center of the follower does not rotate with respect to the center of the cam. Hence,

$$\theta = \phi \qquad f(\theta) = F(\phi)$$
$$f'(\theta) = F'(\phi) \qquad f''(\theta) = F''(\phi)$$

and Eq. (48) becomes

$$\rho = \frac{\{f^2(\theta) + [f'(\theta)]^2\}^{3/2}}{[f(\theta)]^2 + 2[f'(\theta)]^2 - f(\theta)f''(\theta)} \tag{55}$$

Substituting Eqs. (49), (50), and (51) in Eq. (55), we get the radius of curvature of the pitch curve of the cam at a given value of θ. The minimum value of the radius of curvature can be obtained using two approaches:

1. Differentiate Eq. (55) to determine its minimum.
2. Calculate ρ at incremental values of θ and search graphically the ρ minimum. (The appendix presents the design charts to determine ρ_{min} for some of the custom-made motion programs.)

Appendix

The following design charts* shown in Figs. A–D are obtained using Eq. (55) for cycloidal and harmonic motion programs. With the help of these design charts it is possible to determine ρ_{min} for the pitch curve of the cam.

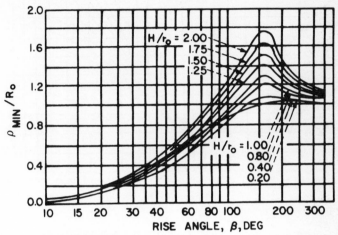

Fig. A Cycloidal motion.

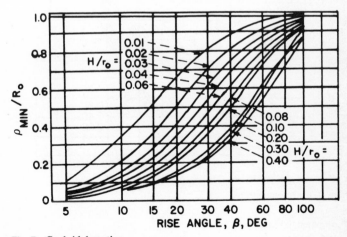

Fig. B Cycloidal motion.

*For a full set of design charts, it is suggested that the reader refer to the paper, "Plate Cam Design... Radius of Curvature," by M. Kloomok and R. Muffler, in *Product Engineering*, September 1955, p. 186.

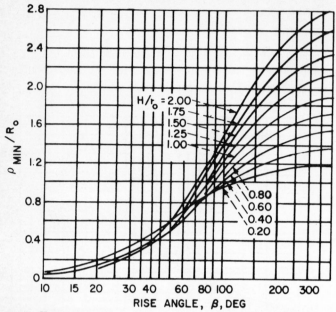

Fig. C Harmonic motion.

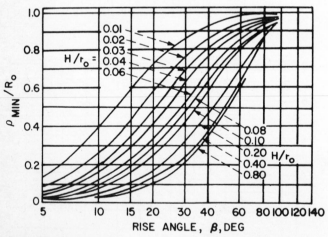

Fig. D Harmonic motion.

Competency Items

- Calculate the size of the roller follower for a cycloidal motion cam having a rise of 1 in. through 60° of cam rotation, if the base circle has a radius of 1 in.
- From the design charts in Figs. A–D determine how the minimum radius of curvature is affected by the base circle radius for a given H and β, also by the angle β for a given H and R.
- Define undercutting.

Objective 3

To calculate the influence of the pressure angle in sizing a cam with a translating roller follower

Activities
- Read the material provided below, and as you read:
  a. Describe the relationship between the pressure angle and the base radius of a cam.
  b. Derive the relationship between the pressure angle, the follower overhang, and the follower length.
  c. Write steps to reduce the pressure angle.
  d. Derive the relationship to study the efficiency of a cam with a translating flat face follower.
- Check your answers to the competency items with your instructor.

Reading Material

In this objective we will study the effects of several parameters on the pressure angle of a radial cam with translating or oscillating roller follower. The magnitude of the pressure angle in such a cam and follower system controls the efficiency of the cam. The smaller the pressure angle, the higher is its efficiency. We will examine the following:

1. Effect of pressure angle on side thrust.
2. Effect of pressure angle on base circle radius.
3. Reduction of pressure angle by offsetting the follower.
4. Sizing a flat face follower.

Effect of pressure angle on side thrust

A radial cam with translating roller follower is shown in Fig. 12. O_1 is the center of rotation of the cam and O_2 is the center of rotation of the roller. Let C and D be the length of the follower overhang and follower bearing. The translating follower is a cylinder with d as its base circle radius. The cam is rotating in a clockwise direction and the follower is moving upward as shown in Fig. 12.

We will adopt the following nomenclature to study the motion of the cam and its follower:

μ = coefficient of friction
N_1, N_2 = forces normal to follower stem

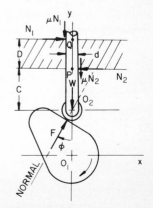

Fig. 12 Radial cam with translating roller follower.

F = force transmitted by cam. The direction of the force is normal to the tangent at the point of contact of follower.

W = external load on the follower

ϕ = pressure angle

ϕ_{max} = maximum pressure angle

Under the static equilibrium condition, the sum of all the forces along the x and y axes must be zero. That is

$$\sum Fy = 0 = -W + F \cos\phi - \mu N_1 - \mu N_2 \tag{56}$$

The moment of all forces about points p and q must also be zero. That is

$$\sum Mp = 0 = -FC \sin\phi + N_1 D - \mu N_1 \frac{d}{2} + \mu N_2 \frac{d}{2} \tag{57}$$

$$\sum Mq = 0 = -F(C + D) \sin\phi + N_2 D - \mu N_1 \frac{d}{2} + \mu N_2 \frac{d}{2} \tag{58}$$

The magnification factor defined by the ratio F/W can be now obtained by solving Eqs. (56), (57), and (58). Thus

$$\frac{F}{W} = \frac{D}{D \cos\phi - (2\mu C + \mu D - \mu^2 d) \sin\phi} \tag{59}$$

With proper lubrication, the friction in the follower bearing is considerably small so that we can assume $\mu^2 = 0$. The ratio W/F provides information about the amount of load that can be received by the follower versus the amount of load that is transmitted by the cam to the follower. When F becomes infinite, W/F becomes small. In limiting situation, $F = \infty$, and $W/F = 0$; and the follower has no motion. Under this condition the follower jams. The ratio W/F becomes zero when

$$D \cos\phi - (2C\mu + \mu D) \sin\phi = 0 \tag{60}$$

When the follower jams, the pressure angle becomes maximum which is obtained from Eq. (60) as

$$\phi_{max} = \tan^{-1} \frac{D}{\mu(2C + D)} \tag{61}$$

From Eq. (61), we can reason that the maximum value of pressure angle ϕ can be held high provided that the coefficient of friction is minimum, the follower overhang C is minimum, and the bearing length D is as large as possible.

The acceptable value of ϕ_{max} is $30°$. The magnification factor F/W becomes infinite at some values of $\phi < 90°$. As ϕ reaches $90°$, the follower stem experiences larger bending moments. With large clearances in follower bearing and elastic follower, the follower stem will dig into the lower corner of the bearing.

Effect of pressure angle on base circle radius

A radial cam with knife edge follower is shown in Fig. 13. The knife edge follower is a special case of roller follower. The radius of roller is zero. Let F be the force transmitted by the cam to the follower. F_T and F_F are the horizontal and vertical components of F. Let V_T be the velocity of the point on the cam at the point of contact V_F be the velocity of the follower. ϕ is the pressure angle. If r_b is the base radius of the cam, then

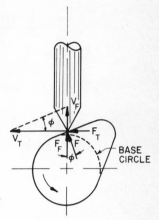

$$V_T = (r_b + y)\omega \qquad (62)$$

where y = displacement of the follower
ω = angular velocity of the cam

Fig. 13 Radial cam with knife edge follower.

From Fig. 13,

$$\tan\phi = \frac{V_F}{V_T} \qquad (63)$$

$$= \frac{V_F}{(r_b + y)\omega} \qquad (64)$$

But $V_F = dy/dt$ and $\omega = d\theta/dt$. Therefore,

$$\tan\phi = \frac{dy/dt}{(r_b + y)\, d\theta/dt} \qquad (65)$$

or

$$\tan\phi = \frac{dy/d\theta}{r_b + y} \qquad (66)$$

From Eq. (66), the pressure angle ϕ is a function of $dy/d\theta$ and y. The maximum value of ϕ occurs when $dy/d\theta$ is maximum and y is minimum. Since $dy/d\theta$ measures the velocity of the follower, the pressure angle becomes large when the velocity is maximum. In order to lower the value of ϕ_{max}, $(V_F)_{max}$ should be low or r_b is large. However, with large r_b, the cam becomes large which is undesirable since the inertial forces become large.

A design chart for cam with a cycloidal or a simple harmonic motion program is shown in Fig. 14. This design chart permits the designer to determine the maximum value of the pressure angle once H and r_b are known.

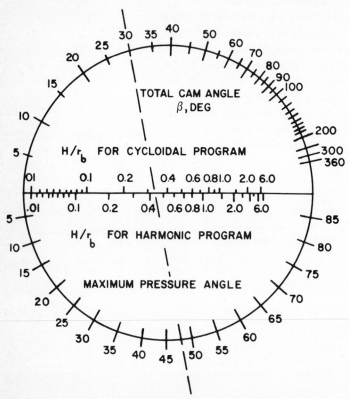

Fig. 14 Nomogram for finding maximum pressure angle for different H/r_b.

Effect of offsetting follower on pressure angle

The maximum value of the pressure angle can be lowered by increasing the base circle radius, lowering the maximum value of the follower velocity, and by offsetting the follower. We will now examine how much offsetting should be done to bring the maximum value of ϕ to a tolerable level.

A radial cam with offset knife edge follower is shown in Fig. 15. Let e be the offset distance, y be the displacement of the follower, $(V_F)_{max}$ be the maximum velocity of the follower, ϕ_{max} the maximum pressure angle, and β be the angle of inclination of V_T with the horizontal line where V_T is the velocity of the point of contact on the cam.

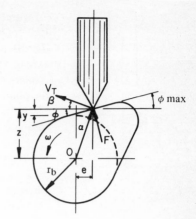

Fig. 15 Radial cam with offset knife edge follower.

From Fig. 15,

$$Z = y + (r_b^2 - e^2)^{1/2} \qquad (67)$$

where r_b = base circle radius.

$$\tan \phi_{max} = \frac{(V_F)_{max} - e\omega}{Z\omega} \qquad (68)$$

where ω = angular velocity of cam.

The maximum pressure angle ϕ_{max} becomes zero when

$$(V_F)_{max} - e\omega \simeq 0 \qquad (69)$$

or

$$e = e_{critical} = e_c \simeq \frac{(V_F)_{max}}{\omega} \qquad (70)$$

Equation (68) shows that it is possible to reduce the pressure angle to zero value. Hence, it is possible to increase the efficiency of transmission when the radial cam with translating follower is offset. There is, however, a limit for offsetting the follower. That is, $e \leq e_c$. For practical purposes, $e/e_c \leq 0.5$.

Analysis of cam with flat face translation follower

The cams with flat face translating follower do not have the problems of controlling the pressure angle. In fact, the concept of pressure angle is not applicable for such cams since the force transmitted by the cam is always parallel to the axis of the follower. This condition makes the pressure angle ϕ to be always zero. However, since the transmitted force acts at a distance from the center, the cam-follower system has a turning moment. When the follower has maximum velocity, the turning moment becomes maximum.

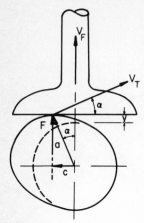

Fig. 16

Figure 16 shows a radial cam with translating flat face follower. Let V_F be the follower velocity, c be the distance at which the force F acts parallel to the axis of the follower, r_b be the base radius, y be the displacement of the follower, V_T be the velocity of the point contact on cam, and α be the angle at which V_T is inclined with the horizontal line. Then, the torque is

$$T = \mu F(r_b + y) + Fc \tag{71}$$

where μ is the coefficient of friction at the point of contact between the cam and its follower.

From Fig. 16,

$$a \cos\alpha = (r_b + y)$$

or

$$a = \frac{r_b + y}{\cos\alpha} \tag{72}$$

Also,

$$V_F = V_T \sin\alpha \tag{73}$$

$$V_T = a\omega \tag{74}$$

Hence,

$$V_F = \frac{(r_b + y)\omega}{\cos\alpha} \sin\alpha$$

or

$$\tan\alpha = \frac{V_F}{(r_b + y)\omega} \tag{75}$$

Also, from Fig. 16

$$\tan\alpha = \frac{c}{r_b + y} \tag{76}$$

Equating Eqs. (75) and (76), we get

$$c = \frac{V_F}{\omega} \tag{77}$$

When V_F becomes maximum, c becomes maximum, which, in turn, increases the turning moment. 'c' can be made smaller if the maximum slope of the follower displacement program is made smaller.

During the rise stroke the flat face follower will experience a lag while $(V_F)_{max}$ is brought to zero value in a short time interval. The deceleration time t_D is usually kept below the critical value $\pi\alpha_{max}/180\omega$.

The angle α plays a role similar to that of a pressure angle in a radial cam with translating roller follower. For higher efficiency, the maximum value of α should be kept as low as possible. From Eq. (76), we observe that

$$\tan\alpha_{max} \simeq \frac{c_{max}}{r_b + y} \tag{78}$$

Hence, to keep low α_{max}, the base circle radius should be large.

Competency Items
- Describe in writing how the base circle radius of a cam affects the pressure angle.
- Describe in writing how offsetting the follower affects the pressure angle.
- State what is undesirable about a very large base circle radius.

Performance Test #1

A cycloidal motion cam has a rise program with $H = 1$ in., $\beta = 60°$. The size of the cam is limited to $r_b = 2$ in., cam speed is 100 rpm. Calculate the maximum pressure angle for this rise with a translating roller follower. Calculate maximum velocity during the rise. Calculate the amount of offset that is necessary to reduce the maximum pressure angle to $25°$. Calculate the minimum radius of curvature of the pitch curve.

Performance Test #2

1. A translating roller follower has a harmonic motion program. The rise of the follower is 0.25 in. through $30°$. The prime circle radius, R_0, to be used is 2.5 in. Find the minimum base radius that can be used so that there will be no undercutting.
2. A harmonic motion cam has a rise of 0.75 in. through an angle of $45°$. The base radius is 2.0 in. The follower is a translating roller and is offset 0.8 in. Find the maximum pressure angle of the follower.

Performance Test #3

1. A plate cam and flat faced follower as shown in Fig. 16, Unit X, has a simple harmonic motion program which occurs during the cam rotation angle of $60°$. If the rise H is 1 in., find the maximum value of c during the program.
2. Develop a design chart that relates maximum velocity with ϕ, H, and r_b for
   a. harmonic motion
   b. cycloidal motion

Supplementary References

1 (9.1–9A.X); **3** (7.21–7.26); **4** (Ch. 4 and 5); **6** (3.1–3.18, 4.16–4.23); **7** (7.7–7.8); **8** (8.7–8.8)

Unit XI. Motion Analysis and Synthesis of Four-bar, Slider-crank, and Inverted Slider-crank Mechanisms

Objectives

1. To state Grashoff criteria for a four-link mechanism
2. To identify the conditions that determine limit positions and dead-center positions of a four-link mechanism
3. To identify the conditions that determine limit positions of a slider-crank mechanism
4. To identify the conditions that determine limit positions of an inverted slider-crank mechanism
5. To identify the conditions that determine minimum and maximum transmission angles of four-link, slider-crank, and inverted slider-crank mechansims
6. To design a drag-link mechanism with an optimum transmission angle
7. To design a crank-rocker mechanism using a minimum transmission angle

Objective 1

To state Grashoff criteria for a four-link mechanism

Activities

- Read the material provided below, and as you read classify four-link mechanisms into one of three categories:
  a. rocker-rocker (double rocker) mechanism
  b. crank rocker mechanism
  c. crank-crank (double-crank or drag-link) mechanism
- Check your answers to the competency items with your instructor.

Reading Material

The practical importance of the four-link mechanisms with Class I kinematic pairs lies in their ability to produce a variety of nonuniform motion and transmit large forces. In addition to design requirements of transmitting nonuniform motion, the four-link mechanisms are also designed so that emphasis is given to the type and quality of motion executed by these mechanisms. This unit examines two points:

types of motion displayed by planar mechanisms
design criteria to control the quality of motion

Types of motion of a four-link mechanism with four revolute pairs

A four-link mechanism with four revolute pairs is shown in Fig. 1. Link $MA = a$ is an input link; link $AB = b$ is a coupler link; link $QB = c$ is a follower link; and link $MQ = d$ is the fixed link. Link MQ is also called a base link. Links MQ, MA, AB and QB are given a vector representation, as shown in Fig. 1.

Angles θ_1, θ_2, θ_3, and θ_4, measured in a counterclockwise direction, describe the relative positions of the links.

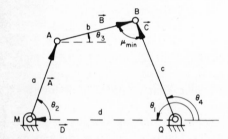

Fig. 1 A four-link mechanism with four revolute pairs.

The different types of motion that a four-link mechanism executes depend upon the proportion of its four links. There are three basic types of motions that a four-link mechanism can execute. The three types of motion can be examined by considering a four-link chain shown in Fig. 2. The three types of mechanisms are described below.

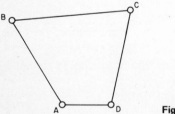

Fig. 2

Rocker-rocker mechanism

Figure 3 shows a rocker-rocker mechanism derived from the kinematic chain shown in Fig. 2 by holding link BC fixed to the ground. Link AB or CD may be considered as input links. Because of the specific link proportions, links BA and CD are able to oscillate through angles ϕ and ψ. Neither link BA nor link CD is able to complete a full rotation. Since links BA and CD are able to rock through finite angles ϕ and ψ, the mechanism shown in Fig. 3 is called a rocker-rocker or double rocker mechanism.

Crank-rocker mechanism

The mechanism shown in Fig. 4 is obtained from Fig. 2 by holding link AB fixed to the ground. If link AD is input and link BC is follower link, then input link AD is able to complete full rotation while follower link BC rocks through angle ϕ. Since the input link is able to rotate 360° and the output link is able to oscillate through angle ϕ, the mechanism in Fig. 4 is called a crank-rocker mechanism.

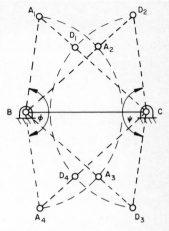

Fig. 3

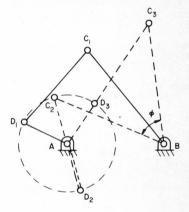

Fig. 4

Crank-crank mechanism

The mechanism shown in Fig. 5 is obtained from Fig. 2 by holding link AD to the ground. Link AB or DC can be considered as input links. Because of the specific link proportions, links AB and DC are able to make full rotation. Hence, the mechanism shown in Fig. 5 is called a crank-crank or double-crank mechanism. A double-crank mechanism is often called a drag-link mechanism.

The type of motion displayed by a four-link mechanism can be predicted by applying Grashoff criteria to this mechanism. The Grashoff criteria can be stated in the following manner:

1. A four-link mechanism belongs to the Class I mechanism provided the sum of the lengths of its shortest and longest links is less than or equal to the sum of the lengths of its other two links. For the Class I mechanisms, the type of motion can be predicted using the following rules:

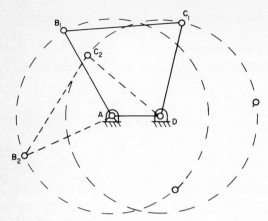

Fig. 5

    a. If the shortest link is the input link, then the four-link mechanism is a crank-rocker mechanism.

    b. If the shortest link is the fixed link, then the four-link mechanism is a crank-crank or a drag-link mechanism.

    c. If the conditions other than those described in (a) or (b) are satisfied, then the four-link mechanism is a rocker-rocker mechanism.

2. A four-link mechanism belongs to the Class II mechanisms provided the sum of the lengths of its shortest and longest links is greater than the sum of the lengths of its other two links. A Class II mechanism is always a rocker-rocker mechanism.

Competency Items

- A four-link kinematic chain has link lengths of $a = 1$ in., $b = 2$ in., $c = 3$ in., and $d = 5$ in. Fix each link, one at a time to the ground, and determine what type of motion will be displayed by each of the four different mechanisms.
- Plot θ_2 vs. θ_4 for the four-link mechanism shown in Fig. 4.
- Plot θ_2 vs. θ_4 for the four-link mechanism shown in Fig. 5.
- Build a cardboard model of the kinematic chain shown in Fig. 2, and demonstrate the three types of motion that the derived mechanisms will display.

Objective 2

To identify the conditions that determine limit positions and dead-center positions of a four-link mechanism

Activities

- Read the material provided below, and involve yourself to:

    a. Name the conditions that define limit positions and dead-center positions of a four-link mechanism.

b. Determine the limit positions and dead-center positions using graphical construction.

c. Calculate the angle of oscillation between the two limit positions.

• Check your answers to the competency items with your instructor.

Reading Material

Part I: Limit positions and dead centers of a four-link mechanism

When a four-link mechanism is designed either for coordinated input and output crank positions or for coordinated input and coupler link positions, a designer is required to check for the continuous mobility of a designed linkage through these positions. While checking for mobility, there are two situations that warrant a designer's attention: the limit positions and the dead-center positions of a designed four-link mechanism.

A limit position for an output link of a four-link mechanism, shown in Fig. 6, is defined as a position in which the interior angle between its coupler link and input link becomes either 360° or 180°. Thus, when a mechanism is in its limit position, the pivot points M, A', and B lie on a straight line. A four-link mechanism can have a maximum of two limit positions.

Because of its motion, a crank-rocker mechanism has two limit positions. Figure 6a describes the first limit position. The angle between input link MA_1 and coupler link A_1B_1 is 180°. Figure 6b illustrates the second limit position. The angle between input link MA_2 and coupler link A_2B_2 is 360°.

If θ_{4L_1} and θ_{4L_2} describe the two limit positions of the output link QB, then ϕ, the angle of oscillation of a crank-rocker mechanism, is given as

$$\phi = \text{angle of oscillation of output link } QB = \theta_{4L_2} - \theta_{4L_1} \qquad (1)$$

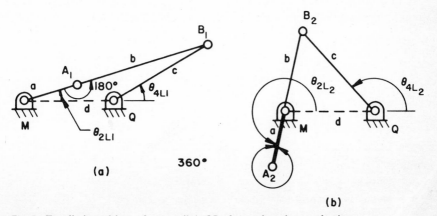

Fig. 6 Two limit positions of output link QB of a crank-rocker mechanism.

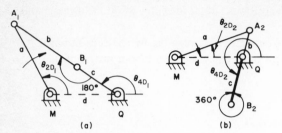

Fig. 7 Two dead-center positions of output link QB of a four-link mechanism.

A dead-center position for an output link of a four-link mechanism is defined as a position in which the interior angle between its coupler link and follower link becomes either $360°$ or $180°$. Thus, when a mechanism is in its dead-center position, pivot points A, B, and Q lie on a straight line. A four-link mechanism with a predefined input link can have a maximum of two dead-center positions.

A rocker-rocker mechanism has dead-center positions because of its motion. In fact, a limit position and a dead-center position can both exist in a rocker-rocker mechanism. Figure 7a and b show a rocker-rocker mechanism in its two dead-center positions. Figure 8a and b illustrates a rocker-rocker mechanism in its limit position and dead-center position.

The existence of a dead center in a designed four-link mechanism could be an undesirable design situation if a continuous motion is required. If input link MA_2 in Fig. 8b attempts to rotate clockwise, then we observe that output link QB_2 will be locked. That is, the mechanism has become an *instantaneous structure*. If the mobility of the mechanism is to be retained at the dead-center position, then link QB_2 must be displaced in the desired direction by an external force. Link QB_2 can be displaced either to position QB_3 or to position QB_4 from dead-center position QB_2.

A drag-link mechanism does not have either a limit position or a dead-center position because both cranks are required to rotate through $360°$.

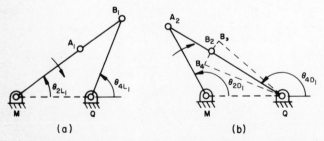

Fig. 8 Motion analysis of a double rocker mechanism: **a.** Limit position of QB; **b.** dead-center position of QB.

*Part II: Determination of limit and dead-center positions–
graphical approach*

Limit positions and dead-center positions can be determined using either a graphical or a mathematical approach. The graphical approach is much simpler than the mathematical approach; however, the mathematical approach is exact and accurate.

Graphical approach—If link lengths $MA = a$, $AB = b$, $QB = c$, and $MQ = d$ of a four-link mechanism of Fig. 1 are given, then its limit positions or dead-center positions can be determined. The graphical approach of determining either the limit positions or the dead-center positions will be illustrated using numerical examples.

Example 1.

A four-link mechanism has the following dimensions: $a = 1$ in., $b = 3$ in., $c = 3$ in., and $d = 4$ in. Draw its limit positions.

Solution.

Repeat the following steps for the first limit positions and compare your results with those in Fig. 9a.

1. Locate fixed centers M and Q at the ends of line $MQ = 4$ in.
2. With M as a center and a radius equal to $MA_1 + A_1B_1 = a + b = 1 + 3 = 4$ in., draw an arc of circle X.
3. With Q as a center and a radius equal to $QB_1 = c = 3$ in., draw an arc of circle Y to intersect circle X in point B_1.
4. Using the conventions of Fig. 6, measure angles θ_{2L_1} and θ_{4L_1}. These angles define the first limit position of the linkage.

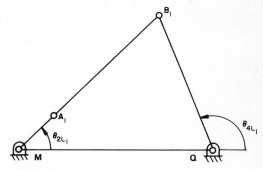

Fig. 9a First limit position.

Repeat the following steps for the second limit position and compare your results with those in Fig. 9b.

1. With M as a center and a radius equal to $A_2B_2 - MA_2 = b - a = 3 - 1 = 2$ in., draw an arc of circle X'.

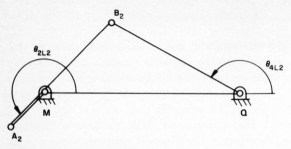

Fig. 9b   Second limit position.

2. With Q as a center and a radius equal to $QB = c = 3$ in., draw an arc of circle Y' to intersect circle X' in point B_2.
3. Measure angles θ_{2L_2} and θ_{4L_2}, as shown in Fig. 9b. These angles define the second limit position of the linkage.

Example 2.
A four-link rocker-rocker mechanism has the following dimensions: $a = 2$ in., $b = 1$ in., $c = 3$ in., and $d = 3$ in. Find its limit position and dead-center position.
Solution.
    Steps for determining the limit position are described in Example 1. The following steps are used for finding the dead-center position:

1. As shown in Fig. 10, with M as a center and a radius equal to $A_2 = a = 2$ in., draw an arc of circle X.
2. With Q as a center and a radius equal to $QB_2 + B_2A_2 = 3 + 1 = 4$ in., draw an arc of a circle Y to intersect the circle X in A_2.
3. Measure angle θ_{2D_1} and θ_{4D_1}, as shown in Fig. 8b. These angles define the dead-center position of the mechanism.

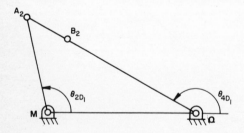

Fig. 10   Construction for the dead-center position.

Part III: Relationships to calculate angles of limit positions and dead-center positions—mathematical approach
    From Fig. 6a, the cosine law for triangle MB_1Q yields:

$$\theta_{2L_1} = \cos^{-1}\left[\frac{(a+b)^2 + d^2 - c^2}{2(a+b)d}\right] \qquad (2)$$

$$\theta_{4L_1} = \cos^{-1}\left[\frac{(a + b)^2 - c^2 - d^2}{2cd}\right] \tag{3}$$

From Fig. 6b, the cosine law for triangle MB_2Q

$$\theta_{2L_2} = \cos^{-1}\left[\frac{(b - a)^2 + d^2 - c^2}{2(b - a)d}\right] + 180° \tag{4}$$

$$\theta_{4L_2} = \cos^{-1}\left[\frac{(b - a)^2 - c^2 - d^2}{2cd}\right] \tag{5}$$

Equations (2) and (3) permit us to calculate angles for the first limit position and Eqs. (4) and (5) permit us to calculate angles for the second limit position. The angles of oscillation ϕ of the output link QB is given by Eq. (6).

$$\phi = \cos^{-1}\left[\frac{(b - a)^2 - d^2 - c^2}{2cd}\right] - \cos^{-1}\left[\frac{(a + b)^2 - c^2 - d^2}{2cd}\right] \tag{6}$$

The angles that define the dead-center positions of a four-link mechanism can be calculated in a similar manner. Thus, from Fig. 7a, the cosine law for triangle MA_1Q yields:

$$\theta_{2D_1} = \cos^{-1}\left[\frac{a^2 + d^2 - (b + c)^2}{2ad}\right] \tag{7}$$

$$\theta_{4D_1} = \cos^{-1}\left[\frac{a^2 - (b + c)^2 - d^2}{2(b + c)d}\right] \tag{8}$$

From Fig. 7b, the cosine law for triangle MA_2Q yields:

$$\theta_{2D_2} = \cos^{-1}\left[\frac{a^2 + d^2 - (b - c)^2}{2ad}\right] \tag{9}$$

$$\theta_{4D_2} = \cos^{-1}\left[\frac{a^2 - (b - c)^2 - d^2}{2ad}\right] + 180° \tag{10}$$

Equations (7) and (8) permit us to calculate angles for the first dead-center positions, and Eqs. (9) and (10) permit us to calculate angles for the second dead-center position.

Competency Items
- A four-link mechanism has link lengths of $a = 4$ in., $b = 2$ in., $c = 3$ in., and $d = 3$ in. Determine all possible limit and dead-center positions. Assume d is fixed and a is input.
- Using Eqs. (2)–(10), calculate the angles that define dead-center positions and limit positions of the mechanism defined above.
- Write two design situations where you will use a crank-rocker mechanism.
- Write two design situations where you will use a rocker-rocker mechanism.

Objective 3

To identify the conditions that determine limit positions of a slider-crank mechanism

Activities
- Read the material provided below, and as you read:
    a. Name the conditions that define limit positions of a slider-crank mechanism.
    b. Determine the two limit positions using graphical construction.
    c. Calculate the length of the stroke of the output sliding block.
- Check your answers to the competency items with your instructor.

Reading Material
   A slider-crank mechanism, shown in Fig. 11, is a four-link mechanism with three revolute pairs and a prism pair. The fourth link, associated with the prism pair and the ground, is of infinite length. Input link $MA = a$ makes a complete rotation while slider block B executes a reciprocating motion. Coupler link $AB = b$ and the offset distance e, defined as an eccentricity of the mechanism, are the other two nonvarying parameters of a slider-crank

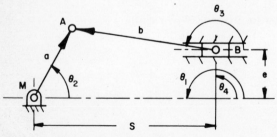

Fig. 11 A slider-crank mechanism.

mechanism. Variable parameters are the angles θ_2, θ_3, and S. Angles θ_2 and θ_3 measure relative displacements of input and coupler links, and S measures the distance traveled by block B.

A slider-crank mechanism is basically a four-link mechanism with its output link of infinite length. For this reason, a slider-crank mechanism has limit positions or dead-center positions.

Figure 12a shows a slider-crank mechanism in its first limit position. The input-crank makes angle θ_{2L_1} with the horizontal line. Slider block B has traveled through its maximum distance, S_{max}.

The second limit position is illustrated in Fig. 12b. The input crank makes angle θ_{2L_2} with the horizontal line. The slider block occupies the initial position, S_{min}. Net distance traveled by the slider block measures the length of the stroke and is given by the difference of S_{max} and S_{min}.

$$\text{Length of the stroke } = S_{max} - S_{min} \tag{11}$$

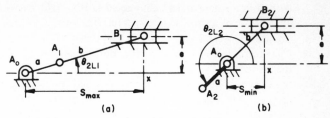

Fig. 12   Limit positions of a slider-crank mechanism.

Figure 13 shows a slider-crank mechanism in its dead-center position. The dead-center position is identified by the geometry that the coupler link is normal to the axis of slider block B.

Fig. 13   Dead-center position of a
slider-crank mechanism.

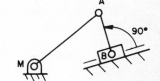

Example 3.

Find the limit positions and the length of the stroke of a slider-crank mechanism with the following dimensions: a = 1 in., b = 3 in., S_{max} = 3.75 in. The axis of the slider block is parallel to the horizontal.

Solution.

Repeat the constructions shown in Fig. 14.

1. Draw horizontal line MX of length 3.75 in.
2. At X draw line XB_1 perpendicular to line MX.
3. With M as a center and a radius equal to $MB_1 = a + b = 1 + 3 = 4$ in., draw an arc of a circle to intersect XB_1 at B_1.

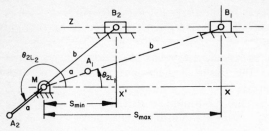

Fig. 14 Limit positions and stroke of a slider-crank mechanism.

4. Connect MB_1. Place revolute pairs at A_1, M, and B_1 and a slider pair at B_1 with its axis parallel to MX.
5. Measure angle θ_{2L_1}. This angle defines the first limit position.
6. Through B_1 draw line B_1Z parallel to line MX.
7. With M as a center and a radius equal to $MB_2 = b - a = 3 - 1 = 2$ in., draw an arc of a circle to intersect line B_1Z in B_2.
8. Connect B_2M and locate A_2 on B_2M so that $B_2A_2 = b$ or $MA_2 = a$.
9. Place revolute pairs at A_2 and B_2. Measure angle θ_{2L_2}. This angle defines the second limit position.
10. Draw line B_2X' perpendicular to MX and intersecting MX in X'. The distance $X'X$ measures the total length of the stroke of output slider block B.

Example 4.

Derive mathematical relationships for determining limit positions of a slider-crank mechanism.

Solution.

The right-angle triangles MB_1X and MB_2X' of Figs. 12a and b yield:

$$\theta_{2L_1} = \sin^{-1}\left(\frac{e}{a + b}\right) \tag{12}$$

$$S_{\max} = [(a + b)^2 - e^2]^{1/2} \tag{13}$$

$$\theta_{2L_2} = \sin^{-1}\left(\frac{e}{b - a}\right) \tag{14}$$

$$S_{\min} = [(b - a)^2 - e^2]^{1/2} \tag{15}$$

$$\text{Length of the stroke} = [(a + b)^2 - e^2]^{1/2} - [(b - a)^2 - e^2]^{1/2} \tag{16}$$

Competency Items

• Define a dead-center position for a slider-crank mechanism.
• Draw the limit position and the dead-center position for a slider-crank–type mechanism if the input were through the slider.
• Derive a mathematical relationship to calculate angles for dead-center positions.

Objective 4

To identify the conditions that determine limit positions of an inverted slider-crank mechanism

Activities

• Read the material provided below, and as you read:
  a. Name the conditions that define limit positions of an inverted slider-crank mechanism.
  b. Determine the two limit positions using graphical construction.
  c. Calculate the angle of oscillation of the output link.
• Check your answers to the competency items with your instructor.

Reading Material

Part I: Inverted slider-crank mechanism, Type I

An inverted slider-crank mechanism, as shown in Fig. 15, has three revolute pairs and a slider pair. The slider pair is placed at the third joint, and output is obtained through the revolute pair at the fourth joint. Because of the slider pair at the third joint, the coupler link AB is of variable length. Link $MA = a$ is the input link and link $QB = e$ is the output link. Note link QB describes the eccentricity of the mechanism and can be of a finite, including zero, length. Let fixed link $MQ = d$. Angles θ_1, θ_2, θ_3, and θ_4 describe the relative motion of the links. Angle ABQ is always $90°$ because of the slider pair at the third joint.

An inverted slider-crank mechanism can function as either a crank-rocker mechanism or as a drag-link mechanism. The type of motion depends upon numerical values of the parameters a, e, and d.

If $a < d + e$, then the inverted slider-crank mechanism is a crank-rocker mechanism. Because of the crank-rocker type of motion, such an inverted slider-crank mechanism has two limit positions which are shown in Fig. 16.

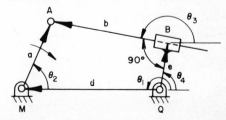

Fig. 15 Inverted slider-crank mechanism, Type I.

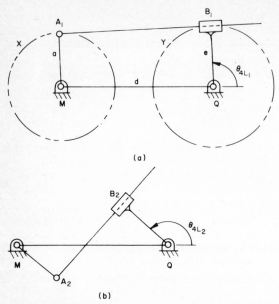

(a)

(b)

Fig. 16 Limit positions of an inverted slider-crank mechanism, Type I.

The two limit positions QB_1 and QB_2 can be determined using the following steps:

1. Draw a circle X with MA_1 as a radius and M as a center.
2. Draw a circle Y with QB_1 as a radius and Q as a center.
3. Draw two lines B_1A_1 and B_2A_2 as tangents to two circles X and Y.
4. The two limit positions are MA_1B_1Q and MA_2B_2Q.
5. The angle of oscillation ϕ of the output link is given by:

$$\phi = \theta_{4L_2} - \theta_{4L_1} \tag{17}$$

Part II: Motion analysis of an inverted slider-crank mechanism, Type II

The Type II inverted slider-crank mechanism has the same type and number of pairs as the Type I inverted slider-crank mechanism. The difference is the way in which the prism pair is placed. In the Type I mechanism, the prism pair is

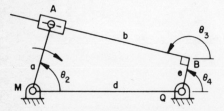

Fig. 17 Inverted slider-crank mechanism, Type II

connected to the output link while in the Type II mechanism, the prism pair is connected to the input link, as shown in Fig. 17. Let link $MA = a$ be the input link. Then $AB = b$ and $QB = e$ constitute the output-link and $MQ = d$ is the fixed link. Angles θ_2 and θ_4 describe angular displacements of input and output links.

Despite the different location of the prism pair, the input-output characteristic of the Type II mechanism is the same as that of the Type I.

Competency Items

- Plot the input-output for an inverted slider-crank mechanism when $a = 1.0$ in., $d = 3.0$ in., $e = 0.5$ in. Draw its two limit positions.
- Plot the input-output for an inverted slider-crank mechanism when $a = 1.0$ in., $d = 0.5$ in., $e = 0.0$ in.
- Describe two practical applications of an inverted slider-crank mechanism.

Objective 5

To identify the conditions that determine minimum and maximum transmission angles of four-link, slider-crank, and inverted slider-crank mechanisms

Activities

- Read the material provided below, and as you read:
  a. Define transmission angle in four-link, slider-crank, and inverted slider-crank mechanisms.
  b. State the importance of the transmission angle in the design of mechanisms.
  c. Calculate the minimum value of the transmission angle.
- Check your answers to the competency items with your instructor.

Reading Material

In addition to taking care of limit positions and dead-center positions in the design of a four-link mechanism, it is equally important to insure that the designed linkage does not have unusually high acceleration and that the motion of the designed linkage is relatively smooth or jerk-free. It is significant that this examination is done while the mechanism is in the design stage. Fortunately, the quality of the motion of a four-link mechanism can be examined using the criteria of a transmission angle.

Part I. Minimum transmission angle in a four-link mechanism

In Fig. 18, four-link mechanism $MABQ$ is shown with input link MA coupler link AB and output link QB. Transmission angle μ is the angle between the direction of the absolute motion T_a of output link QB and the direction of relative motion T_r of coupler link AB with respect to input link MA. The direction of relative motion T_r is perpendicular to the coupler link and the

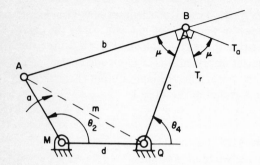

Fig. 18 Definition of transmission angle in a four-link mechanism.

direction of absolute motion T_a of output link MB is perpendicular to the output link.

Since

Angle $T_r B T_a = \mu$

and

Angle $QBT_a = 90°$

then

Angle $QBT_r = 90° - \mu$

Also since,

Angle $ABT_r = 90°$

and

Angle $QBT_r = 90° - \mu$

then

Angle $ABQ = \mu$

Thus, the transmission angle can also be defined as the included angle between coupler link AB and output link QB.

The quality of motion of a four-link mechanism is dependent upon the minimum value of transmission angle μ. The force transmission from the coupler link to the output link is most effective when the transmission angle is $90°$. As the input link rotates, the value of the transmission angle changes. It is desirable that the deviation of the transmission angle from the ideal value of $90°$ is kept to a minimum. Lower values of the transmission angle ($\mu < 15°$) create unusually higher acceleration, objectionable noise and jerk at high speed. A mechanism with minimum transmission angle less than $40°$ becomes unacceptable for high speed design.

Since the minimum value of the transmission angle is a critical design value in a four-link mechanism, it is important to know the technique to determine the positions of a four-link mechanism at which the transmission angle becomes minimum. For this purpose, examine Fig. 18. Application of the cosine law to triangle MAQ yields:

$$a^2 + d^2 - 2ad \cos \theta_2 = m^2 \tag{18}$$

Similarly, for triangle ABQ the cosine law yields:

$$b^2 + c^2 - 2bc \cos \mu = m^2 \tag{19}$$

Combining Eqs. (18) and (19), the result is:

$$a^2 + d^2 - 2ad \cos \theta_2 = b^2 + c^2 - 2bc \cos \mu \tag{20}$$

For the determination of a maximum or a minimum value of the transmission angle, Eq. (20) should be differentiated with respect to θ_2 and the resultant derivative should be set to zero. Thus,

$$\frac{d\mu}{d\theta_2} = \frac{ad \sin \theta_2}{bc \sin \mu} = 0 \tag{21}$$

Since the parameters a and d are never zero, Eq. (21) is satisfied when:

$$\sin \theta_2 = 0 \tag{22}$$

or

$$\theta_2 = 0° \text{ or } 180°$$

Thus, when $\theta_2 = 0°$ or $180°$, the transmission angle is either maximum or minimum. It can be seen from Figs. 19 and 20 that the transmission angle becomes maximum when $\theta_2 = 180°$ and minimum when $\theta_2 = 0°$. Figure 19a and b describes a drag-link mechanism with a maximum and a minimum transmission

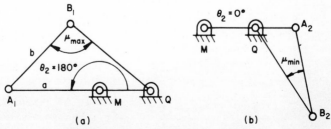

Fig. 19 Drag-link mechanism. **a.** Maximum transmission angle position; **b.** minimum transmission angle position.

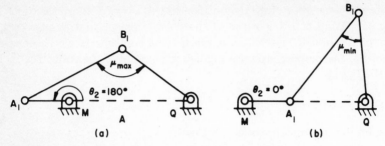

Fig. 20 Crank-rocker mechanism. **a.** Maximum transmission angle position: **b.** minimum transmission angle position.

angle. Figure 20a and b illustrates a crank-rocker mechanism with a maximum and a minimum transmission angle.

Part II: Minimum transmission angle in a slider-crank mechanism

Figure 21 describes transmission angle μ of a slider-crank mechanism. The transmission angle is obtained by erecting normal BY to the straight-line path of the slider and measuring angle ABY.

From Fig. 21,

$$AX = AZ - e = a \sin\theta_2 - e$$

and

$$AX = b \sin(90° - \mu) = b \cos\mu$$

Hence,

$$a \sin\theta_2 - e = b \cos\mu$$

Differentiating the above equation with respect to θ_2 yields:

$$a \cos\theta_2 = -b \sin\mu \frac{d\mu}{d\theta_2}$$

or

$$\frac{d\mu}{d\theta_2} = - \frac{a \cos\theta_2}{b \sin\mu}$$

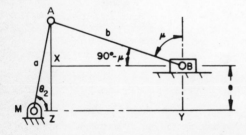

Fig. 21 Definition of a transmission angle in a slider-crank mechanism.

For maximum or minimum transmission angle,

$$\frac{d\mu}{d\theta_2} = 0$$

That is,

$$\cos\theta_2 = 0, \text{ or } \theta_2 = 90°, 270°$$

Therefore, the transmission angle becomes maximum or minimum when $\theta_2 = 270°$ or $90°$.

Figure 22a and b describes positions of a slider-crank mechanism in which the transmission angle is maximum and minimum.

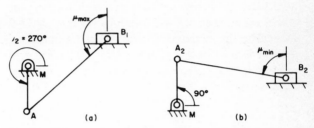

Fig. 22 Slider-crank mechanism. **a.** Maximum transmission angle position: **b.** minimum transmission angle position.

Part III: Minimum transmission angle in an inverted slider-crank mechanism

As discussed in the previous objective, there are two types of inverted slider-crank mechanisms. The Type I inverted slider-crank mechanism with slider pair at the output link always maintains the ideal maximum value of the transmission angle. In other words, μ is always $90°$.

Transmission angle μ for the inverted slider-crank mechanism Type II is defined in Fig. 23. The minimum value of the transmission angle is determined using the graphical procedure described in Fig. 24.

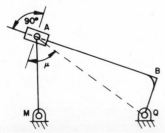

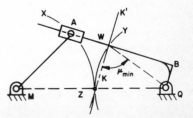

Fig. 23 Definition of transmission angle μ in an inverted slider-crank mechanism, Type II.

Fig. 24 Determination of μ_{min} of an inverted slider-crank mechanism, Type II.

From Fig. 24,

1. With M as a center and a radius equal to MA, draw an arc of circle X to intersect fixed link MQ at Z.
2. With M as a center and a radius equal to MZ, draw an arc of circle Y to intersect BA in W.
3. At W draw a line KK' perpendicular to AB.
4. Connect M and W.
5. Measure angle $KWQ = \mu_{min}$.

Competency Items

- Calculate the minimum and maximum transmission angles for a four-link mechanism when $MA = 1.0$ in., $AB = 2.0$ in., $BQ = 2.0$ in., and $MQ = 2.0$ in.
- Calculate the minimum and maximum transmission angles for a four-link mechanism when $MA = 2.0$ in., $AB = 2.0$ in., $QB = 2.0$ in., and $MQ = 1.0$ in.
- Calculate the minimum and maximum transmission angles for a slider-crank mechanism when $MA = 1.0$ in., $AB = 3.0$ in., and $e = 0$ in.
- Calculate the minimum transmission angle for an inverted slider-crank mechanism Type II when $MA = 2.0$ in., $MQ = 3.0$ in., $QB = 0.5$ in.

Objective 6

To design a drag-link mechanism with an optimum transmission angle

Activities

- Read the material provided below, and as you read:
  a. Design a drag-link mechanism using the graphical technique.
  b. Design a drag-link mechanism using the mathematical technique.
  c. Design a drag-link mechanism using a design chart.
- Check your answers to the competency items with your instructor.

Reading Material

A drag-link mechanism is designed to transform uniformly rotating motions into irregularly rotating motions. For this reason, a drag-link mechanism is placed in series with another mechanism to alter motion characterisitcs such as velocity, acceleration, etc. For a drag-link mechanism to provide a good quality of motion or a motion which does not have unusually high accelerations, the design of the drag-link mechanism must be such that the transmission angle is optimized.

Figure 25 shows a drag-link mechanism in its two positions QMA_4B_4 and QMA_3B_3, having maximum and minimum transmission angles. Corresponding to $180°$ rotation of input link MA output link QB rotates through angle ψ'_μ. Let angles B_4QM and A_3QB_3 be equal to β_1 and β_2'. Then,

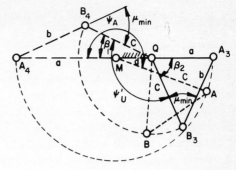

Fig. 25 A drag-link mechanism in its two positions, QMA_4B_4 and QA_3B_3, having maximum and minimum transmission angles.

$$\psi_\mu' = \beta_1 + (180 - \beta_2) \tag{23}$$

For the purpose of optimum transmission conditions let

$$\text{Angle } A_4B_4Q = 180 - \mu_{\min} \tag{24}$$

where

$$\text{Angle } QB_3A_3 = \mu_{\min}$$

A drag-link mechanism with optimum transmission angle conditions defined by Eq. (24) can be designed using either a graphical or an analytical approach. The graphical approach presented below is described using Fig. 26.

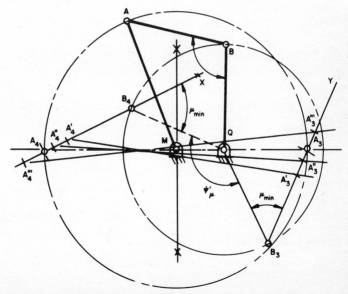

Fig. 26 Graphical approach for a design of a drag-link mechanism with optimum transmission angle.

1. Select the position of fixed pivot point Q and draw sector B_4QB_3 so that $QB_4 = QB_3$ and angle $B_4QB_3 = \psi'_\mu$.
2. At B_4 draw line X, making angle $QB_4X = \mu_{min}$ with QB_4 as shown in Fig. 26.
3. At B_3 draw line Y, making angle $QB_3Y = \mu_{min}$ with line QB_3.
4. With B_3 and B_4 as centers and any arbitrary radius, draw arcs to cut lines X and Y in points A'_3, A''_3, A'''_3, and A'_4, A''_4, A'''_4, etc.
5. Join A'_3 and A'_4, A''_3 and A''_4, A'''_3 and A'''_4, etc. The fixed pivot lies on the line that passes through fixed pivot Q. A_3A_4 is such a line.
6. Bisect A_3A_4. The point of intersection of the perpendicular bisector with line A_3A_4 locates fixed pivot M. MA_3B_3Q is the designed drag-link mechanism with prescribed ψ'_μ and μ_{min}.

A drag-link with optimum transmission angle is designed more accurately by using a mathematical approach which requires solving three simultaneous equations.

These equations are obtained by applying sine and cosine laws to triangles MQA_4B_4 and QA_3B_3. The law of cosines yields:

$$b^2 + c^2 + 2bc \cos \mu_{min} = (a + d)^2 \tag{25}$$

$$b^2 + c^2 - 2bc \cos \mu_{min} = (a - d)^2 \tag{26}$$

$$c^2 + (a + d)^2 - 2c(a + d) \cos \beta_1 = b^2 \tag{27}$$

$$c^2 + (a + d)^2 - 2c(a - d) \cos \beta_2 = b^2 \tag{28}$$

The law of sines yields:

$$\frac{a + d}{\sin \mu_{min}} = \frac{b}{\sin \beta_1} \tag{29}$$

$$\frac{a - d}{\sin \mu_{min}} = \frac{b}{\sin \beta_1} \tag{30}$$

Also,

$$\psi'_\mu = \beta_1 + (180° - \beta_2) \tag{23}$$

Adding Eqs. (25) and (26), the result is:

$$b^2 + c^2 = a^2 + d^2 \tag{31}$$

Subtracting Eq. (26) from Eq. (25) yields:

$$\cos \mu_{min} = \frac{ad}{bc} \tag{32}$$

Taking cosines of both sides of Eq. (23), the result is:

$$\cos\psi'_\mu = \sin\beta_1 \sin\beta_2 - \cos\beta_1 \cos\beta_2 \tag{33}$$

Substituting for $\cos\beta_1$, $\cos\beta_2$, $\sin\beta_1$, and $\sin\beta_2$ from Eqs. (27)–(30) in Eq. (33) and simplifying the resultant equation by using Eqs. (31) and (32), we obtain:

$$\frac{a}{d} = \left[-\frac{q}{2p} + \left(\frac{r}{p} + \frac{q^2}{4p^2} \right) \right]^{1/2} \tag{34}$$

$$\frac{b}{d} = \left[\frac{(a/d)^2(1 + \cos\psi'_\mu) + (1 - \cos\psi'_\mu)}{2\cos^2\mu_{min}} \right]^{1/2} \tag{35}$$

$$\frac{c}{d} = \frac{a/d}{(b/d)\cos\mu_{min}} \tag{36}$$

where

$$p = (1 + \cos\psi'_\mu)\left(1 - \frac{1 + \cos\psi'_\mu}{2\cos^2\mu_{min}} \right)$$

$$q = (\cos\psi'_\mu - 1)\left(1 + \frac{\cos\psi'_\mu + 1}{\cos^2\mu_{min}} \right)$$

$$r = (\cos\psi'_\mu - 1)\left(1 + \frac{\cos\psi'_\mu - 1}{2\cos^2\mu_{min}} \right)$$

Equations (34), (35), and (36) are used to obtain design charts shown in Figs. 27–29. A drag-link mechanism with prescribed values of ψ'_μ and μ_{min} can be designed using either Eqs. (34)–(36) or the design charts shown in Figs. 27–29.

Competency Items
- Design a drag-link mechanism for $\psi_\mu = 100°$ and $\mu_{min} = 35°$.
  Check your answers using the following three methods:
  a. graphical synthesis
  b. design Eqs. (25)–(27)
  c. design charts
- Design an inverted slider-crank mechanism that functions as a drag-link mechanism.

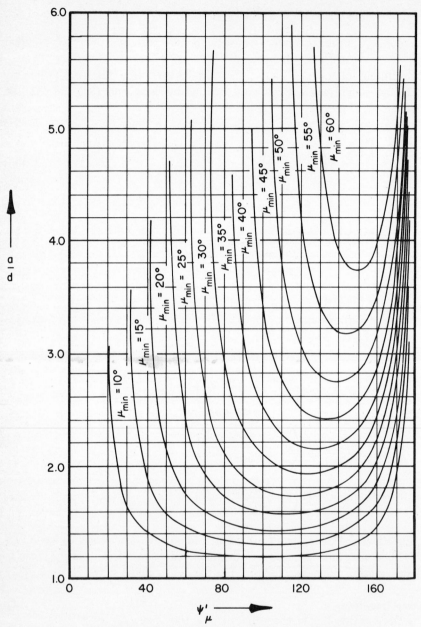

Fig. 27 Design of drag-link mechanism with most favorable transmission angle.

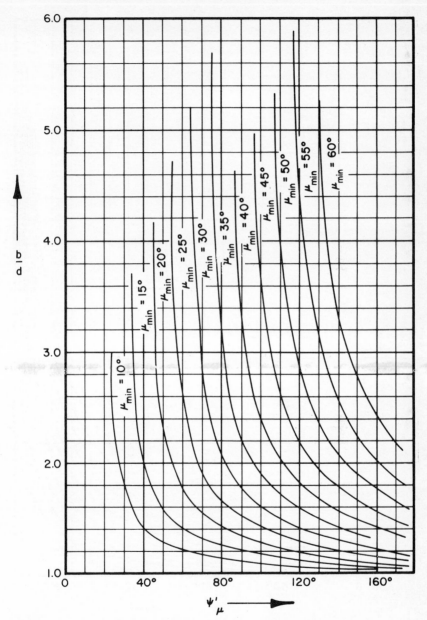

Fig. 28 Design of drag-link mechanism with most favorable transmission angle.

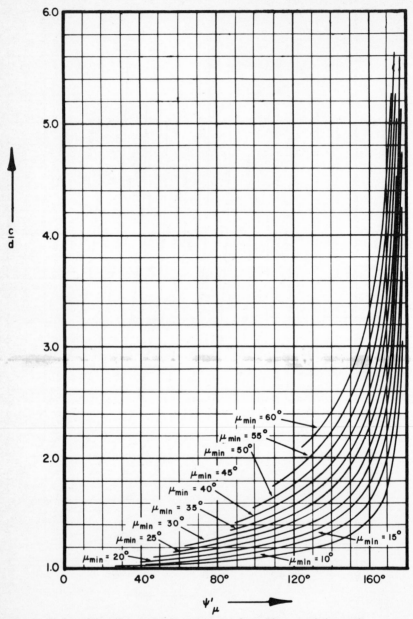

Fig. 29 Design of drag-link mechanism with most favorable transmission angle.

Objective 7

To design a crank-rocker mechanism using a minimum transmission angle

Activities
- Read the material provided below, and as you read:
  a. Design a crank-rocker mechanism with a minimum transmission angle using the graphical technique.
  b. Design a crank-rocker mechanism with a minimum transmission angle using the mathematical technique.
  c. Design a crank-rocker mechanism with a minimum transmission angle using a design chart.
- Check your answers to the competency items with your instructor.

Reading Material

A crank-rocker mechanism is shown in Fig. 30. While input link MA rotates in a clockwise direction, position MA_1 to position MA_2, the output link rocks through angle ϕ. This rocking is described as the forward stroke of the output link. As the input link continues to rotate from position MA_2 to occupy initial position MA_1 the output link executes the reverse stroke to occupy position QB_1. Let angle A_2MB_1 be equal to α. Then the forward and the reverse strokes take place in $(180° + \alpha)$ and $(180° - \alpha)$ rotation of input link MA. The time ratio of a crank-rocker mechanism is defined as:

$$TR = \frac{\text{time for forward stroke}}{\text{time for reverse stroke}} = \frac{180 + \alpha}{180 - \alpha} \qquad (37)$$

The time ratio of a crank-rocker mechanism indicates how fast the reverse stroke is executed. If $TR > 1$, then a rocker has a quick return program. If $TR < 1$, then a rocker has a quick forward program.

In addition to parameters α and ϕ, a crank-rocker mechanism also has another parameter, β, which measures the relative orientation of the input link from the fixed link and helps locate the first limit position of a crank-rocker mechanism.

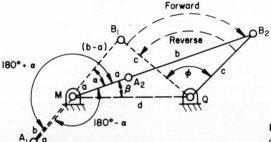

Fig. 30 Design parameters for a crank-rocker mechanism.

A crank-rocker mechanism can be designed much more quickly using a graphical approach when it needs to satisfy two design conditions described by α and ϕ. The graphical approach described in Fig. 31 requires the following steps:

1. Choose $QB_2 = c$ and draw output sector B_1QB_2, so that angle $B_1QB_2 = \phi$.
2. At B_2 draw any line X.
3. At B_1 draw line Y, making angle α with line X. The point of intersection of lines X and Y locates fixed pivot M.
4. With M as a center and MB_2 as a radius, draw an arc of a circle to intersect line Y in point Z. Note $B_1Z = 2a$ where $MA_1 = a$.
5. Bisect B_1Z to determine length a of the input link. Determine coupler $A_2B_2 = b = MB_2 - a$.

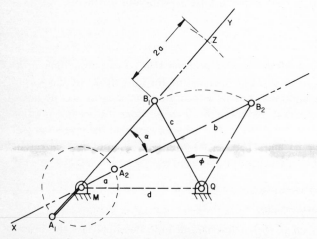

Fig. 31 Design of a crank-rocker mechanism.

This graphical approach is further extended to design a crank-rocker mechanism with maximum value of minimum transmission angle. This extended graphical approach is based on the application of similarly varying triangle.

The necessary steps leading to a design of a crank-rocker mechanism with maximum value of minimum transmission angle are described below:

1. Select an arbitrary length of fixed link MQ. See Fig. 32.
2. On MQ construct triangle MDQ so that angle $QMD = \phi/2 - \alpha$ and angle $MQD = \phi/2$.
3. On DQ construct triangle DQC so that angle $DQC = \phi$ and $QD = QC$.
4. With D and C as centers and a radius equal to MD, draw two arcs of circles X and Y, as shown in Fig. 32. Arcs X and Y describe the locus of points B_1 and B_2 of output link QB.
5. With Q as a center and different radii, draw arcs x_1y_1, x_2y_2, x_3y_3, etc. to intersect x and y in points x_1, x_2, x_3, etc., and y_1, y_2, y_3, etc.

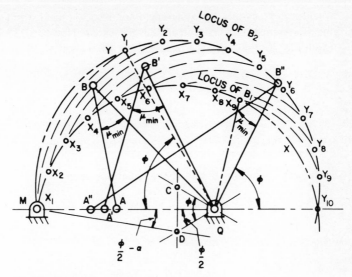

Fig. 32 Design of a crank-rocker mechanism.

6. Points M, y_1, and Q describe a crank-rocker mechanism in its first limit position. Similarly, points M, x_1, and Q describe the same crank-rocker mechanism in its second limit position.

7. Repeat the steps described in Fig. 31, to determine the lengths of input and coupler links of all possible crank-rocker mechanisms from loci arcs X and Y.

8. Determine the minimum transmission angles of all the designed crank-rocker mechanisms and make a plot of these values against values β_1, β_2, β_3, etc. where β describes the position of input link MA corresponding to the first limit position of output link QB. Figure 33 shows an example of such a chart.

9. Select the mechanism which has the maximum value of the minimum transmission angle.

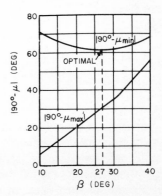

Fig. 33 Plot of μ_{min} versus positions of β.

Analytical procedure for designing a crank-rocker mechanism with maximum value of minimum transmission angle is presented below.

From Fig. 30, the following relationships can be written:

$$c^2 = (a + b)^2 + d^2 - 2(a + b)d \cos \beta \tag{38}$$

$$c^2 = (b - a)^2 + d^2 - 2(b - a)d \cos(\alpha + \beta) \tag{39}$$

$$c^2(1 - \cos \phi) = b^2(1 - \cos \alpha) + a^2(1 + \cos \alpha) \tag{40}$$

Simultaneous solutions of the above equations yield

$$A_1 b^4 + A_2 b^3 + A_3 b^2 + A_4 b + A_5 = 0 \tag{41}$$

Where

$$A_1 = (\cos \phi - \cos \alpha)[\cos \beta - \cos(\alpha + \beta)]^2$$

$$A_2 = 2a(1 - \cos \alpha)[\cos^2 \beta - \cos^2(\alpha + \beta)]$$

$$A_3 = 2a^2\{(2 - \cos \phi)[\cos^2 \beta + \cos^2(\alpha + \beta)]$$
$$- 2 \cos \alpha \cos \beta \cos(\alpha + \beta) - 2(1 - \cos \phi)\}$$

$$A_4 = 2a^3(1 + \cos \alpha)[\cos^2 \beta - \cos^2(\alpha + \beta)]$$

and

$$A_5 = a^4(\cos \phi + \cos \alpha)[(\cos \beta + \cos(\alpha + \beta)]^2$$

$$d = \frac{2ab}{b[\cos \beta - \cos(\alpha + \beta)] + a[\cos \beta + \cos(\alpha + \beta)]} \tag{42}$$

$$c = [(b + a)^2 + d^2 - 2(a + b)d \cos \beta]^{1/2} \tag{43}$$

Also,

$$\mu_{min} = \cos^{-1}\left[\frac{b^2 + c^2 - (d - a)^2}{2bc}\right] \tag{44}$$

An iterative procedure is adopted to obtain the link lengths of a crank-rocker mechanism with maximum value of minimum transmission angle when the time ratio and the angle of oscillation are the required design parameters. The procedure includes the following steps:

For the specified values of α and ϕ:

1. Let $a = 1$.
2. Set $\beta = 0$ and select increment $\Delta\beta$.
3. Set $\beta = \beta + \Delta\beta$.
4. Compute link dimension b from the polynomial Eq. (41).
5. Compute link dimensions d and c from Eqs. (42) and (43).
6. Compute μ_{min} from Eq. (44).
7. Repeat Steps (3)–(5) until β reaches a value of $90°$.
8. Select the mechanism with maximum vlaue of μ_{min}.

A design chart, prepared using the above procedure, for a crank-rocker mechanism with $\alpha = 10°$ is shown in Fig. 34.

The iterative procedure described above, however, is not needed when a crank-rocker mechanism is designed for unit time ratio. If $TR = 1$, then $\alpha = 0$ and $\mu_{max} = 180° - \mu_{min}$. Under these conditions, the simultaneous solution of the equations

$$\cos\theta_1 = \frac{c^2 + d^2 - (a + b)^2}{2cd} \tag{45}$$

$$\cos\theta_2 = \frac{c^2 + d^2 - (b - a)^2}{2cd} \tag{46}$$

$$\phi = \theta_1 - \theta_2 \tag{47}$$

$$\cos\mu_{max} = \frac{b^2 + c^2 - (d + a)^2}{2bc} \tag{48}$$

$$\cos\mu_{min} = \frac{b^2 + c^2 - (d - a)^2}{2bc} \tag{49}$$

yield

$$\frac{b}{d} = \left(\frac{1 - \cos\phi}{2\cos^2\mu_{min}}\right)^{1/2} \tag{50}$$

$$\frac{c}{d} = \left[\frac{1 - (b/a)^2}{1 - (b/d)^2 \cos^2\mu_{min}}\right]^{1/2} \tag{51}$$

and

$$\frac{a}{d} = \left[\left(\frac{b}{d}\right)^2 + \left(\frac{c}{d}\right)^2 - 1.0\right]^{1/2} \tag{52}$$

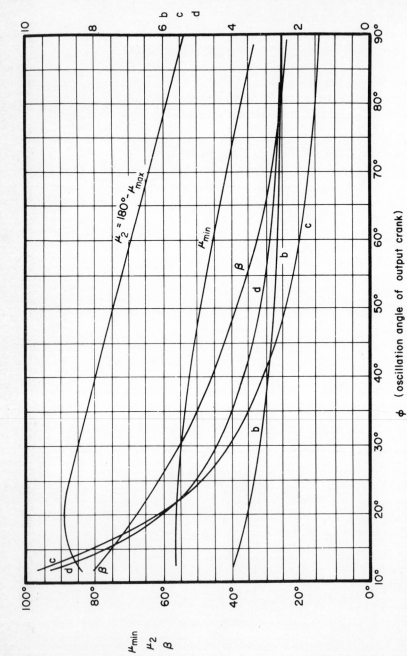

Fig. 34 Design chart for a crank-rocker mechanism ($\alpha = 10°$, $a = 1.0$).

A design chart for a crank-rocker mechanism with optimum transmission angle and unit time ratio ($\alpha = 0$) is shown in Fig. 35.

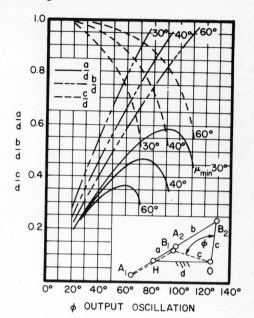

Fig. 35 Design chart for a crank-rocker mechanism with optimum transmission angle and unit time ratio ($\alpha = 0$).

ϕ OUTPUT OSCILLATION

Competency Items

- Design a crank-rocker mechanism in which $\alpha = 10°$, $\phi = 60°$, and $\mu_{min} = 40°$ using:
  a. graphical technique
  b. analytical iterative technique
  c. design charts
- Design a slider-crank mechanism with 2.0-in. output stroke. The forward stroke is executed in $210°$ rotation of the input crank. Illustrate your design techniques using:
  a. analytical technique
  b. graphical technique
- Draw a chart to design an optimum crank-rocker mechanism with $\alpha = 20°$, $30°, \ldots 60°$, and varying values of ϕ.

Performance Test #1

1. Complete the chart for the linkage shown below.

Grounded link	Input link	Type of motion
AB	AD	
AD	AB	
AD	CD	
CD	BC	

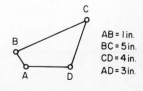

AB = 1 in.
BC = 5 in.
CD = 4 in.
AD = 3 in.

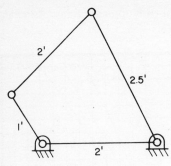

2. For the mechanism shown determine the time ratio and output angle ϕ.

3. Design a drag-link mechanism with a minimum transmission angle of $45°$ and an output angle of $100°$. Show your mechanism in its two critical positions.

Performance Test #2

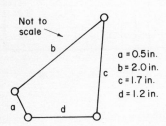

Not to scale

b

a = 0.5 in.
b = 2.0 in.
c = 1.7 in.
d = 1.2 in.

1. State which link(s) could be grounded:
   a. if it is possible to make a crank-rocker mechanism from the four-link chain shown
   b. if it is possible to make a drag link from the chain

2. On the figure show the limit positions of the output link e.

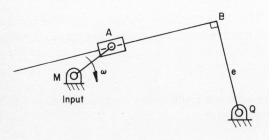

3. On the figure show the maximum and minimum transmission angles.

4. Design a drag-link mechanism for $\psi_\mu' = 120°$ if $\mu_{min} = 40°$. Verify on a drawing that your design satisfies the requirements.

Performance Test #3

1. For the four-link mechanism in the figure draw:
   a. find the maximum value of a if the mechanism is a crank rocker with MA as the input
   b. find the minimum value of a if the mechanism is a drag link

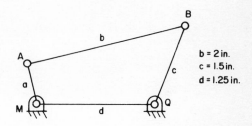

b = 2 in.
c = 1.5 in.
d = 1.25 in.

2. Find the minimum transmission angle μ_{min} of the inverted slider-crank mechanism shown.
   At what value of θ does μ_{min} occur?

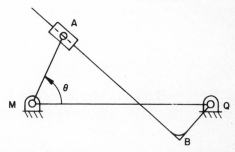

3. A crank-rocker mechanism as shown must be designed. M is the input shaft and Q, the output shaft. Output QB must oscillate through an angle of 50° and the minimum transmission angle is to be 35°. The time ratio is unity.
   a. Find the lengths of links MA, AB, and QB.
   b. Verify on a drawing that your design satisfies the requirements.

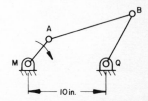

Supplementary References

2 (2.7, 2.13, 3.3); 3 (2.18); 7 (6.2, 6.14, 11.4); 8 (9.1-9.2); 9 1.1-2.3)

Unit XII. Linkage Synthesis Coordinating Input and Output Motions

Objectives
1. To synthesize a four-link and a slider-crank mechanism to coordinate their input and output displacements
2. To synthesize a four-link and a slider-crank mechanism for two positions of input and output links
3. To synthesize a four-link and a slider-crank mechanism for three positions using pole techniques
4. To synthesize a four-link and a slider-crank mechanism for three positions using the inversion technique
5. To synthesize a four-link mechanism to coordinate for four positions the motions of its input and output links using the point position reduction technique
6. To synthesize a four-link mechanism using the overlay technique

Objective 1

To synthesize a four-link and a slider-crank mechanism to coordinate their input and output displacements

Activities
- Read the material provided below, and as you read:
  a. Construct poles.
  b. Name properties of poles.
  c. Synthesize a four-link mechanism for its two positions of input and output links.
  d. Synthesize a slider-crank mechanism for its two positions of input link and slider block.
- Check your answers to the competency items with your instructor.

Reading Material

A number of problems in design involve the synthesis of a mechanism with input and output members rotating in a prescribed relationship. Typical examples demonstrating the practical applicabilities of a four-link mechanism include:

mechanical linearizer
radio tuning mechanisms
nonlinear rotary valve actuators
controlled acceleration drives (as shuttles on a textile loom)
computing components of analog computers
constant angular velocity ratio shaft coupling
quick return mechanisms and mechanisms with constant velocity drive motion
linkages and mechanisms having momentary dwells
linakges to replace cams with rotary input and output, especially in heavy duty
    and high speed applications
linkages to replace gears

In this unit we will examine different graphical techniques for synthesizing a mechanism for different prescribed positions of its input and output links. These techniques will be used to synthesize (a) a four-link mechanism with four revolute pairs, and (b) a slider-crank mechanism. The problem of position synthesis of input and output links will be considered in four parts:

two-position synthesis
three-position synthesis
four-position synthesis
five-or-more–position synthesis

Objective 1 presents synthesis technique for two position-synthesis using the properties of a pole. In Fig. 1, two positions MA_1B_1Q and MA_2B_2Q of a four-link mechanism are described. Pole P_{12} is obtained by locating the point of intersection of the bisectors of chords A_1A_2 and B_1B_2.

The subscript "12" with pole point P_{12} describes the pole point for the two positions MA_1 and MA_2 of input link MA, and QB_1 and QB_2 of output link QB. The following properties are associated with pole point P_{12}:

1. Since pole point P_{12} lies on the bisectors of A_1A_2 and B_1B_2, P_{12} becomes the common center of rotation of points A_1 and A_2 and B_1 and B_2.

2. Since P_{12} is a common center of rotation, points A_1 and A_2 and B_1 and B_2 lie on two concentric circles with radii equal to $P_{12}A_1$ and $P_{12}B_1$, with P_{12} as a center.

3. Since $A_1B_1 = A_2B_2$, the angles intercepted by chords A_1A_2 and B_1B_2 at center P_{12} are equal. That is

    angle $A_1P_{12}A_2$ = angle $B_1P_{12}B_2$

4. $\frac{1}{2}$ angle $A_1P_{12}A_2$ = $\frac{1}{2}$ angle $B_1P_{12}B_2$

That is angle $A_1P_{12}M$ = angle $B_2P_{12}Q$. But, angles $A_1P_{12}M$, $B_2P_{12}Q$ are the angles intercepted by the input and the output links. Thus, the input and the output links intercept equal angles at pole P_{12}.

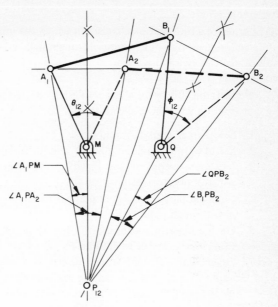

Fig. 1 Basic geometry of a pole.

5. It is proved that angle $A_1 P_{12} A_2$ = angle $B_1 P_{12} B_2$. We add angle $A_2 P_{12} B_1$ to both sides of the above equality. Thus,

$$\text{angle } A_1 P_{12} A_2 + \text{angle } A_2 P_{12} B_1$$
$$= \text{angle } B_1 P_{12} B_2 + \text{angle } A_2 P_{12} B_1$$

that is angle $A_1 P_{12} B_1$ = angle $A_2 P_{12} B_2$. The above relationship states that coupler link AB in its two positions intercepts equal angles at pole P_{12}.

6. Let us examine angle $A_1 P_{12} B_1$:

$$\text{angle } A_1 P_{12} B_1 = \text{angle } A_1 P_{12} M + \text{angle } M P_{12} A_2$$
$$+ \text{angle } A_2 P_{12} B_1$$

But angle $A_1 P_{12} M$ = angle $B_1 P_{12} Q$. Therefore,

$$\text{angle } A_1 P_{12} B_1 = \text{angle } B_1 P_{12} Q + \text{angle } M P_{12} A_2$$
$$+ \text{angle } A_2 P_{12} B_1 = \text{angle } M P_{12} Q$$

That is, coupler link AB and fixed link MQ intercept equal angles at pole P_{12}.

Using the properties of a pole point, we can now proceed to synthesize a four-link mechanism for its two positions of input and output links.

Example 1.

Synthesize a four-link mechanism in which the input link rotates through angle θ_{12} and the output link rotates through angle ϕ_{12}. Both links are required to move in a clockwise direction.

Solution.

Repeat the following steps to design the required mechanism. Check your construction with those shown in Fig. 2.

1. Select fixed centers M and Q.
2. Select output crank length QB, and its initial position QB_1.
3. Lay out output sector B_1QB_2 such that angle $B_1QB_2 = \phi_{12}$. This construction locates position QB_2.
4. Find the perpendicular bisector b of B_1B_2. Note that the bisector will pass through point Q.
5. On the bisector locate arbitrarily any point P_{12}. Measure angle $B_1P_{12}Q$.
6. Connect P_{12} and M to get line $P_{12}Z$.
7. Draw line $P_{12}X$ from point P_{12} such that angle $MP_{12}X =$ angle $QP_{12}B_1$. These angles are measured in a counterclockwise sense from lines QP_{12} and MP_{12}, since it is required for input and output links to move in a clockwise direction.
8. Draw line MY from point M such that angle $ZMY = \frac{1}{2}\theta_{12}$. Angle $\frac{1}{2}\theta_{12}$ is measured in a counterclockwise sense from the reference line $P_{12}Z$. Line MY intersects $P_{12}X$ in A_1.

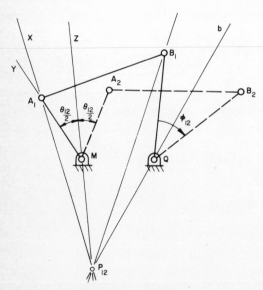

Fig. 2 Two-position design technique.

9. Connect M, A_1, B_1, and Q. Place revolute pairs at joints M, A_1, B_1, and Q. Then, the four-link mechanism MA_1B_1Q will satisfy the design requirement.

10. There are two important points that must be examined.
    a. Check for the existence of dead centers.
    b. Check for the minimum transmission angle. Select the mechanism with a maximum value of minimum transmission angle.

11. If it is required that link MA should move in the direction opposite to that of link QB, then the pivot point M is located on any line P_1Z so that coupler link AB crosses the fixed link MQ.

Example 2.

Synthesize a slider-crank mechanism with eccentricity e for the two positions of input link θ_{12} and output displacement S_{12} of the slider. The slider should move away from the fixed center while input crank moves in a clockwise sense.

Solution.

Repeat the following steps to design the slider-crank mechanism shown in Fig. 3.

1. Draw two parallel lines Q_1 and Q_2 a distance e apart.
2. On line Q_2, select arbitrarily the fixed center M. Measure distance $MC = \frac{1}{2}S_{12}$ in the direction opposite to the desired motion of the slider.
3. Draw two parallel lines Y_1 and Y_2 passing through the points M and C, and normal to line Q_2.
4. At M, draw line KM making angle $\frac{1}{2}\theta_{12}$ with line Y_1M. Angle θ_{12} is measured in a counterclockwise direction from reference line Y_1M. Line KM intersects line Y_2C in pole P_{12}.
5. Draw any line $P_{12}Z_B$ intersecting line Q_1 in B_1.
6. Draw line $P_{12}Z_A$ through point P_{12} so that angle $B_1P_{12}Z_A = \frac{1}{2}\theta_{12}$ measured in a counterclockwise sense.
7. Select point A_1 arbitrarily on line Z_A.

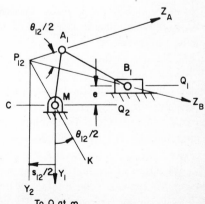

Fig. 3 Design of a slider-crank mechanism for two positions of input link and two positions of slider block.

8. Connect points M, A_1, and B_1. Place revolute pairs at M, A_1, and B_1, and a slider pair at B_1, with its axis coincident with line Q_1. Then, MA_1B_1 is the desired slider-crank mechanism.
9. Since there are an infinite number of solutions, as a designer you are required to select the mechanism which has the optimum design characteristics.

Competency Items
- Design a four-link mechanism so that $\theta_{12} = 45°$ and $\phi_{12} = 60°$. Both input and output links should move in a counterclockwise direction.
- Design a slider-crank mechanism so that $\theta_{12} = 60°$ and $s_{12} = 3$ in. The input link should move in a counterclockwise direction while the output slider block should move away from fixed point M.

Objective 2

To synthesize a four-link and a slider-crank mechanism for two positions of input and output links

Activities
- Read the material provided below, and as you read:
  a. Demonstrate the procedure to get inversion of a linkage.
  b. Define the principle of inversion using an inversion technique.
  c. Design a four-link mechanism to coordinate input and output links for two positions using an inversion technique.
  d. Design a slider-crank mechanism.
- Check your answers to the competency items with your instructor.

Reading Material
In order to design a four-link mechanism for two-position synthesis using the inversion technique, it is necessary to know the underlying principle of the method of inversion.

Fig. 4a describes a four-link mechanism in its two positions MA_1B_1Q and MA_2B_2Q. The input and the output links MA and QB rotate through angle θ_{12} and ϕ_{12} in a clockwise direction. In Fig. 4a, the fixed link is MQ.

The principle of inversion first considers input link MA_1 fixed to the ground, instead of link MQ. Then move the link MQ through angle θ_{12} in the counterclockwise sense to obtain position $MA_1B_2'Q'$, shown in Fig. 4b.

Now compare positions $MA_1B_2'Q'$ (Fig. 4b), with position MA_2B_2Q (Fig. 4a). We observe that the two configurations are congruent. But the configuration in Fig. 4b appears to be rotated, as if MA_1B_1Q is a structure, through an angle θ_{12} in a counterclockwise sense. The configuration of Fig. 4b is superimposed in Fig. 4a.

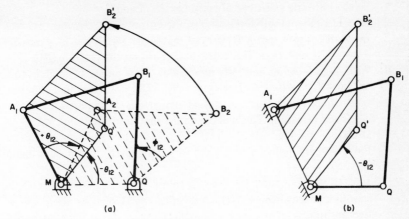

Fig. 4 Inversion of a four-link mechanism.

Thus, the inversion technique permits us to locate B_2' by rotating B_2 about M through angle θ_{12}, in a counterclockwise sense. Furthermore, B_2' lies on a circle with radius $A_1 B_1$ and center A_1. This property is utilized in synthesizing a four-link mechanism for its two positions. The following example demonstrates the application of the inversion principle in synthesizing a four-link mechanism to coordinate its motion through two positions.

Example 3.

Synthesize a four-link mechanism for its two positions of its input and output links rotating through angles θ_{12} and ϕ_{12} in a clockwise direction.

Solution.

Repeat the following steps to design the four-link mechanism shown in Fig. 5.

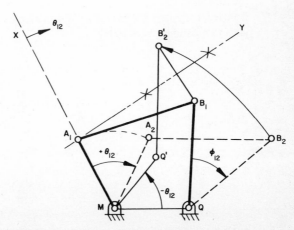

Fig. 5 Two-position design by inversion technique.

1. Select arbitrarly positions of fixed centers M and Q.
2. Select length and initial position of the output link QB_1.
3. Draw the output sector B_1QB_2 so that angle $B_1QB_2 = \phi_{12}$ measured in a clockwise direction. This construction locates the position QB_2.
4. With M as a center and MB_2 as a radius, draw an arc of a circle to rotate MB_2 to MB_2' so that $\angle B_2'MB_2 = -\theta_{12}$. QB_2 is rotated in a counterclockwise direction because in actual design QB is required to move in a clockwise direction. This will locate the first position of link MA.
5. Draw any line MX.
6. Find the perpendicular bisector Y of chord B_1B_2' to intersect line MX in A_1.
7. Join M, A_1, B_1, and Q. Place revolute pairs at these joints. Then the four-link mechanism MA_1B_1Q satisfies the design requirements of the two-position synthesis.

Example 4.

Synthesize a slider-crank mechanism with eccentricity e for the two positions of its input link, θ_{12}, and the two positions S_{12}, of its slider block. The input link should rotate in a clockwise direction while the slider block slides away from fixed center M.

Solution.

Repeat the following steps to obtain the mechanism shown in Fig. 6a.

1. Draw two lines Q_1 and Q_2 a distance e apart.
2. Locate arbitrarily a center M on line Q_2.
3. Locate arbitrarily a point B_1 on line Q_1.
4. Locate B_2 from B_1 so that $B_1B_2 = S_{12}$.
5. Select the initial position and the length of MA_1'.
6. Rotate MA_1' about center M through angle θ_{12} in a clockwise direction to locate MA_2'.
7. Construct triangle $A_1'MB_1$. Rotate this triangle about center M, so that A_1' goes to A_2' and B_1 goes to B_1'.

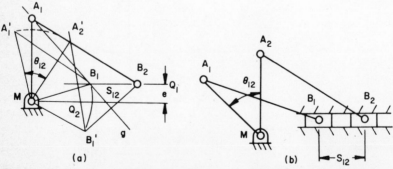

Fig. 6 a. Two-position synthesis of a slider-crank mechanism, using an inversion technique: **b.** one of the many possible solutions of the designed mechanism.

8. Locate A_1 anywhere on perpendicular bisector g of the chord $B_1'B_2$.
9. Join M, A_1, and B_1. Place revolute pairs at M, A_1, B_1, and a slider pair at B with its axis coincident with line Q_1. Then MA_1B_1 is the mechanism that will satisfy the given design requirement.
10. There are an infinite number of solutions. One of the many solutions is shown in Fig. 6b.

Competency Items

- Design a four-bar mechanism and a slider-crank mechanism to meet the requirements for the competency items of Objective 1: use the inversion technique. Compare your results.
- Describe the inversion technique to synthesize a four-link mechanism for two positions of its input and output links which are required to move in opposite directions.

Objective 3

To synthesize a four-link and a slider-crank mechanism for three positions using pole techniques

Activities

- Read the material provided below, and as you read:
  a. Demonstrate the application of the pole technique to synthesize a four-link mechanism for its three positions of input and output links.
  b. Demonstrate the application of the pole technique to synthesize a slider-crank mechanism for three positions of its input link and slider crank.
- Check your answers to the competency items with your instructor.

Reading Material

In the two-position synthesis, using the pole technique, several choices exist in location of the pole and the initial position of the input and output links. One of these choices is utilized to synthesize a four-link mechanism for three-position synthesis. The synthesis technique is demonstrated in the example presented below.

Example 5.

Synthesize a four-link mechanism for its three positions of input and output links given by θ_{12}, θ_{13}, ϕ_{12}, and ϕ_{13}, as shown in Fig. 7a.
Solution.

Repeat the following construction that leads to the design of a four-link mechanism as shown in Fig. 7b.

1. Locate fixed centers M and Q.
2. Locate arbitrarily the initial position of the output link QB_1 of arbitrary length.

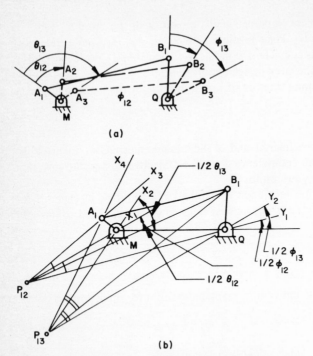

(a)

(b)

Fig. 7　**a.** Three-position synthesis problem statement: **b.** construction for synthesizing a four-link mechanism for three-position synthesis.

3. At point Q draw lines Y_1 and Y_2 making angles $\frac{1}{2}\phi_{12}$ and $\frac{1}{2}\phi_{13}$ with MQ. The angles are measured in a counterclockwise direction.
4. At point M draw lines X_1 and X_2 making angles $\frac{1}{2}\theta_{12}$ and $\frac{1}{2}\theta_{13}$ with MQ. The angles are measured in a counterclockwise direction.
5. Poles P_{12} and P_{13} are the points of intersection of lines X_1 and Y_1, and X_2 and Y_2.
6. Join P_{12} and B_1, and P_{13} and B_1.
7. At P_{12} draw line $P_{12}X_3$ making angle $X_3P_{12}B_1$ = angle $MP_{12}Q$.
8. At P_{13} draw line $P_{13}X_4$ making angle $X_4P_{13}B_1$ = angle $MP_{13}Q$.
9. Lines $P_{12}X_3$ and $P_{13}X_4$ intersect in the point A_1.
10. Join M, A_1, B_1, and Q. Place revolute pairs at these joints. The four-link mechanism MA_1B_1Q shown in Fig. 7b is the designed linkage. The mechanism, however, must be examined for the existence of dead centers and the limit positions within the region of interest.

Example 6.

Using the pole technique, design a slider-crank mechanism with eccentricity e for three positions θ_{12} and θ_{13} of the input crank, and three positions S_{12} and S_{13} of the output displacement of the slider block. The input crank rotates in a

clockwise direction and the slider block moves away from the fixed center.
Solution.

The following constructions, shown in Fig. 8, yield the required mechanism:

1. Draw two parallel lines q_1 and q_2 a distance e apart.
2. Select arbitrarily center M on line q_2. Draw line n normal to line q_2 at M.
3. Measure distance $MN_1 = \frac{1}{2}S_{12}$ and $MN_2 = \frac{1}{2}S_{13}$ on line q_2 as shown in Fig. 8a.
4. Erect normals p and q to line q_2 at N_1 and N_2.
5. At M draw line l and m, making angles $\frac{1}{2}\theta_{12}$ and $\frac{1}{2}\theta_{13}$ with line n. Lines l and m intersect p and q to yield pole points P_{12} and P_{13}.
6. Select any point B_1 on line q_1. Join points P_{12} and B_1, and P_{13} and B_1.
7. At pole P_{12}, draw line $P_{12}Z_A$ making angle $\frac{1}{2}\theta_{12}$ with line $P_{12}B_1$. $\frac{1}{2}\theta_{12}$ is measured in a counterclockwise direction.
8. At pole P_{13}, draw line $P_{13}Z'_A$ making angle $\frac{1}{2}\theta_{13}$ with line $P_{13}B_1$. $\frac{1}{2}\theta_{13}$ is measured in a counterclockwise direction.
9. Point A_1 is obtained as a point of intersection of lines $P_{12}Z_A$ and $P_{13}Z'_A$.

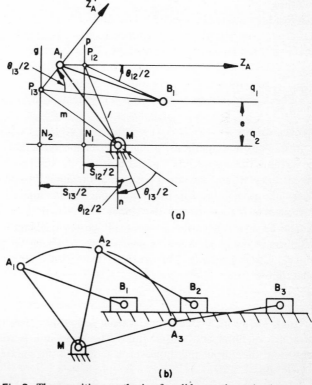

Fig. 8 Three-position synthesis of a slider-crank mechanism using the pole technique

10. Join M, A_1, and B_1. Place revolute pairs at M, A_1, B_1, and a slider pair at B_1, with its axis coincident with the line q_1. Then, MA_1B_1 is the slider-crank mechanism that meets the given design requirements. There are an infinite number of solutions. Fig. 8b describes one such solution.

Competency Items
- Design a four-bar mechanism so that $\theta_{12} = 30°$, $\theta_{23} = 30°$, and $\phi_{12} = 45°$, $\phi_{23} = 60°$. Input moves counterclockwise, and output moves clockwise.
- Design a slider-crank mechanism so that $\theta_{12} = 30°$, $\theta_{23} = 45°$, and $S_{12} = 2$ in., $S_{23} = 2$ in. Input moves clockwise and output moves away from a fixed pivot.

Objective 4

To synthesize a four-link and a slider-crank mechanism for three positions using an inversion technique.

Activities
- Read the material provided below, and as you read:
  a. Design a four-link mechanism to coordinate the motions of input and output links for three positions.
  b. Design a slider-crank mechanism to coordinate the motions of input link and output slider block for three positions.
- Check your answers to the competency items with your instructor.

Reading Material
In two-position synthesis, using an inversion technique, several choices exist concerning the output link's initial position, its link length, the location of the fixed center M, and the initial position of the input link. In order to synthesize a four-link mechanism for three positions of its input and output links the inversion technique utilizes the choice of initial position of the input link. It should be noted that other choices still exist and, therefore, there exist infinite numbers of design solutions of mechanisms satisfying the given requirements of three-position synthesis. The synthesis technique is illustrated in the example that follows.

Example 7.
Using an inversion technique, design a four-link mechanism for the three positions of input and output links. The input link is rotating in a clockwise direction, the output link however, is moving in the counterclockwise direction.
Solution.
Repeat the following constructions, shown in Fig. 9, that yield the required mechanism.

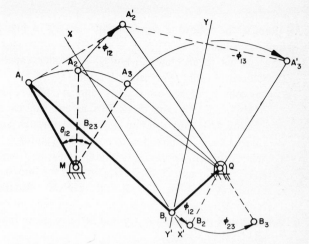

Fig. 9 Inversion technique for three-position synthesis.

1. Select the position of fixed center M and construct an input sector by selecting an appropriate length of the input link. Angles A_1MA_2 and A_1MA_3 are θ_{12} and θ_{13}. Positions A_2 and A_3 are obtained by rotating the input link MA_1 through angles θ_{12} and θ_{13} in a clockwise direction.
2. Select arbitrarily the position of fixed center Q.
3. Join QA_2 and rotate it about Q through angle $-\phi_{12}$ in a clockwise direction to obtain QA_2'.
4. Join QA_3 and rotate it about Q through angle $-\phi_{13}$ in a clockwise direction to obtain QA_3'.
5. Find the perpendicular bisector XX' and YY' of chords A_1A_2' and $A_2'A_3'$.
6. The point of intersection of the two bisectors XX' and YY' gives point B_1.
7. Join M, A_1, B_1, and Q. Place revolute pairs at these joints. Then, the four-link mechanism MA_1B_1Q is the mechanism that satisfies the given requirement. The mechanism must be examined, however, for the existence of dead centers or limit positions in the region of interest.
8. The mechanism can also be designed by selecting the output link length, and by working with the output sector to obtain A_1. There exist infinite numbers of mechanisms that can satisfy the given requirement of three-position synthesis. If the input link QB is required to rotate in a clockwise direction, QA_3' and QA_2', are obtained by rotating QA_3 and QA_2 in a counterclockwise direction.

Example 8.

Design, using an inversion technique, a slider-crank mechanism for its three positions θ_{12} and θ_{13} of the input link, and three positions S_{12} and S_{13} of the output slider block. The input link is required to rotate in the counterclockwise direction while the slider block moves toward fixed center M.

Solution.

Repeat the following constructions, shown in Fig. 10a, that yield the required mechanism.

1. Draw two parallel lines Q_1 and Q_2 a distance e apart.
2. Select arbitrarily fixed center M on line Q_2.
3. Lay out input sectors $A_1'MA_2'$ and $A_1'MA_3'$ so that angles $A_1'MA_2'$ and $A_1'MA_3'$ are θ_{12} and θ_{13}. This layout can be achieved by arbitrarily locating MA_1' and rotating it about M through angles θ_{12} and θ_{13} in a counterclockwise direction to obtain MA_2' and MA_3'.
4. Select arbitrarily point B_1 on line Q_1. Measure from B_1 on line Q_1 distances $B_1B_2 = S_{12}$ and $B_1B_3 = S_{13}$ toward the direction of M.
5. Construct triangles $A_2'B_2M$ and $A_3'B_3M$. Rotate these triangles about center M through angles θ_{12} and θ_{13} in the clockwise direction to obtain their new positions $MA_1'B_2'$ and $MA_1'B_3'$. Note that angle $B_2MB_2' = -\theta_{12}$ and $B_3MB_3' = -\theta_{13}$.
6. Find the perpendicular bisectors X and Y of chords B_1B_2' and B_1B_3'. These bisectors intersect at point A_1.
7. Join M, A_1, and B_1. Place revolute pairs at M, A_1, and B_1, and a slider pair at B_1 with its axis coincident with line Q_1. Then MA_1B_1 is the slider-crank

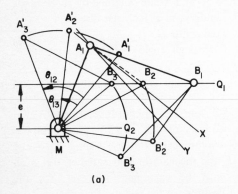

(a)

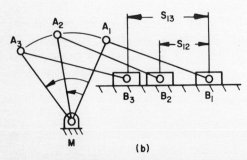

(b)

Fig. 10 Three-position synthesis of a slider-crank mechanism using an inversion technique.

mechanism satisfying the given design requirements of three-position synthesis.

8. Here again, there exist an infinite number of mechanisms that will satisfy the design requirements. One of the many solutions is presented in Fig. 10b.

Competency Item

- Synthesize the mechanisms described in Unit XII, Objective 3 using the inversion technique. Compare the two sets of results.

Objective 5

To synthesize a four-link mechanism to coordinate for four positions the motions of its input and output links using the point position reduction technique

Activities

- Read the material provided below, and as you read:
  a. Demonstrate the application of the principle of point position reduction technique.
  b. Design a four-link mechanism to coordinate for four positions the motions of input and output links.
- Check your answers to the competency items with your instructor.

Reading Material

The principle of point position reduction technique is based on the condition that one of the six poles, defining the four positions of the input and the output links, coincides with one of the two fixed centers M and Q of a four-link mechanism. Figure 11 shows a four-link mechanism MA_1B_1Q. The second position MA_2B_2Q of the four-link mechanism (shown in the dotted line), is obtained by rotating the input link MA_1 through angle θ_{12}, the output link MB_1 correspondingly rotates through angle ϕ_{12}. For the four-link mechanism under consideration, pole P_{12}, obtained as a point of intersection of two perpendicular bisectors of chords A_1A_2 and B_1B_2, coincides with the fixed center Q. The inversion of this four-link mechanism about QB_1 (that is, QB_1 is fixed to the ground), will make QB_2 coincide with QB_1, QA_2 will coincide with QA_1, and QM will be displaced to QM' so that angle MQM' is equal to $-\phi_{12}$, where ϕ_{12} is the angular displacement of output link QB_1 to occupy the position QB_2. Since QA_2 coincides with QA_1, the two-position problem is reduced to a one-position problem. If we have three positions of a linkage, and if pole P_{12} coincides with Q, then the three-position synthesis problem will be reduced to a problem of two-position synthesis.

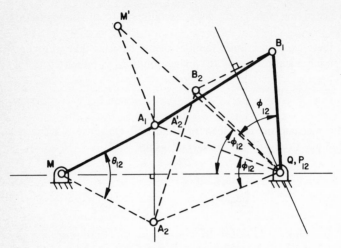

Fig. 11 Principle of point position reduction technique.

If it is desired to synthesize a mechanism for four positions of input and output links, then it is possible to reduce the four-position synthesis problem to a three-position synthesis problem by forcing the condition of coinciding one of the six poles, say pole P_{12}, with fixed center Q. The point position reduction technique can be better understood by repeating the steps presented in the illustrative example that follows.

Example 9.

Synthesize a four-link mechanism for four positions θ_{12}, θ_{23}, and θ_{34} of input link moving in a clockwise direction and four positions ϕ_{12}, ϕ_{23}, and ϕ_{34} of output link moving in a counterclockwise direction.

Solution.

Repeat the following construction, shown in Fig. 12, that yields the mechanism satisfying the given design requirements.

1. Select arbitrarily fixed center M.
2. Select the length of input link MA_1 and its initial position.
3. Rotate MA_1 about center M through angles θ_{12}, θ_{13}, and θ_{14} in a clockwise direction to locate positions MA_2, MA_3, and MA_4.
4. Find the perpendicular bisector X of chord $A_3 A_4$.
5. On the bisector X locate point Q so that angle $A_3 Q A_4 = \phi_{34}$. An overlay can be made to locate the point Q. An alternate procedure is to compute the distance MQ using the relationship

$$
MQ = MA \left[\frac{\sin\left(\dfrac{\theta_{34} + \phi_{34}}{2}\right)}{\sin(\phi_{34}/2)} \right] \tag{1}
$$

Note that pole P_{34} will coincide with the fixed center Q.

6. Join QA_2 and rotate it about Q in a clockwise direction through angle ϕ_{12} to locate position QA_2'. We are taking inversion about QB_1.

7. Join QA_3 and rotate it about Q in a clockwise direction through angle ϕ_{13} to locate position QA_3'.

8. Join QA_4 and rotate it about Q in a clockwise direction through angle ϕ_{14} to locate position QA_4'. Since the pole P_{34} is coincident with fixed center Q, QA_3' and QA_4' will coincide with each other.

9. Find the perpendicular bisectors Y and Z of the two chords A_1A_2' and $A_2'A_3'$. The point of intersection of these bisectors X and Y is the point B_1.

10. Join points M, A_1, B_1, and Q. Place revolute pairs at these joints. Then, the four-link mechanism MA_1B_1Q is the mechanism that satisfies the given design requirements.

11. Other poles P_{12}, P_{23}, P_{13}, P_{14}, and P_{24} can be made to coincide with fixed center Q. These poles provide opportunities to obtain other possible solutions.

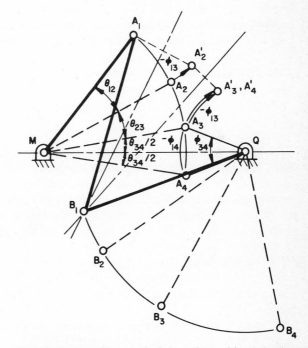

Fig. 12 Four-position synthesis by point position reduction technique.

Competency Items

• Using point position reduction, design a four-bar mechanism so that

$$\theta_{12} = 20° \quad \phi_{12} = 40°$$
$$\theta_{23} = 30° \quad \phi_{23} = 30°$$
$$\theta_{34} = 20° \quad \phi_{34} = 20°$$

Input and output moves clockwise. Design your mechanism so that pole P_{14} coincides with Q.

- Repeat the design procedures by letting poles P_{12}, P_{13}, P_{14}, P_{23}, P_{24}, one at a time, coincide with Q.
- Write steps to design a slider-crank mechanism using point position reduction technique.

Objective 6

To synthesize a four-link mechanism using the overlay technique

Activities
- Read the material provided below, and as you read:
  a. Demonstrate the application of the principle of overlay technique in synthesizing a four-link mechanism.
  b. Synthesize a four-link mechanism to generate a specific function.
  c. Synthesize six- and eight-link mechanisms for function generation.
- Check your answers to the competency items with your instructor.

Reading Material
The overlay technique is a graphical technique for the synthesis of four-, six-, eight-, etc., link mechanisms to generate a sequence of specified positions of the output link. Because of the visual display, the overlay technique is extremely versatile and widely applicable for the design of mechanisms. Depending upon the severity of design requirements, the method is capable of synthesizing mechanisms with an extreme ease for more than five and less than ten positions well within practical engineering accuracy.

The design procedure involve three simple steps.

The first step is to assume any convenient output link length and lay out an output link sector showing all the required positions of the output link in one cycle.

The second step is to assume a convenient coupler link length usually in a range of 75–125% of output link's length. With the coupler length as a radius, draw coupler loci circles with centers at each moving end of the output link.

Finally, the third step is to draw the input link sector showing all corresponding positions of the input link on a piece of transparency—this is the overlay. Fit the overlay so that all the moving ends of the input link lie on their respective coupler loci circles. The overlay sector should be drawn for several input link lengths so that a wide variety of solutions can be investigated.

Advantages of the overlay technique include:

1. For the typical simple design problem involving two- to five-position synthesis of the input and the output links, the method is simple, accurate for all practical purposes, and quick.
2. It is possible to easily find several alternate linkage configurations and visually check for desirable geometric proportions, dead center, limit positions, and transmission angle.
3. For more difficult design situations, the method is adaptable to dividing the required output among two or more four-link mechanisms in series. Thus, the method provides a synthesis technique for the series six-, eight-, ten-, etc., link mechanisms.
4. The series six-, eight-, ten-, etc., link mechanisms have the additional advantages of allowing arbitrary location of the output crank axis including collinearity with the input axis.
5. The overlay technique is equally adaptable to the slider-crank and inverted slider-crank mechanisms.

The overlay synthesis technique is illustrated using three examples.

Example 10.

Design, using an overlay technique, a four-link mechanism for three-position synthesis θ_{12}, θ_{13} of the input crank and ϕ_{12}, ϕ_{13} of the output crank.

Solution.

Repeat the following construction, shown in Fig. 13, that yield the desired linkage.

1. On a transparent drawing paper, lay out the output sector of the output link QB to any convenient scale.
2. Select a coupler length AB. $(1.25 > AB/QB > 0.75)$. Draw loci of coupler link A for each position of coupler link B.
3. On a separate paper, construct the input sector with a series of arcs representing several lengths of input link MA.
4. Fit the construction of Steps 1 and 2, on the construction of the loci circles of coupler point A until all the corresponding points A on the overlay lie on their respective loci circles. The length MA and location of the fixed center M is automatically determined by the "fit." Search for at least one other solution. If the output link is to have opposite rotation to that of the input link, the overlay is merely laid out in reverse or turned over.
5. Examine the solutions for limit positions, dead centers, and minimum transmission angle.

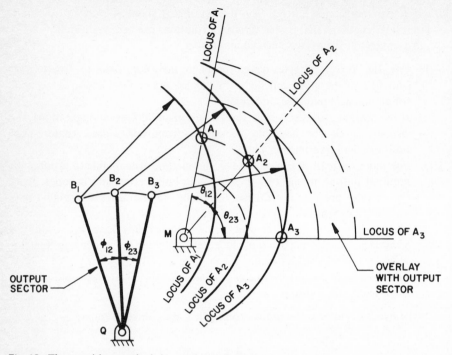

Fig. 13 Three-position synthesis by overlay technique.

Example 11.

Design, using the overlay technique, a four-link mechanism in which the input displacement θ_2 and the output displacement θ_4 are describing the function $y = x^2$ in the range $1 \le x \le 5$. That is, for the designed linkage, when the input crank displacements measure x on a linear scale, the output crank displacements measure to the corresponding values of $y = x^2$ on a linear scale.

Solution.

Repeat the following constructions, shown in Fig. 14, that yield the desired mechanism.

1. Tabulate, as shown in Table 1, the results of the computation of different values of y for chosen increments of x.
2. Compute the percentage of the total allowable link rotation corresponding to each value of x and y. For example, for $x = 2$, the percentage of total allowable input link rotation is obtained from

$$\frac{2.0 - 1}{5 - 1} \times 100\% = 25\%$$

The percentage of total allowable output link rotation for $x = 2$, is obtained from

$$\frac{4-1}{25-1} \times 100\% = 12.5\%$$

3. Select a range for input link rotation $(\theta_2)_x$, and $(\theta_4)_y$, and compute angular positions of links corresponding to each value of x.

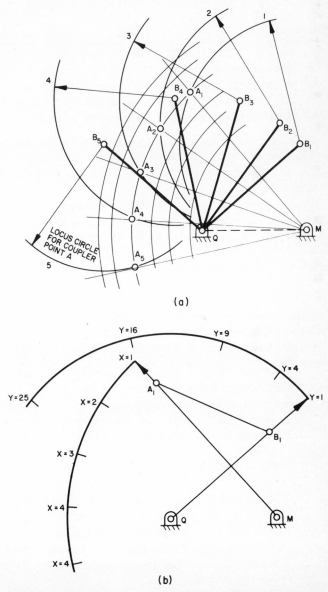

(a)

(b)

Fig. 14 a. Synthesis of $y = x^2$ by overlay method: **b.** a function generator $y = x^2$.

TABLE 1 Design Data to Generate Function $y = x^2$

x	y	$\%(\theta_2)_x$	$\%(\theta_4)_y$	$(\theta_2^\circ)_x$	$(\theta_4^\circ)_y$
1	1	0	0	0°	0°
2	4	25	12.5	15°	12.5°
3	9	50	33.33	30°	33.33°
4	16	75	62.5	45°	62.5°
5	25	100	100	60°	100°

4. The last two columns in Table 1 are based on the choice that total rotation of the input and the output links are 60° and 100°, respectively. These two columns provide the input and output sectors. Use the procedure described in example 10 to synthesize the mechanism.

We observe that there are many possible solutions. One solution is shown in Fig. 14b.

Example 12.

Design, using the overlay technique, a six-link mechanism (two four-link mechanisms in series) for generating a function given by $y = 3x^3 + 2x^2 + x$ for the range $0 \leq x \leq 4$. Design the linkage for five distinct, finitely separated points. Also, investigate the relative accuracy between the four-link and six-link mechanisms synthesized to generate this function.

Solution.

Study carefully the logic of this procedure and repeat each step by performing the tasks set out in this solution. Several good points of the overlay technique in designing multiloop mechanisms will be stressed. The design procedure requires careful considerations of the following steps.

1. Construct an input-output diagram by computing percentage input displacement and percentage output displacement. For the computation, see Table 2. The input-output diagram is shown in Fig. 15. From Fig. 15, or from Table 2, compute the total percentage deviation of the output by subtracting the percentage input from the percentage output. That is,

$$D_T = \text{total deviation} = \%(\theta_4)_y - \%(\theta_2)_x \qquad (2)$$

2. The function $y = 3x^3 + 2x^2 + x$ is a nonsymmetric function. It is possible to design a four-link mechanism to generate this function. However, such a four-link mechanism has a larger percentage error than a six-link mechanism generating this function. Therefore, it is highly desirable that a six-link mechanism is designed to generate this non-symmetric, skewed function.

The six-link mechanism is designed by dividing the total deviation into two parts A and B. The deviation part A will be used to design the first

TABLE 2 Design Data to Generate Function $y = 3x^3 + 2x^2 + x$

1	2	3	4	5	6	7	8	9
		$\%(\theta_2)x$	$\%(\theta_6)y$	$D_T{}^*$	$D_A{}^\dagger$	$D_B{}^\ddagger$	$I_B\S$	$O_B\P$
x	y			Total deviation	% Deviation of four-bar A	% Deviation of four-bar B	% Input displ. of four-bar B	% Output displ. of four-bar B
0	0	0	0	0	0	0	0	0
0.2	0.304	5	0.133	− 4.867	− 3.8	− 1.067	1.2	0.133
0.4	0.912	10	0.400	− 9.600	− 7.2	− 2.4	2.8	0.400
0.6	1.968	15	0.863	− 14.137	− 10.2	− 3.937	4.8	0.863
0.8	3.616	20	1.585	− 18.414	− 12.8	− 5.614	7.2	1.585
1.0	6.000	25	2.631	− 22.369	− 15.0	− 7.369	10.0	2.631
1.2	9.264	30	4.063	− 25.937	− 16.8	− 9.137	13.2	4.063
1.4	13.5522	35	5.943	− 29.047	− 18.2	− 10.847	16.8	5.943
1.6	19.008	40	8.336	− 31.664	− 19.2	− 12.464	20.8	8.336
1.8	25.776	45	11.305	− 33.695	− 19.8	− 13.895	25.2	11.305
2.0	34.000	50	14.912	− 35.088	− 20.0	− 15.088	30.0	14.912
2.2	43.824	55	19.221	− 35.779	− 19.8	− 15.979	35.2	19.221
2.4	55.392	60	24.294	− 35.706	− 19.2	− 16.506	40.8	24.294
2.6	68.848	65	30.196	− 34.804	− 18.2	16.604	46.8	30.196
2.8	84.336	70	36.989	− 33.011	− 16.8	−16.211	53.2	36.989
3.0	102.000	75	44.736	− 30.264	− 15.0	−15.264	60.0	44.736
3.2	121.984	80	53.501	− 26.499	− 12.8	−14.699	67.2	53.501
3.4	144.432	85	63.347	− 21.653	− 10.2	−11.453	74.8	63.347
3.6	169.488	90	74.336	− 15.664	− 7.2	− 8.464	82.8	74.336
3.8	197.296	95	86.533	− 8.467	− 3.8	− 4.667	91.2	86.533
4.0	228.000	100	100.000	0	0	0	100.0	100.000

*Total deviation = $\%(\theta_6)y - \%(\theta_2)x$.
$\dagger D_A$ is obtained by using a symmetric curve.
$\ddagger D_B = D_T - D_A$.

$\S$Output of four-bar A = Input of four-bar B, $I_B = (\theta_2)x + D_A$.
$\P$Output of four-bar $B = I_B + D_B = (\theta_2)x + D_A + D_B = (\theta_6)y$.

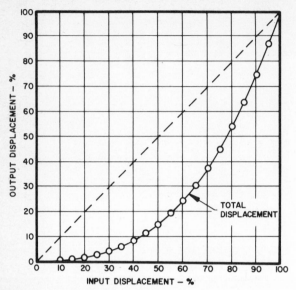

Fig. 15 The input-output displacement diagram for example 12.

four-link mechanism, and deviation part B will be used to design the second four-link mechanism. Then, these two mechanisms will be conencted in a series to form a six-link mechanism.

3. The trick of designing a six-link mechanism to generate the given function with a minimum structural error is in dividing the total deviation into two parts. One must try, however, that both parts adding up to the total deviation are as symmetric as possible.

   The total deviation in Column 5 is divided into two parts D_A and D_B as listed in Columns 6 and 7. Thus,

$$D_T = D_A + D_B \tag{3}$$

4. The input for the first four-link mechanism, named as "mechanism A," is the value of $\%(\theta_2)_x$. The output of mechanism A is given by

$$\%(\theta_4)_A = \%\left[(\theta_2)_x\right]_A + D_A \tag{4}$$

5. Since the second four-link mechanism named as "mechanism B," will be connected in series with mechanism A, the output of mechanism A must be the input of mechanism B. That is

$$\%(\theta_2)_B = \%(\theta_4)_A = \%\left[(\theta_2)_x\right]_A + D_A \tag{5}$$

Column 8 in Table 2 is computed using Eq. (5).

6. Finally, the output of the mechanism B is given by

$$(\theta_4)_B = (\theta_2)_B + D_B \tag{6}$$

Substituting Eq. (5) into Eq. (6), we get

$$(\theta_4)_B = \left[(\theta_2)_x\right]_A + D_A + D_B \tag{7}$$

$$= \left[(\theta_2)_x\right]_A + D_T \tag{8}$$

That is,

$$(\theta_4)_B = (\theta_4)_Y \tag{9}$$

or, output of mechanism $B = y = 3x^3 + 2x^2 + x$.

Thus, the six-link mechanism, obtained by placing mechanism A and mechanism B, generates the desired function.

In Fig. 16, the percentage deviation D_A is plotted versus $[(\theta_2)_x]_A$. The percentage deviation D_B is plotted versus $(\theta_2)_B$.

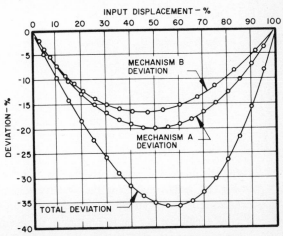

Fig. 16 Plots of displacements and deviations.

7. Using the numerical values of Columns 3 and 8, mechanism A, as shown in Fig. 17, is designed using the overlay technique described in example 10. The input link has an oscillation of $160°$ and the output link has an oscillation of $80°$. The overlay technique is used to synthesize for five positions of the input and output links. These positions are listed in Table 3.

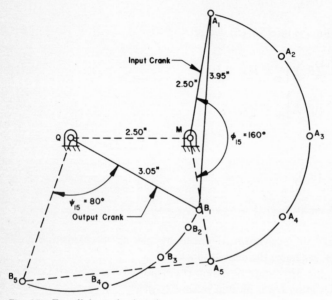

Fig. 17   Four-link mechanism A.

TABLE 3   Design Data for Mechanism A

Position	x	$\%(\theta_2)_x$	$\%(\theta)_y$	θ_2	θ_6
1	0	0	0	0	$0°$
2	1	25	10	$40°$	$8°$
3	2	50	30	$80°$	$24°$
4	3	75	60	$120°$	$48°$
5	4	100	100	$160°$	$80°$

8. Using the numerical values of Columns 8 and 9, mechanism B, as shown in Fig. 18, is designed using the overlay technique. The input link has an oscillation of $80°$ and the output link has an oscillation of $110°$. The overlay technique is used to synthesize for five positions of the input and output links. These positions are listed in Table 4.

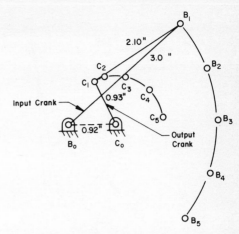

Fig. 18   Four-link mechanism B.

TABLE 4   Design Data for Mechanism B

Position	x	$\%(\theta_4)$	$\%(\theta_6)$	θ_4 (deg)	θ_6 (deg)
1	0	0	0	0	0
2	1	25	11.10	20	12.21
3	2	50	33.60	40	36.96
4	3	75	63.51	60	69.86
5	4	100	100.00	80	110.00

9. Mechanisms A and B are connected in series, as shown in Fig. 19, to obtain a six-link mechanism generating the desired function.

10. Since the technique is graphical, the designed linkage is expected to have some error. Table 5 shows the percentage error of the different positions of the designed linkage. The percentage error is computed using Eq. (10).

$$\text{Percentage error} = \frac{|\text{ theoretical output } - \text{ observed output }|}{\text{theoretical output}} \times 100\%$$

$$(10)$$

11. It is important to note that the designed six-link mechanism has a smaller percentage error than a four-link mechanism also generating the given function. Using the values of Columns 3 and 4, a four-link mechanism, as shown in Fig. 20, is designed using the overlay technique. From the results of Table 6, it is clear that the four-link mechanism of Fig. 20 has a larger percentage error than the six-link mechanism of Fig. 19.

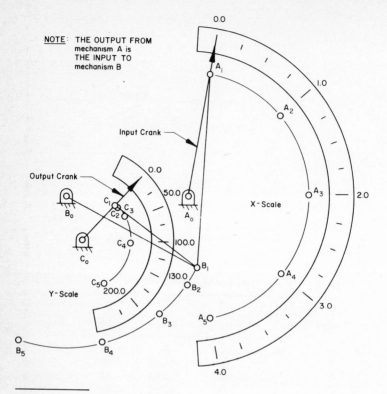

Fig. 19 Six-link function generator.

TABLE 5

Position	Theoretical output (deg)	Actual output (deg)	Difference (deg)	% Error
1	0.00	0.00	0.00	0.00
2	2.89	3.00	0.11	3.81
3	16.40	16.75	0.35	2.13
4	49.75	49.21	0.54	1.09
5	110.00	110.00	0.00	0.00

12. The two mechanisms *A* and *B* can be connected using a ternary link of any proportion. This choice makes it possible to specify the exact location of the output fixed center of the six-link mechanism.

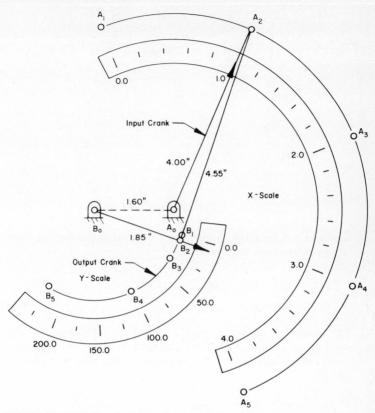

Fig. 20 Four-link function generator.

TABLE 6

x Input	Four-link error (%)	Six-link error (%)
0	0.00	0.00
1	29.60	3.81
2	1.11	2.13
3	6.38	1.09
4	0.00	0.00

Competency Items

- Design a four-link mechanism to generate function $y = x^3$ for $0 \leq x \leq 3$. Use five precision positions for the overlay technique.
- Design a slider-crank mechanism to generate $y = x^2$ for $-1 < x < 1$.
- Design a drag-link mechanism to drive the slider-crank mechanism. The output slider block should have approximately constant velocity for 60% of its forward stroke.

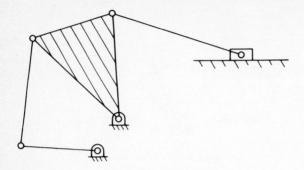

Performance Test #1

1. Synthesize a four-link mechanism to match the following three positions, using the inversion technique:

$$\theta_{12} = 30° \quad \phi_{12} = 30°$$
$$\theta_{23} = 45° \quad \phi_{23} = 30°$$

The input θ is clockwise and the output ϕ is clockwise. Be certain that no dead-center position is in the operating range of the mechanism.

2. Synthesize a slider-crank mechanism using the pole technique for the three positions.

$$\theta_{12} = 30° \quad S_{12} = 0.5 \text{ in.}$$
$$\theta_{23} = 45° \quad S_{23} = 0.5 \text{ in.}$$

The input θ moves counterclockwise and the slider moves away from the fixed center.

Performance Test #2

1. Synthesize a slider-crank mechanism using the inversion technique for the three positions:

$$\theta_{12} = 55° \quad S_{12} = 1 \text{ in.} \quad e = \frac{1}{2} \text{ in.}$$
$$\theta_{23} = 35° \quad S_{23} = \frac{3}{4} \text{ in.}$$

The input θ moves clockwise and the slider moves toward the fixed center. State whether your transmission angle is satisfactory in the range of action.

2. Synthesize a four-link mechanism to match the following three positions using the overlay technique:

$$\theta_{12} = 30° \quad \phi_{12} = 25°$$
$$\theta_{23} = 30° \quad \phi_{23} = 35°$$

The input θ moves clockwise while the output is counterclockwise. Be sure that there is no dead-center position in the range of operation.

Performance Test #3

1. Design a slider-crank mechanism (using the pole technique) so that $\theta_{12} = 75°$, $s_{12} = 2.5$ in., and $e = 1$ in. The input moves counterclockwise and the slider moves toward the fixed pivot. Determine whether your transmission angle is satisfactory in the range of action.

2. Synthesize a four-link mechanism to match the following three positions using the overlay technique:

$$\theta_{12} = 25° \quad \phi_{12} = 30°$$
$$\theta_{23} = 30° \quad \phi_{23} = 35°$$

The input θ moves counterclockwise while the output moves counterclockwise. Be sure that there is no dead-center position in the range of action.

Supplementary References

1 (11.1-11.3, 11.4); **3** (12.5-12.6); **7** (11.1, 11.4-11.7); **8** (9.3-9.5); **9** (3.1-3.6, 5.1)

Unit XIII. Synthesis of Four-link, Slider-crank, and Inverted Slider-crank Mechanisms Using the Analytical Approach

Objectives

1. To synthesize a four-link mechanism to coordinate three positions of output and input links
2. To synthesize a four-link mechanism to coordinate three positions of input link and three positions of a coupler point
3. To synthesize a four-link mechanism to guide a rigid body through its three finitely separated positions
4. To synthesize a function generator mechanism using the least-square technique
5. To synthesize a four-link mechanism for prescribed values of displacement, velocity, and acceleration

Objective 1

To synthesize a four-link mechanism to coordinate three positions of input and output links

Activities

- Read the material provided below, and as you read:
  a. Synthesize a four-link mechanism.
- Check your answers to the competency items with your instructor.

Reading Material

In Unit IV, Objective 1, we have derived Eq. (19) which relates rotation angles θ_2 and θ_4 of input and output links of a four-link mechanism shown in Fig. 1.

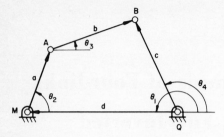

Fig. 1

This relationship, called *Freudenstein's equation*, is written as

$$k_1 \cos \theta_4 - k_2 \cos \theta_2 + k_3 = \cos(\theta_2 - \theta_4) \tag{1}$$

where

$$k_1 = d/a$$
$$k_2 = d/c$$
$$k_3 = (a^2 - b^2 + c^2 + d^2)/(2ac)$$

Suppose we are given three values θ_2 and θ_4 as $\theta_{21}, \theta_{22}, \theta_{23}$ and $\theta_{41}, \theta_{42}, \theta_{43}$, and we are asked to calculate dimensions a, b, c, and d of the four-link mechanism whose input link MA and output link QB will exhibit such displacements. For such a synthesis problem, we can utilize Eq. (1), treating it as a linear equation with constant coefficients k_1, k_2, and k_3. Since we will know numerical values of $\theta_{21}, \theta_{22}, \theta_{23}$ and $\theta_{41}, \theta_{42}, \theta_{43}$, we can write Eq. (1) three times. Thus,

$$k_1 \cos \theta_{41} - k_2 \cos \theta_{21} + k_3 = \cos(\theta_{21} - \theta_{41}) \tag{2}$$

$$k_1 \cos \theta_{42} - k_2 \cos \theta_{22} + k_3 = \cos(\theta_{22} - \theta_{42}) \tag{3}$$

$$k_1 \cos \theta_{43} - k_2 \cos \theta_{23} + k_3 = \cos(\theta_{23} - \theta_{43}) \tag{4}$$

We can solve these equations using Cramer's rule. Let us define four determinants:

$$\Delta = \begin{vmatrix} \cos \theta_{41} & -\cos \theta_{21} & 1 \\ \cos \theta_{42} & -\cos \theta_{22} & 1 \\ \cos \theta_{43} & -\cos \theta_{23} & 1 \end{vmatrix}$$

$$\Delta_1 = \begin{vmatrix} \cos(\theta_{21} - \theta_{41}) & -\cos \theta_{21} & 1 \\ \cos(\theta_{22} - \theta_{42}) & -\cos \theta_{22} & 1 \\ \cos(\theta_{23} - \theta_{43}) & -\cos \theta_{23} & 1 \end{vmatrix}$$

$$\Delta_2 = \begin{vmatrix} \cos\theta_{41} & \cos(\theta_{21} - \theta_{41}) & 1 \\ \cos\theta_{42} & \cos(\theta_{22} - \theta_{42}) & 1 \\ \cos\theta_{43} & \cos(\theta_{23} - \theta_{43}) & 1 \end{vmatrix}$$

$$\Delta_3 = \begin{vmatrix} \cos\theta_{41} & -\cos\theta_{21} & \cos(\theta_{21} - \theta_{41}) \\ \cos\theta_{42} & -\cos\theta_{22} & \cos(\theta_{22} - \theta_{42}) \\ \cos\theta_{43} & -\cos\theta_{23} & \cos(\theta_{23} - \theta_{43}) \end{vmatrix}$$

According to Cramer's rule,

$$k_1 = \frac{\Delta_1}{\Delta} \tag{5}$$

$$k_2 = \frac{\Delta_2}{\Delta} \tag{6}$$

$$k_3 = \frac{\Delta_3}{\Delta} \tag{7}$$

Once numerical values of k_1, k_2, and k_3 are known, link lengths a, b, c, and d can be computed using the following steps:

Assume $a = 1$.
Calculate d from $k_1 = d/a$.
Calculate c from $k_2 = d/c$.
Calculate b from $k_3 = (a^2 - b^2 + c^2 + d^2)/(2\ ac)$.

The use of the Freudenstein equation permits us to design a four-link mechanism for three precision positions of input and output links. A four-link mechanism can be designed at most for five precision positions. The five-position synthesis can be accomplished provided θ_2 and θ_4 are measured from some arbitrary reference rather than from the reference fixed link MQ.

For such a case, however, the synthesis equations become nonlinear and other approaches are required to solve such synthesis equations.

The approach demonstrated above can be applied to synthesize a slider-crank mechanism to coordinate three positions of its input link and three positions of its slider block.

Competency Items
- Design a four-link mechanism so that:

$$\theta_{21} = 15° \qquad \theta_{41} = 30°$$

$$\theta_{22} = 30° \qquad \theta_{42} = 40°$$

$$\theta_{23} = 45° \qquad \theta_{43} = 55°$$

• Write steps to design a slider-crank mechanism for three positions: θ_{21}, θ_{22}, and θ_{23} of input link MA and s_1, s_2, and s_3 of slider block B.

Objective 2

To synthesize a four-link mechanism to coordinate three positions of input link and three positions of a coupler point

Activities
• Read the material provided below, and as you read:
   a. State the synthesis problem.
   b. Derive synthesis equations.
   c. Demonstrate applications of the technique in the synthesis of a four-link mechanism.
• Check your answers to the competency items with your instructor.

Reading Material

Part I: Definition of synthesis problem

Four-link mechanism $MABQ$ with coupler point C is shown in Fig. 2, in its three positions MA_1B_1Q, MA_2B_2Q, and MA_3B_3Q in an xy frame of reference. The positions of coupler point C are located using vector $R = re^{i\delta}$. Let the coordinates of pivots M, A, B, and Q be (x_m, y_m), (x_{a1}, y_{a1}), (x_{b1}, y_{b1}), and (x_q, y_q).

For the synthesis of a four-link mechanism to coordinate three positions of input link MA with three positions of coupler point, we will give three values of r and δ to describe three positions of point C and θ_{21}, θ_{22}, and θ_{23} to describe three corresponding positions of input link MA. We will be asked to calculate $(x_m y_m)$, $(x_q y_q)$, MA, AB, QB, MQ, AC, and BC.

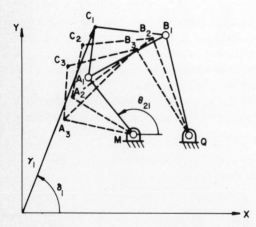

Y

Fig. 2 Definition of synthesis problem.

X

Part II: Derivation of synthesis equations

Figure 3 shows the vector notations that are necessary to describe a four-link mechanism for our synthesis problem.

From Fig. 3, loops *OMAC* and *OQBC* will give the following vector relationship:

$$OM + MA + AC = OC$$
$$OQ + QB + BC = OC$$

Using complex number notation, the above relationships will yield:

$$r_1 e^{j\alpha} + r_2 e^{j\theta_2} + r_3 e^{j\psi} = r e^{j\delta}$$
$$r_4 e^{j\beta} + r_5 e^{j\theta_4} + r_6 e^{j\phi} = r e^{j\delta}$$

Substituting $e^{j\theta} = \cos\theta + j\sin\theta$, etc., and separating the resultant into their real and complex parts, we get

$$r_1 \cos\alpha + r_2 \cos\theta_2 + r_3 \cos\psi = r\cos\delta \tag{8}$$

$$r_1 \sin\alpha + r_2 \sin\theta_2 + r_3 \sin\psi = r\sin\delta \tag{9}$$

$$r_4 \cos\beta + r_5 \cos\theta_4 + r_6 \cos\phi = r\cos\delta \tag{10}$$

$$r_4 \sin\beta + r_5 \sin\theta_4 + r_6 \sin\phi = r\sin\delta \tag{11}$$

Using the identities $\cos^2\psi + \sin^2\psi = 1$ and $\cos^2\theta_4 + \sin^2\theta_4 = 1$, we can eliminate unwanted angles ψ and θ_4 from Eqs. (8) and (9) and Eqs. (10) and (11). The resultant equations can be written as

$$r_1[2r\cos(\alpha - \delta)] + r_2[2r\cos(\theta_2 - \delta)] + (r_3^2 - r_2^2 - r_1^2)$$
$$= r^2 + r_1 r_2[2\cos(\theta_2 - \alpha)] \tag{12}$$

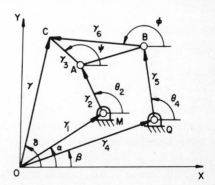

Fig. 3   Vector notations.

$$r_4[2r \cos(\delta - \beta)] + r_6[2r \cos(\delta - \phi)] + (r_5^2 - r_4^2 - r_6^2) \tag{12}$$
$$= r^2 + r_4 r_6[2 \cos(\phi - \beta)] \tag{(Cont.)}$$

Let

$$
\begin{aligned}
k_1 &= r_1 & k_5 &= r_4 \\
k_2 &= r_2 & k_6 &= r_6 \\
k_3 &= r_3^2 - r_2^2 - r_1^2 \quad & k_7 &= r_5^2 - r_4^2 - r_6^2 \\
k_4 &= r_1 r_2 & k_8 &= r_4 r_6
\end{aligned}
\tag{13}
$$

Substituting Eq. (13), in Eqs. (11) and (12) we get

$$k_1[2r \cos(\alpha - \delta)] + k_2[2r \cos(\theta_2 - \delta)] + k_3$$
$$= r^2 + k_4[2 \cos(\theta_2 - \alpha)] \tag{14}$$

$$k_5[2r \cos(\delta - \beta)] + k_6[2r \cos(\delta - \phi)] + k_7$$
$$= r^2 + k_8[2 \cos(\phi - \beta)] \tag{15}$$

From the above we observe that $k_4 = k_1 k_2$ and $k_8 = k_5 k_6$.

In order to synthesize a mechanism we will first solve for r_1, r_2, and r_3 from Eq. (14), and then solve for r_4, r_5, and r_6 from Eq. (15). Because of the relationships $k_4 = k_1 k_2$ and $k_8 = k_5 k_6$, Eqs. (14) and (15) cannot be solved in the same manner as in the previous objective. We can adapt the following procedure to Eq. (14).

Let $k_4 = \lambda = k_1 k_2$ and define

$$k_p = l_p + \lambda m_p \quad \text{for} \quad p = 1, \ldots 3 \tag{16}$$

Equation (16) is similar to defining a complex number. Parameter λ however has to satisfy the compatibility condition

$$\lambda = k_1 k_2 = (l_1 + \lambda m_1)(l_2 + \lambda m_2) \tag{17}$$

Substituting $k_p = l_p + \lambda m_p$ for $p = 1, 3$ in Eq. (14), and separating the components into two groups—one with, and the other without λ—we get two linear equations in l_1, l_2, l_3, and m_1, m_2, m_3. These are:

$$l_1[2r \cos(\alpha - \delta)] + l_2[2r \cos(\theta_2 - \delta)] + l_3 = r^2 \tag{18}$$

$$m_1[2r \cos(\alpha - \delta)] + m_2[2r \cos(\theta_2 - \delta)] + m_3 = [2 \cos(\theta_2 - \alpha)]$$
$$(19)$$

For three-position synthesis, variable angles δ and θ_2 will take values $\delta_1, \delta_2,$ $\delta_3, \theta_{21}, \theta_{22},$ and θ_{23}. Hence, we can write Eqs. (18) and (19) three times and solve the linear equations in l_1, l_2, l_3 and m_1, m_2, m_3 as unknown in a manner similar to that demonstrated in Unit XIII, Objective 1. Numerical values of λ are calculated from Eq. (17). This will lead us to calculate k_1, k_2, k_3, from which we can calculate r_1, r_2, r_3 and ψ_1, ψ_2, ψ_3.

From the geometry of Fig. 2, we observe that

$$\phi_2 = \phi_1 + (\psi_2 - \psi_1)$$

$$\phi_3 = \phi_1 + (\psi_3 - \psi_1)$$

Equation (15) can now be utilized to solve for unknowns $r_4, r_5,$ and r_6. The procedure is similar to that utilized for solving Eq. (14).

Part III: Stepwise procedure for synthesis
1. Given: Three values of $r, \delta,$ and θ_2
2. Assume: $\alpha, \beta,$ and ϕ_1
3. Calculate:
   a. r_1, r_2, r_3 using Eq. (14)
   b. $\psi_1, \psi_2,$ and ψ_3 with known values of r_1, r_2, r_3
   c. $\phi_2 = \phi_1 + (\psi_2 - \psi_1), \phi_3 = \phi_1 + (\psi_3 - \psi_1)$
   d. r_4, r_5, r_6, using Eq. (15)
4. Number of solutions: There are two compatibility conditions defined by the conditions $k_4 = k_1 k_2$ and $k_8 = k_5 k_6$. Each of these conditions will yield two sets of values for $r_1, r_2, r_3, r_4, r_5,$ and r_6. Hence, we can expect a maximum of four possible solutions.

Competency Items
- Derive the synthesis equations to design a slider-crank mechanism to coordinate three positions of its coupler point with three positions of its input link.
- Derive the synthesis equations to design an inverted slider-crank mechanism to coordinate three positions of its coupler point with three positions of its input link.

Objective 3

To synthesize a four-link mechanism to guide a rigid body through its three finitely separated positions

Activities

- Read the material provided below, and as you read:
  a. Define the synthesis problem.
  b. Derive synthesis equations.
  c. Demonstrate the synthesis technique.
- Check your answers to the competency items with your instructor.

Reading Material

Part I: Definition of the synthesis problem

Figure 4 shows rigid body $EFCD$ in its three positions $E_1F_1C_1D_1$, $E_2F_2C_2D_2$, and $E_3F_3C_3D_3$. We wish to design four-link mechanism $MABQ$ whose couple link AB will carry rigid body $EFCD$ and guide it through its three finitely separated positions.

In order to synthesize this equation, the specification of rigid body $EFCD$ in its three positions requires us to consider point P and specify its coordinates using polar coordinates r and δ, and specify angles ψ_1, ψ_2, and ψ_3, which are measured with reference to the x axis. The synthesis technique will locate two points A and B whose three positions A_1, A_2, A_3, and B_1, B_2, B_3, will lie on circles with M and Q as centers and MA and QB as radii.

Referring to Fig. 4 we can state our synthesis problem in the following manner:

1. Given: $(r_1\delta_1)$, $(r_2\delta_2)$, $(r_3\delta_3)$ and ψ_1, ψ_2, ψ_3 for a rigid body $EFCD$
2. Required: r_1, r_2, r_3, r_4, r_5, and r_6

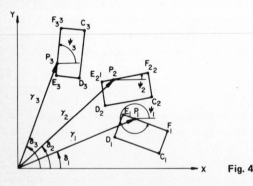

Fig. 4

Part II: Derivation of synthesis equations

Figure 3 is redrawn as Fig. 5. The synthesis equations are derived, hence, that will lead us to design a four-link mechanism to guide a rigid body through its three finitely separated positions.

From the geometry of Fig. 2, loops $OMAP$ and $OQBP$ will yield:

$$OM + MA + AP = OP$$
$$OQ + QB + BP = OP$$

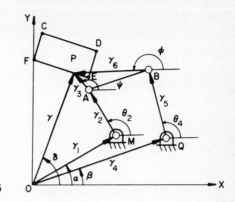

Fig. 5

Using complex number notations, the above relationship will yield

$$r_1 e^{j\alpha} + r_2 e^{j\theta_2} + r_3 e^{j\psi} = r e^{j\delta}$$

$$r_4 e^{j\beta} + r_5 e^{j\theta_4} + r_6 e^{j\phi} = r e^{j\delta}$$

Substituting $e^{j\theta} = \cos\theta + j\sin\theta$, etc., and separating the real and the complex parts, we get from the above equations

$$r_1 \cos\alpha + r_2 \cos\theta_2 + r_3 \cos\psi = r\cos\delta \tag{20}$$

$$r_1 \sin\alpha + r_2 \sin\theta_2 + r_3 \sin\psi = r\sin\delta \tag{21}$$

$$r_4 \cos\beta + r_5 \cos\theta_4 + r_6 \cos\phi = r\cos\delta \tag{22}$$

$$r_4 \sin\beta + r_5 \sin\theta_4 + r_6 \sin\phi = r\sin\delta \tag{23}$$

These equations are the same as Eqs. (9)–(11). Because of the rigid body guidance problems, we will eliminate unknown angle θ_2 from Eqs. (20) and (21) using the identity $\cos^2\theta_2 + \sin^2\theta_2 = 1$.

Also, from Eqs. (22) and (23), we will eliminate angle θ_4 using the identity $\cos^2\theta_4 + \sin^2\theta_4 = 1$. The resultant equations can be written as

$$[r_2^2 - r_1^2 - r_3^2] + r_1[2r\cos(\delta - \alpha)] + r_3[2r\cos(\delta - \psi)]$$
$$= r^2 + r_3 r_1[2\cos(\psi - \alpha)] \tag{24}$$

$$[r_5^2 - r_4^2 - r_6^2] + r_4[2r\cos(\delta - \beta)] + r_6[2r\cos(\delta - \phi)]$$
$$+ [r_5^2 - r_4^2 - r_6^2] = r^2 + r_4 r_6[2\cos(\phi - \beta)] \tag{25}$$

Let

$$k_1 = r_1 \qquad k_3 = r_2^2 - r_1^2 - r_3^2 \qquad k_5 = r_4 \qquad k_7 = r_5^2 - r_4^2 - r_6^2$$

$$k_2 = r_3 \qquad k_4 = r_3 r_1 \qquad\qquad\qquad k_6 = r_6 \qquad k_8 = r_4 r_6$$

Hence Eqs. (24) and (25) will become

$$k_1[2r \cos(\delta - \alpha)] + k_2[2r \cos(\delta - \psi)] + k_3$$
$$= r^2 + k_4[2 \cos(\psi - \alpha)] \tag{26}$$

$$k_5[2r \cos(\delta - \beta)] + k_6[2r \cos(\delta - \phi)] + k_7$$
$$= r^2 + k_8[2 \cos(\phi - \beta)] \tag{27}$$

Equations (26) and (27) are similar to Eqs. (14) and (15) in Objective 2 and can be solved in a similar manner for k_1, k_2, k_3, k_5, k_6, and k_7. Once all the numerical values of k_1, k_2, k_3, k_5, k_6, and k_7 are known, *linkage* parameters r_1, r_2, r_3, r_4, r_5, and r_6, can be calculated.

Part III: Stepwise procedure for synthesis
1. Given: Three values of r, δ, and ψ
2. Assume: α, β, and ϕ_1
3. Calculate:
   a. r_1, r_2, r_3 using Eq. (26)
   b. θ_{21}, θ_{22}, θ_{23} with known values of r_1, r_2, r_3,
   c. $(\phi_2 = \phi_1 + (\psi_2 - \psi_1)$

   $\phi_3 = \phi_1 + (\psi_3 - \psi_1)$
   d. r_4, r_5, r_6 using Eq. (27)
4. Number of solutions: There are two compatibility conditions: $k_4 = k_1 k_2$ and $k_8 = k_5 k_6$. Each of these conditions will yield two sets of values for r_1, r_2, r_3, r_5, and r_6. Hence, we can expect a maximum of four possible solutions.

Competency Items
- Derive the synthesis equations to design a slider-crank mechanism that will guide a rigid body through its finitely separated three positions.
- Derive the synthesis equations to design an inverted slider-crank mechanism that will guide a rigid body through its finitely separated three positions.

Objective 4

To synthesize a function generator mechanism using the least-square technique

Activities
- Read the material provided below, and as you read:
   a. Demonstrate the application of the least-square technique in curve fitting.

b. Demonstrate the application of the least-square technique in synthesizing a function generator mechanism.
- Check your answers to the competency items with your instructor.

Reading Material

In Objective 1, we examined a synthesis technique that permits us to synthesize a mechanism to coordinate motions of its input and output links for three finitely separated positions. Quite often, we are required to obtain an average performance over a region which may be divided into more than five positions. For such a design situation, the least-square technique can be considered to synthesize a mechanism.

Figure 6 shows a four-link mechanism. We are specified values θ_{2i} and for $i = 1, \ldots, n$, and we wish to synthesize this mechanism so that with calculated values of $a, b, c,$ and d, the quantity

$$E = \frac{1}{n-1} \sum_{i=1}^{n} (\theta_{4i} - \phi_{4i})^2$$

is minimum, where ϕ_{4i} are the actual angular displacements of the output link of the synthesized mechanism. Since θ_{4i} is the displacement that we wish to obtain and ϕ_{4i} is the displacement that we obtain in reality, the quantity E defines the mean squared error. The smaller this error, the better is the fit and more satisfactory is the designed mechanism.

In order to obtain the fit, we will consider Freudenstein's equation

$$k_1 \cos \theta_4 - k_2 \cos \theta_2 + k_3 = \cos(\theta_2 - \theta_4)$$

where $k_1 = d/a$

$k_2 = d/c$

$k_3 = (a^2 - b^2 + c^2 + d^2)/(2ac)$

To apply the least-square technique, let us define

$$D = \sum_{i=1}^{n} [k_1 \cos \theta_{4i} - k_2 \cos \theta_{2i} + k_3 - \cos(\theta_{2i} - \theta_{4i})]^2 \tag{28}$$

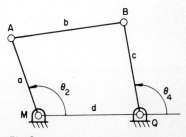

Fig. 6

For the minimum error between θ_{4i} and ϕ_{4i}, based on the least-square technique, we are required to take the partial derivative of D with respect to k_1, k_2, and k_3 and set these derivatives to zero. Thus,

$$\frac{\partial D}{\partial k_1} = \sum_{i=1}^{n} [(k_1 \cos \theta_{4i} - k_2 \cos \theta_{2i} + k_3 - \cos (\theta_{2i} - \theta_{4i})] \cos \theta_{4i}$$
$$= 0 \quad (29)$$

$$\frac{\partial D}{\partial k_2} = \sum_{i=1}^{n} [(k_1 \cos \theta_{4i} - k_2 \cos \theta_{2i} + k_3 - \cos (\theta_{2i} - \theta_{4i})] \cos \theta_{2i}$$
$$= 0 \quad (30)$$

$$\frac{\partial D}{\partial k_3} = \sum_{i=1}^{n} [(k_1 \cos \theta_{4i} - k_2 \cos \theta_{2i} + k_3 - \cos (\theta_{2i} - \theta_{4i})] = 0$$
$$(31)$$

Equations (29), (30), and (31) can be simplified to obtain

$$k_1 \sum \cos^2 \theta_{4i} - k_2 \sum \cos \theta_{2i} \cos \theta_{4i} + k_3 \sum \cos \theta_{4i}$$
$$= \sum \cos (\theta_{2i} - \theta_{4i}) \cos \theta_{4i} \quad (32)$$

$$k_1 \sum \cos \theta_{4i} \cos \theta_{2i} - k_2 \sum \cos^2 \theta_{2i} + k_3 \sum \cos \theta_{2i}$$
$$= \sum \cos (\theta_{2i} - \theta_{4i}) \cos \theta_{2i} \quad (33)$$

$$k_1 \sum \cos \theta_{4i} - k_2 \sum \cos \theta_{2i} + k_3 = \sum \cos (\theta_{2i} - \theta_{4i}) \quad (34)$$

Equation (32), (33), and (34) describe three simultaneous linear, nonhomogeneous equations in three unknowns k_1, k_2, and k_3. These equations are similar to Eqs. (2), (3), and (4) and are solved using the technique demonstrated in Objective 1.

Competency Items

- Derive the synthesis equations to design a function generator slider-crank mechanism.
- Derive the synthesis equations to design a function generator inverted slider-crank mechanism.

Objective 5

To synthesize a four-link mechanism for prescribed values of displacement, velocity, and acceleration

Activities

- Read the material provided below, and as you read:
  a. Synthesize a four-link mechanism with prescribed displacement, velocity, and acceleration.
  b. Demonstrate the application of Bloch's synthesis technique to other types of mechanisms.
- Check your answers to the competency items with your instructor.

Reading Material

A four-link mechanism is shown in Fig. 7.

For synthesis purpose we are given for one particular position of the mechanism, θ_2, $\dot{\theta}_2$, and $\ddot{\theta}_2$ and corresponding values for θ_4, $\dot{\theta}_4$, and $\ddot{\theta}_4$. We are required to obtain link lengths MA, AB, QB, and MQ of the four-link mechanism that will satisfy the design requirements. For this purpose let us examine the vector equation that describes vector polygon $MABQ$ in Fig. 7.

$$\mathbf{R}_1 + \mathbf{R}_2 + \mathbf{R}_3 = \mathbf{R}_4 \tag{35}$$

Where

$$\mathbf{R}_1 = de^{j\theta_1} \quad \mathbf{R}_3 = be^{j\theta_3}$$
$$\mathbf{R}_2 = ae^{j\theta_2} \quad \mathbf{R}_4 = ce^{j\theta_4}$$

Using complex number notations, Eq. (35) can be written as

$$de^{j\theta_1} + ae^{j\theta_2} + be^{j\theta_3} = ce^{j\theta_4} \tag{36}$$

Taking the successive time derivative of Eq. (36) and substituting

$$\dot{\theta}_i = \omega_i$$
$$\ddot{\theta}_i = \alpha_i$$

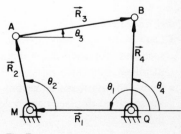

Fig. 7

for $i = 2, \ldots, 4$, we get

$$ja\omega_2 e^{j\theta_2} + jb\omega_3 e^{j\theta_3} = jc\omega_4 e^{j\theta_4} \tag{37}$$

$$ja(\omega_2^2 + j\alpha_2)e^{j\theta_2} + jb(\omega_3^2 + j\alpha_3)e^{j\theta_3} = j(\omega_4^2 + j\alpha_4)e^{j\theta_4} \tag{38}$$

Transforming Eqs. (36), (37), and (38) back into vector equations we get

$$\mathbf{R}_2 + \mathbf{R}_3 - \mathbf{R}_4 = -\mathbf{R}_1 \tag{39}$$

$$\omega_2 \mathbf{R}_2 + \omega_3 \mathbf{R}_3 - \omega_4 \mathbf{R}_4 = 0 \tag{40}$$

$$(\omega_2^2 + j\alpha_2)\mathbf{R}_2 + (\omega_3^2 + j\alpha_3)\mathbf{R}_3 - (\omega_4^2 + j\alpha_4)\mathbf{R}_4 = 0 \tag{41}$$

Equation (39), (40), and (41) can be solved using the technique shown in Objective 1. For example,

$$\mathbf{R}_2 = \frac{\begin{vmatrix} -\mathbf{R}_1 & 1 & -1 \\ 0 & \omega_3 & -\omega_4 \\ 0 & (\omega_3^2 + j\alpha_3) & -(\omega_4^2 + j\alpha_4) \end{vmatrix}}{\begin{vmatrix} 1 & 1 & -1 \\ \omega_2 & \omega_3 & -\omega_4 \\ (\omega_2^2 + j\alpha_2) & (\omega_3^2 + j\alpha_3) & -(\omega_4^2 + j\alpha_4) \end{vmatrix}}$$

Similarly we can calculate $\mathbf{R}_3$ and $\mathbf{R}_4$. The vector $\mathbf{R}_1$ can be calculated from Equation (39).

The synthesis technique demonstrated above is called *Bloch's synthesis technique*. Vector $\mathbf{R}_1 = r_1 e^{j(\pi)}$. In order to calculate $\mathbf{R}_2$, assume $r_1 = 1$ unit. Similarly, calculate $\mathbf{R}_3$ and $\mathbf{R}_4$.

Competency Items

- Derive the synthesis equations to design a slider-crank mechanism for specified displacement, velocity, and acceleration of its input and output members.
- Derive the synthesis equations to design an inverted slider-crank mechanism for specified displacement, velocity, and acceleration of its input and output links.

Performance Test #1

1. Design a four-link mechanism to coordinate motions of input and output links for three positions governed by a function $y = x^2$ for $0 \le x \le 2$. Assume $\Delta x = 1$.
2. Design a four-link mechanism to guide a rigid body through three finitely separated positions which are described using three sets of values for (r, δ, ψ) as $(4 \cdot 5, 30°, 80°), (1 \cdot 98, 52°, 91°)$, and $(3 \cdot 8, 68°, 98°)$. Assume $\alpha = 40°, \beta = 34°, \phi_1 = 95°$.

Performance Test #2

1. Using the least-square technique, design a four-link mechanism to coordinate motions of input and output links for ten positions governed by a function $y = \log x$ for $0 < x \le 10$. Assume $\Delta x = 1$; the ranges for θ_2 and θ_4 are from $0°$ to $100°$.
2. Design a four-link mechanism to coordinate three positions of the coupler point with three positions of the input link. The three sets of values for (r, δ) are $(4 \cdot 5, 30°)$, $(1 \cdot 98, 52°)$, and $(3 \cdot 8, 68°)$. The corresponding values of θ_2 are $\theta_{21} = 80°, \theta_{22} = 91°, \theta_{23} = 98°$. Assume $\alpha = 40°, \beta = 34°$, and $\phi_1 = 95°$.

Performance Test #3

Design a crank-rocker mechanism with a coupler point drawing a coupler curve which is approximated by twelve positions having the following coordinates.

Position, i	x_i (in)	y_i (in)	Input-crank rotation, θ_{2i}
1	1.22	3.66	161°
2	2.04	4.45	131°
3	3.24	4.9	101°
4	4.57	4.8	71°
5	5.62	4.06	41°
6	6.12	3.2	11°
7	5.8	2.36	−19°
8	4.58	1.85	−49°
9	3.1	1.46	−79°
10	1.93	1.48	−109°
11	1.2	1.93	−139°
12	0.9	2.7	−169°

Unit XIV. Coupler Curves of Four-link, Slider-crank, and Inverted Slider-crank Mechanisms

Objectives
1. To identify the coupler curves of a four-link mechanism
2. To derive relationships to calculate coordinates of a coupler point of a four-link mechanism
3. To derive relationships to calculate coordinates of a coupler point of a slider-crank mechanism
4. To derive relationships to calculate coordinates of a coupler point of an inverted slider-crank mechanism

Objective 1

To identify the coupler curves of a four-link mechanism

Activities
- Read the material provided below, and as you read:
  a. Define a coupler curve of a four-link mechanism.
  b. Build a four-link adjustable mechanism with a coupler point.
  c. Plot the coupler curves of a four-link mechanism.
- Check your answers to the competency items with your instructor.

Reading Material
We observed earlier that a four-link mechanism is capable of providing two different types of output motion. These output motions are derived from the follower link and from the coupler link. In previous units, we designed a four-link mechanism for the motion coordination of input, output, and coupler links. The present unit is devoted to the study of the path traced by a point lying in the plane of the coupler link. The study of coupler curves will be utilized in synthesizing a six-link mechanism for programmed motion of its output link.

Figure 1 illustrates crank-rocker mechanism MA_1B_1Q with coupler points C_1, D_1, E_1, etc. As the input crank rotates through $360°$ to complete one cycle, output link QB, oscillates between the two limit positions marked as B_1 and B_2, and coupler points C_1, D_1, E_1, etc., trace curves 1, 2, 3, etc., as shown in Fig. 1. Note that the different positions of coupler points C_1, D_1, etc., draw different shapes and sizes of coupler curves.

In the sections that follow, we will examine for the synthesis purpose, the coupler curves of

1. crank-rocker mechanisms
2. drag-link mechanisms
3. slider-crank mechanisms
4. inverted slider-crank mechanism, Type I

Competency Items

- After examining Fig. 1, explain in writing how the family of coupler curves was obtained.
- Write two applications for these coupler curves.
- Plot the coupler curves of a slider-crank mechanism.
- Plot the coupler curves of an inverted slider-crank mechanism.

Objective 2

To derive relationships to calculate coordinates of a coupler point of a four-link mechanism

Activities

- Read the material provided below, and as you read:
  a. Develop an algorithm for plotting coupler curves of a four-link mechanism.
  b. Plot the coupler curves of a crank-rocker mechanism.
  c. Plot the coupler curves of a drag-link mechanism.
  d. Classify the coupler curves of four-link mechanisms.
- Check your answers to the competency items with your instructor.

Reading Material

Part I: Derivation of relationships to calculate coordinates of a coupler point of a four-link mechanism

Figure 2 shows four-link mechanism MA_1B_1Q with coupler point C. Let $MA_1 = a$, $A_1B_1 = b$, $B_1Q = c$, and $MQ = d$. Let input link MA_1 make angle θ_2 with fixed link MQ. Let MQ represent the X axis and MY normal to MX represent the Y axis. Join A_1Q. Let angle $A_1QM = \gamma$, and angle $B_1A_1Q = \beta$.

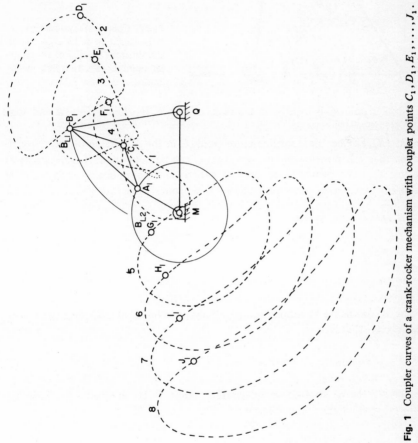

Fig. 1 Coupler curves of a crank-rocker mechanism with coupler points $C_1, D_1, E_1, \ldots, J_1$.

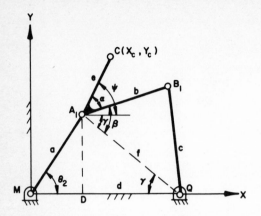

Fig. 2 Coordinate system and notations used to derive equation of coordinates (X_c, Y_c) of the coupler point C of a four-link mechanism.

Coupler point C is located in the plane of link A_1B_1. Let $A_1C = e$ and angle $CA_1B_1 = \alpha$.

Let (X_c, Y_c) be the coordinates of point C in the XY frame of reference. In order that the position of coupler point C can be computed, it is necessary to relate (X_c, Y_c) coordinates in terms of parameters a, b, c, d, e, α, and θ_2 of the four-link mechanism. For this purpose, draw from A_1 line l parallel to MQ. Then

$$\angle QA_1l = \angle A_1QM = \gamma$$

Let angle $CA_1l = \psi$, then

$$\psi = \beta - \gamma + \alpha \tag{1}$$

The x coordinate of coupler point C is obtained by taking the projection along the X axis. That is

$$X_c = a\cos\theta_2 + e\cos\psi \tag{2}$$

Similarly, the y coordinate of coupler point C is obtained by taking the projection along the y axis. That is

$$Y_c = a\sin\theta + e\sin\psi \tag{3}$$

Substituting Eq. (1) in Eqs. (2) and (3), we get

$$X_c = a\cos\theta_2 + e\cos(\beta - \gamma + \alpha) \tag{4}$$

and

$$Y_c = a\sin\theta_2 + e\sin(\beta - \gamma + \alpha) \tag{5}$$

If angles β and γ are known, then it is possible to compute coordinates X_c and Y_c for each position of the input crank.

The application of the cosine law to triangle B_1A_1Q yields for

$$\cos\beta = \frac{b^2 + f^2 - c^2}{2bf} \tag{6}$$

Where $A_1Q = f$ is given by

$$f = (a^2 + d^2 - 2ad\cos\theta_2)^{1/2} \tag{7}$$

Using the trigonometric identity

$$\tan^2\beta = \sec^2\beta - 1 \tag{8}$$

Equation (6) can be rewritten as

$$\beta = \tan^{-1}\left\{\frac{[4b^2f^2 - (b^2 + f^2 - c^2)^2]^{1/2}}{b^2 + f^2 - c^2}\right\} \tag{9}$$

From triangle MA_1D, and QA_1D

$$\tan\gamma = \frac{A_1D}{DQ} = \frac{a\sin\theta_2}{d - a\cos\theta_2} \tag{10}$$

That is

$$\gamma = \tan^{-1}\left(\frac{a\sin\theta_2}{d - a\cos\theta_2}\right) \tag{11}$$

Thus, the X_c and Y_c of point C can be computed using Eqs. (4), (5), (9), and (11) if parameters a, b, c, d, e, α, and θ_2 are known.

Part II: Coupler curves of a crank-rocker mechanism

Using Eqs. (4) and (5), coupler curves of crank-rocker mechanisms are plotted for $5°$ increment of input crank rotation. See Figs. 3–8. The location of the coupler point is decided by angle α and length AC. The coupler curves, shown in Figs. 3–8, are plotted for $30°$ incremental values of α and coupler point A is located from point C at distances given by the ratio $AC/AB = 0.5, 1.0, 1.5$, and 2.0. The crank-rocker mechanism used in plotting the coupler curves has the following kinematic parameters:

$a = 1.0$ unit

$b = 2.02$ units   angle of oscillation, $\phi = 60°$

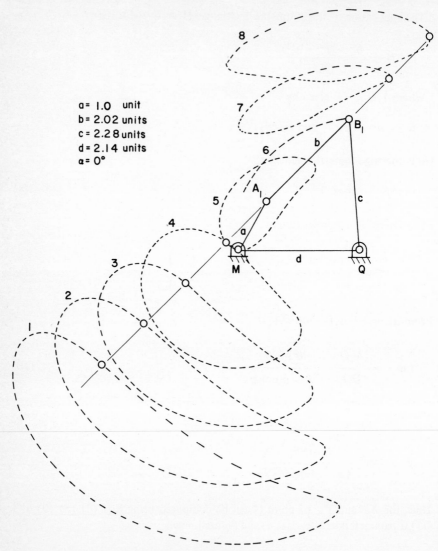

a = 1.0 unit
b = 2.02 units
c = 2.28 units
d = 2.14 units
α = 0°

Fig. 3 Coupler curves of a crank-rocker mechanism.

c = 2.28 units	time ratio	=	1.5 : 1
d = 2.14 units	maximum μ_{min}	=	29.7°

We observe that there appear to be basically six different types of coupler curves of a crank-rocker mechanism.

1. Flattened curves with circular arcs. See coupler curves 1 and 2 of Figs. 3, 4, 6, 7, and 8.

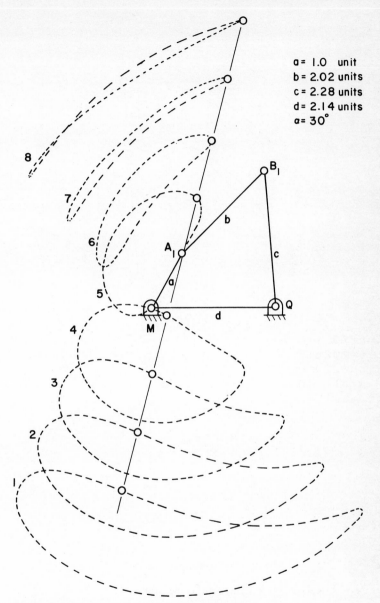

a = 1.0 unit
b = 2.02 units
c = 2.28 units
d = 2.14 units
α = 30°

Fig. 4 Coupler curves of a crank-rocker mechanism.

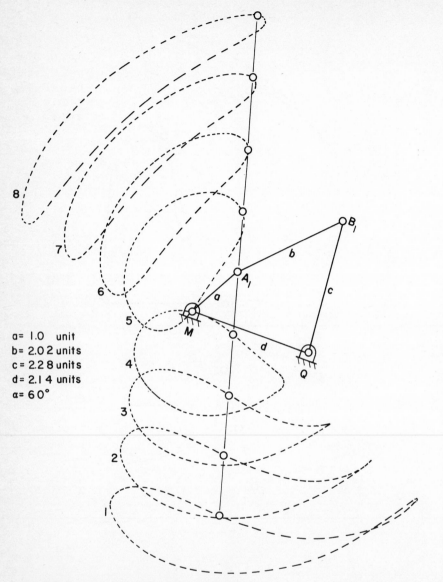

a= 1.0   unit
b= 2.02 units
c= 2.28 units
d= 2.14 units
α= 60°

Fig. 5 Coupler curves of a crank-rocker mechanism.

2. Flattened curves with circular arc and straight line segment. See curves 3 and 4 of Figs. 3, 4, 6, 7, and 8.

3. Curves with double points or figure-eight curves. See curve 8 of Fig. 4; curve 1 of Fig. 5; curves 6, 7, and 8 of Fig. 6; curves 7 and 8 of Fig. 7; curve 6 of Fig. 8. Whenever a coupler point traces the same point twice, the coupler is said to have "double points" at that point.

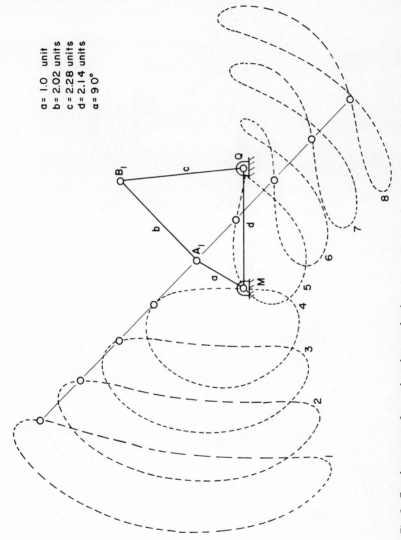

a = 1.0 unit
b = 2.02 units
c = 2.28 units
d = 2.14 units
α = 90°

Fig. 6 Coupler curves of a crank-rocker mechanism.

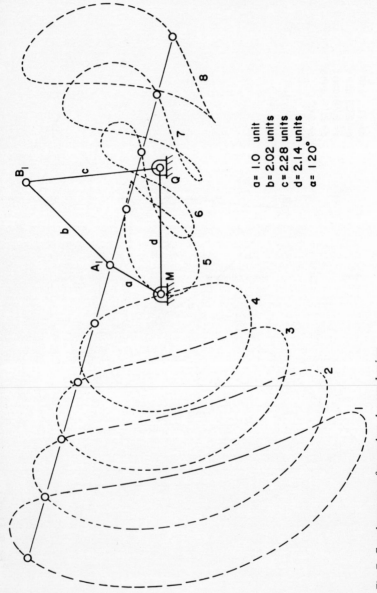

a = 1.0 unit
b = 2.02 units
c = 2.28 units
d = 2.14 units
α = 120°

Fig. 7 Coupler curves of a crank-rocker mechanism.

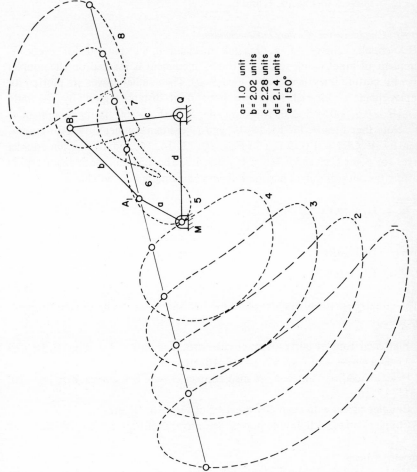

a = 1.0   unit
b = 2.02 units
c = 2.28 units
d = 2.14 units
α = 150°

Fig. 8 Coupler curves of a crank-rocker mechanism.

4. Coupler curves with two intersecting straight line segments. See curve 8 of Fig. 3, and curves 7 and 8 of Fig. 8.
5. Coupler curves with cusp. Whenever two branches of a coupler curve are tangent to each other, the coupler curve is said to have a cusp at that point. See curve 8 of Fig. 4, curve 2 of Fig. 5, and curve 8 of Fig. 7.
6. Coupler curves with "air foil" shape. See curves 7 of Fig. 3; curves 3 and 4 of Fig. 5; curve 5 of Figs. 6, 7, and 8.

Part III: Coupler curves of a drag-link mechanism

The coupler curves of a drag-link mechanism are plotted for every $5°$ increment of input crank rotation. The $5°$ increment is marked on the coupler curve by cutting lines as shown in Figs. 9–20. The coupler curves are plotted for coupler point C located from pin joint A at distances given by the ratio $AC/AB = 0.5, 1.0, 1.5, 2.0$. Angle α varies from $0°$ to $360°$ in the increment of $30°$. Note that Figs. 9, 11, 13, 15, 17, and 19 contain coupler curves for point C given by $AC/AB = 1.0, 2.0$, and Figs. 10, 12, 14, 16, and 18 contain coupler curves for point C given by $AC/AB = 0.5, 1.5$. The drag-link mechanism used in plotting the coupler curves has the following kinematic parameters.

$$a = 1.73 \text{ units}$$
$$b = 1.41 \text{ units} \qquad \psi = 120°$$
$$c = 1.41 \text{ units} \qquad \mu_{min} = 180° - \mu_{max} = 30°$$
$$d = 1.0 \text{ unit}$$

The coupler curves of the drag-link mechanism appear to be of four different types.

1. Flattened coupler curves with circular arcs. See curves 2 of Figs. 9, 10, and 11; and curves 4 of Figs. 13, 14, 15, 16, etc.
2. Heart-shaped or "limacon"-shaped coupler curves. See curves 4 of Figs. 10, 11, 12, etc.
3. Coupler curves with cusp. See curves 2 of Figs. 16, 18, etc.
4. Coupler curves with double points. See curves 2 of Figs. 15, 17, 19, etc.

Competency Items
- Make a cardboard model of a crank-rocker mechanism with a series of small holes in the coupler link. Generate coupler curves by placing a pencil point in the hole, allowing a curve to be traced as the crank is rotated $360°$.
- Sketch a coupler curve of any of the shapes described in this unit. Show a mechanism that will generate a curve like the one you have sketched.

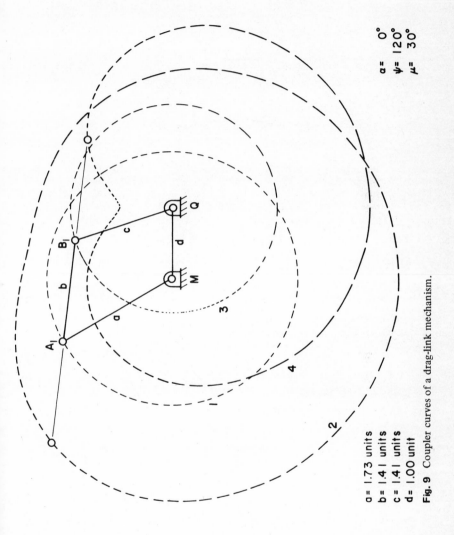

a = 0°
ψ = 120°
μ = 30°

a = 1.73 units
b = 1.41 units
c = 1.41 units
d = 1.00 unit

Fig. 9 Coupler curves of a drag-link mechanism.

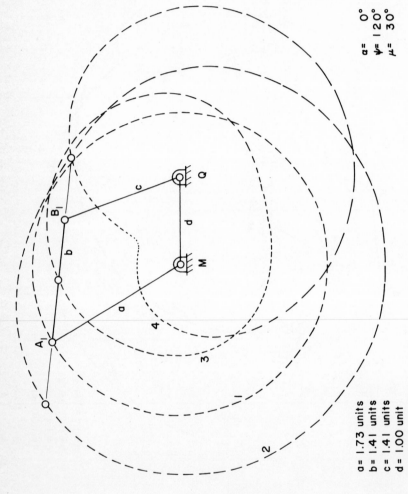

a = 0°
ψ = 120°
μ = 30°

a = 1.73 units
b = 1.41 units
c = 1.41 units
d = 1.00 unit

Fig. 10 Coupler curves of a drag-link mechanism.

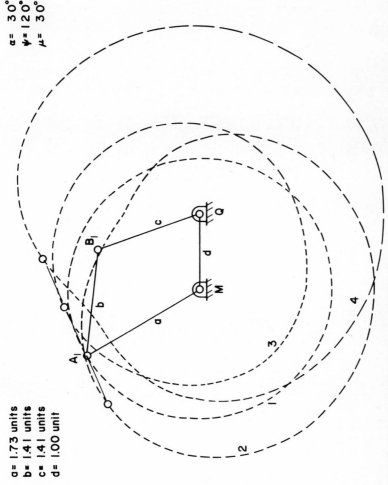

a = 1.73 units
b = 1.41 units
c = 1.41 units
d = 1.00 unit

α = 30°
ψ = 120°
μ = 30°

Fig. 11 Coupler curves of a drag-link mechanism.

329

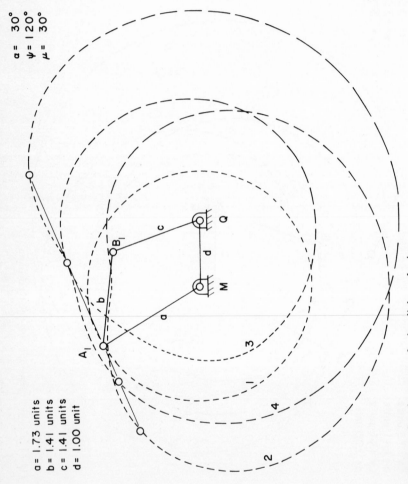

Fig. 12 Coupler curves of a drag-link mechanism.

a = 1.73 units
b = 1.41 units
c = 1.41 units
d = 1.00 unit

α = 30°
ψ = 120°
μ = 30°

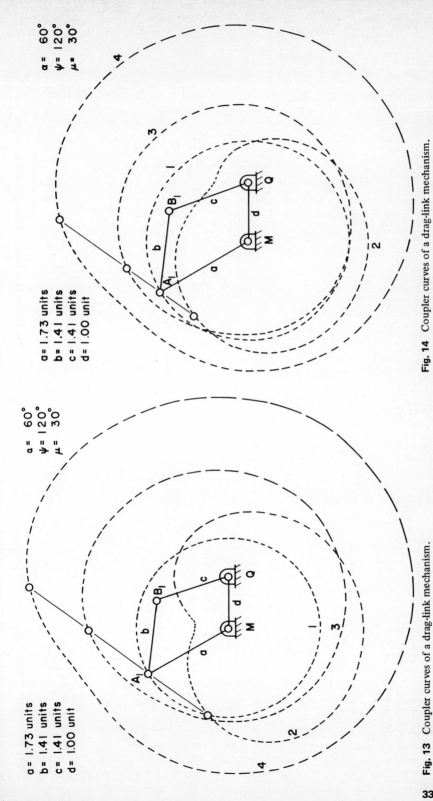

a = 1.73 units
b = 1.41 units
c = 1.41 units
d = 1.00 unit

α = 60°
ψ = 120°
μ = 30°

Fig. 13 Coupler curves of a drag-link mechanism.

a = 1.73 units
b = 1.41 units
c = 1.41 units
d = 1.00 unit

α = 60°
ψ = 120°
μ = 30°

Fig. 14 Coupler curves of a drag-link mechanism.

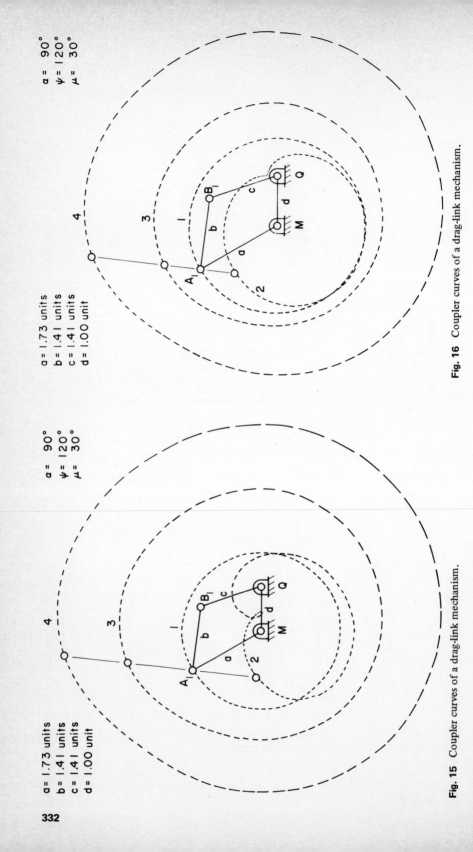

a = 1.73 units
b = 1.41 units
c = 1.41 units
d = 1.00 unit

α = 90°
ψ = 120°
μ = 30°

Fig. 15 Coupler curves of a drag-link mechanism.

a = 1.73 units
b = 1.41 units
c = 1.41 units
d = 1.00 unit

α = 90°
ψ = 120°
μ = 30°

Fig. 16 Coupler curves of a drag-link mechanism.

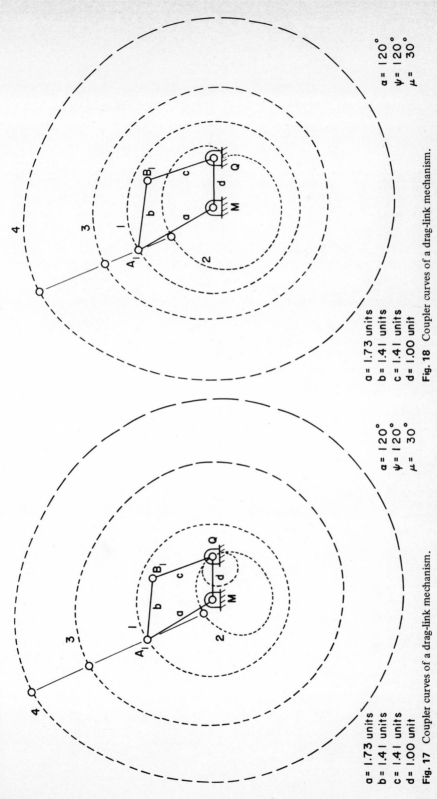

a = 1.73 units
b = 1.41 units
c = 1.41 units
d = 1.00 unit

α = 120°
ψ = 120°
μ = 30°

Fig. 17 Coupler curves of a drag-link mechanism.

a = 1.73 units
b = 1.41 units
c = 1.41 units
d = 1.00 unit

α = 120°
ψ = 120°
μ = 30°

Fig. 18 Coupler curves of a drag-link mechanism.

333

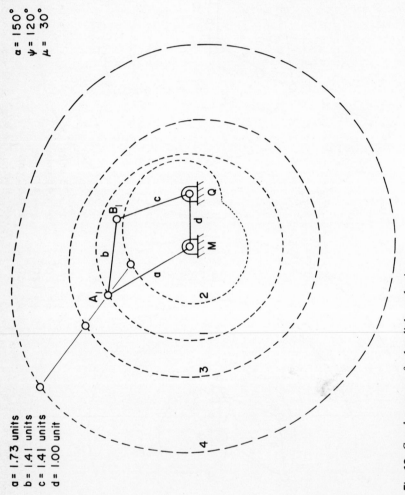

a = 150°
ψ = 120°
μ = 30°

a = 1.73 units
b = 1.41 units
c = 1.41 units
d = 1.00 unit

Fig. 19 Coupler curves of a drag-link mechanism.

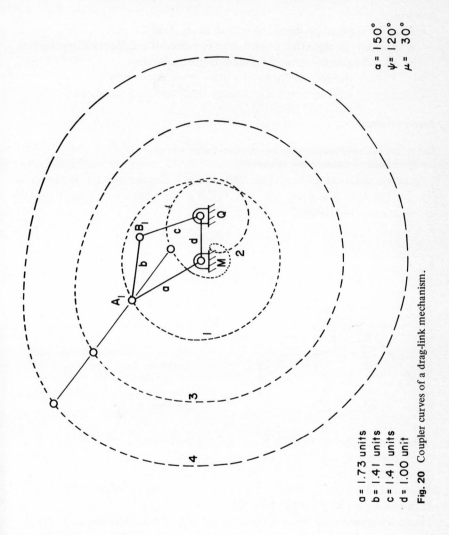

$a = 150°$
$\psi = 120°$
$\mu = 30°$

$a = 1.73$ units
$b = 1.41$ units
$c = 1.41$ units
$d = 1.00$ unit

Fig. 20 Coupler curves of a drag-link mechanism.

335

Objective 3

To derive relationships to calculate coordinates of a coupler point of a slider-crank mechanism

Activities

- Read the material provided below, and as you read:
  a. Develop an algorithm to plot coupler curves of a slider-crank mechanism.
  b. Plot the coupler curves of a slider-crank mechanism.
  c. Classify the coupler curves of a slider-crank mechanism.
- Check your answers to the competency items with your instructor.

Reading Material

Part I: Derivation of relationships to calculate coordinates of a coupler point of a slider-crank mechanism

Eccentric slider-crank mechanism MA_1B_1 with coupler point C is shown in Fig. 21. Let $MA_1 = a$, $A_1B_1 = b$, and $A_1C = c$. Let e denote the eccentricity of the slider-crank mechanism.

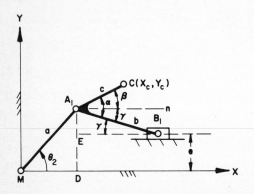

Fig. 21 Coordinate system and notations used to derive equation of coordinates (X_c, Y_c) of the coupler point C of a slider-crank mechanism.

Draw line X passing through M and parallel to the axis of the prism pair. Draw line Y passing through M and perpendicular to line X. Then XMY describes the XY frame of reference for the slider-crank mechanism. Let input crank MA_1 make angle θ_2. Coupler point C is located in the plane of coupler link AB. Let AC make angle α with coupler link AB.

Drop a perpendicular from A_1 cutting the axis of the slider pair in E and line MX at D. Let angle $A_1B_1E = \gamma$. At A_1 draw line n parallel to MX. Let angle $CA_1n = \beta$. Then, since angle $nA_1B_1 = \gamma$,

$$\beta = \alpha - \gamma \tag{12}$$

The (X, Y) coordiantes of coupler point C are obtained by taking projection along MX and MY. Thus,

$$X_c = a \cos \theta_2 + c \cos \beta \tag{13}$$

and

$$Y_c = a \sin \theta_2 + c \sin \beta \tag{14}$$

Substituting Eq. (12) in Eqs. (13) and (14), we get

$$X_c = a \cos \theta_2 + c \cos (\alpha - \gamma) \tag{15}$$

$$Y_c = a \sin \theta_2 + c \sin (\alpha - \gamma) \tag{16}$$

In order to compute the coordinates, we must express angle γ in terms of the kinematic parameters of the mechanism. From triangle AEB, of Fig. 21, we have

$$\sin \gamma = \frac{A_1 E}{A_1 B_1}$$

But

$$A_1 E = AD - DE = a \sin \theta_2 - e \tag{17}$$

and

$$A_1 B_1 = b$$

Therefore,

$$\sin \gamma = \frac{a \sin \theta_2 - e}{b} \tag{18}$$

or

$$\gamma = \sin^{-1} \left(\frac{a \sin \theta_2 - e}{b} \right) \tag{19}$$

Thus, Eqs. (15), (16), and (19) can be used to compute the coordinates of a coupler point provided the numerical value of the kinematic parameters a, b, c, e, θ_2, and γ are known.

Part II: Coupler curves of a slider-crank mechanism

Using Eqs. (15), (16), and (19), coupler curves of a slider-crank mechanism with zero eccentricity are plotted, as shown in Figs. 22–27, for a 5° increment of the input crank rotation. The location of coupler point C is dependent upon the values of α and distance AC.

For the coupler curves in Figs. 22–27, the angle varies from 0° to 360° in the increment of 30° and length AC is computed using ratio $AC/AB = 0.5, 1.0, 1.5,$ 2.0.

The slider-crank mechanism used for plotting the coupler curves has the following kinematic parameters: $a = 1$ unit, $b = 2$ units, $e = 0$. The coupler curves of slider-crank mechanisms are similar in appearance to those of crank-rocker mechanisms, and so, the coupler curves of slider-crank mechanisms can be classified into six different types. When the eccentricity of the slider-crank mechanism is zero and α becomes zero, the coupler curves are symmetric about the axis of the slider pair. See Fig. 22.

Competency Items

- Explain in writing how the coupler curve of a slider-crank mechanism would be affected by using the output of a drag-link mechanism as the input to the slider crank.
- List several coupler curves which have a straight line segment which corresponds to a rotation of 180° of the input crank.
- List several coupler curves which have a straight line segment with approximately constant velocity. Determine what rotational speed the input crank must have to result in a constant velocity of 10 in./sec for the coupler point over the straight line segment.

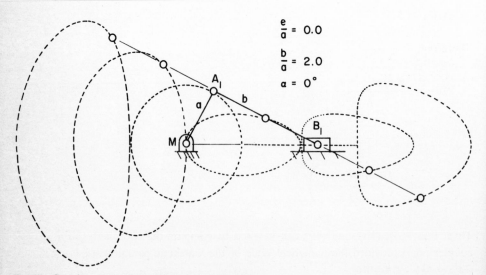

Fig. 22 Coupler curves of the slider-crank mechanism.

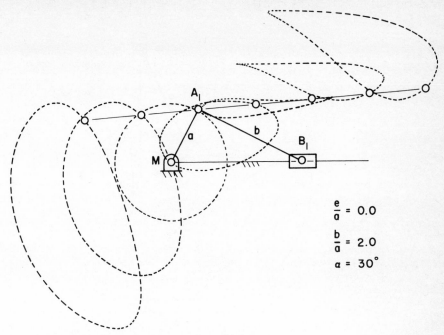

$$\frac{e}{a} = 0.0$$

$$\frac{b}{a} = 2.0$$

$$a = 30°$$

Fig. 23 Coupler curves of the slider-crank mechanism.

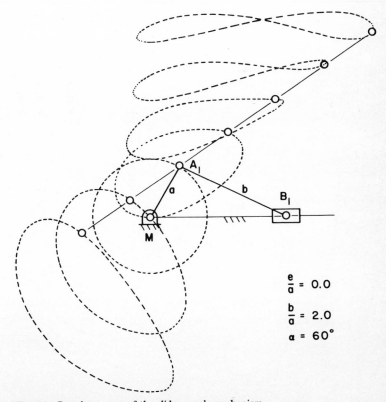

$$\frac{e}{a} = 0.0$$

$$\frac{b}{a} = 2.0$$

$$a = 60°$$

Fig. 24 Coupler curves of the slider-crank mechanism.

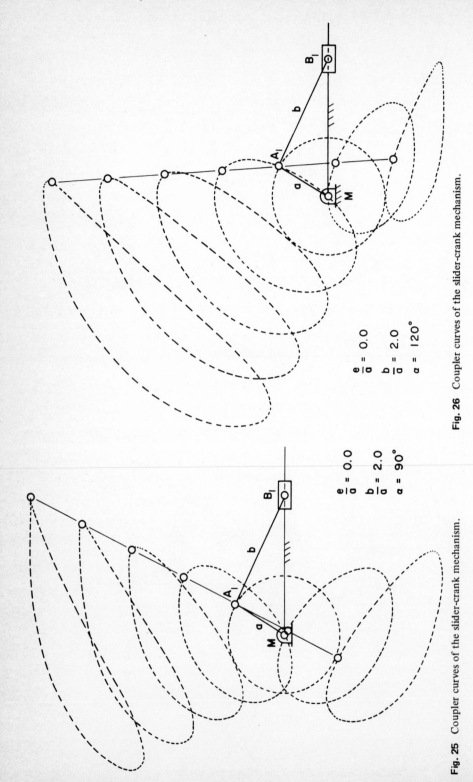

$\dfrac{e}{a} = 0.0$

$\dfrac{b}{a} = 2.0$

$\alpha = 120°$

Fig. 26 Coupler curves of the slider-crank mechanism.

$\dfrac{e}{a} = 0.0$

$\dfrac{b}{a} = 2.0$

$\alpha = 90°$

Fig. 25 Coupler curves of the slider-crank mechanism.

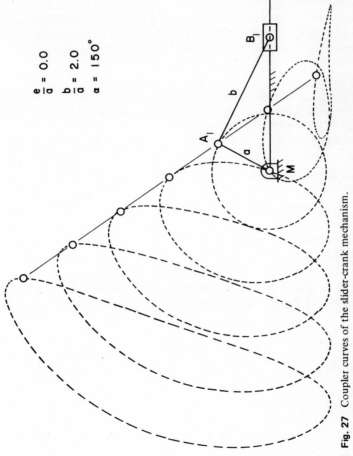

$\dfrac{e}{a} = 0.0$

$\dfrac{b}{a} = 2.0$

$\alpha = 150°$

Fig. 27 Coupler curves of the slider-crank mechanism.

Objective 4

To derive coordinates of a coupler point of an inverted slider-crank mechanism

Activities
- Read the material provided below, and as you read:
  a. Develop an algorithm to plot the coupler curves of an inverted slider-crank mechanism.
  b. Plot the coupler curves of an inverted slider-crank mechanism.
  c. Classify the coupler curves of an inverted slider-crank mechanism.
- Check your answers to the competency items with your instructor.

Reading Material

Part I: Derivation of relationships to calculate coordinates of a coupler point of an inverted slider-crank mechanism, Type I

Inverted slider-crank mechanism MA_1B_1Q with coupler point C in the plane of variable coupler link A_1B_1 is shown in Fig. 28. Let fixed link $MQ = d$ represent the X axis and line Y at M perpendicular to MX be the Y axis. Input link $MA_1 = a$ and output link QB_1 make angles θ_2 and θ with the X axis as shown in the figure. Join A_1 and Q using a dotted line. Let $\angle A_1QM = \beta_1$ and $\angle A_1QB_1 = \beta_2$.

Coupler point C is located in the plane of the coupler link using parameters $A_1C = c$ and angle $CA_1B = \alpha$. Drop a perpendicular from C to the X axis, cutting A_1B_1 at E. Let $\angle A_1CE = \gamma$ and $\angle A_1EC = \beta_3$.

The (X, Y) coordinates of coupler point C are obtained by taking projection along the X and Y axis. Thus,

$$X_c = a\cos\theta_2 + c\sin\gamma \tag{20}$$

and

$$Y_c = a\sin\theta_2 + c\cos\gamma \tag{21}$$

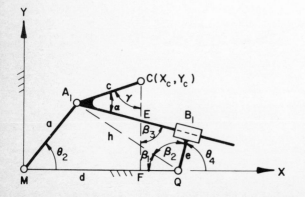

Fig. 28 Coordinate system and notations used to derive equation of the coordinates (X_c, Y_c) of the coupler point C of an inverted slider-crank mechanism.

From triangle ACE

$$\gamma = 180° - \beta_3 - \alpha \tag{22}$$

From quadrilateral $EFQB_1$

$$\beta_3 + \text{angle } EFQ + \beta_1 + \beta_2 + \text{angle } EB_1Q = 360° \tag{23}$$

Because of the slider pair at B_1, angle $EB_1Q = 90°$. Because of the construction, angle $EFQ = 90°$. Therefore,

$$\beta_1 + \beta_2 + \beta_3 = 180° \tag{24}$$

But

$$\theta_4 = 180° - (\beta_1 + \beta_2) \tag{25}$$

Therefore,

$$\beta_3 = \theta_4 \tag{26}$$

and

$$\gamma = 180° - \alpha - \theta_4 \tag{27}$$

Thus, Eqs. (20) and (21) can be written as

$$X_c = a \cos \theta_2 + c \sin(180° - \alpha - \theta_4) \tag{28}$$
$$Y_c = a \sin \theta_2 + c \cos(180° - \alpha - \theta_4) \tag{29}$$

where angle θ_4 can be computed using Eqs. (28) and (29). Thus, Eqs. (28) and (29) can be used to compute (X, Y) coordinates of coupler point C.

Part II: Coupler curves of inverted slider-crank mechanism, Type I

Using Eqs. (28) and (29), coupler curves of an inverted slider-crank mechanism are plotted for 5° of increment of input crank rotation. See Figs. 29–34. The location of coupler point C is decided by angle α and length AC. The coupler curves, shown in Figs. 29–34, are plotted for 30° incremental values of α and coupler point C is located from point A at distance given by $AC = MA, 2MA, 3MA$, etc.

The inverted slider-crank mechanism used in plotting the coupler curves has the following kinematic parameters: $a = 1.0$ units, $d = 2.0$ units, $e = 0$.

We observe that the coupler curves, shown in Figs. 29–39, are similar in appearance to those of crank-rocker mechanisms. Also, the input-output motion of the inverted slider-crank mechanism with the above link proportion will be of the crank-rocker type.

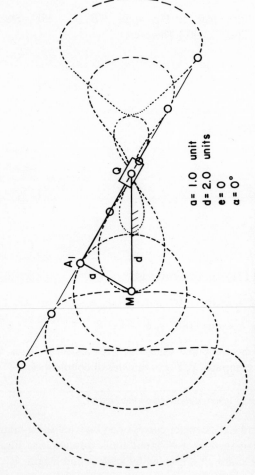

Fig. 29 Coupler curves of an inverted slider-crank mechanism.

a = 1.0 unit
d = 2.0 units
e = 0
α = 0°

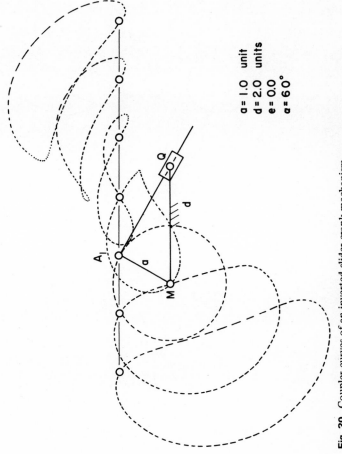

a = 1.0 unit
d = 2.0 units
e = 0.0
α = 60°

Fig. 30 Coupler curves of an inverted slider-crank mechanism.

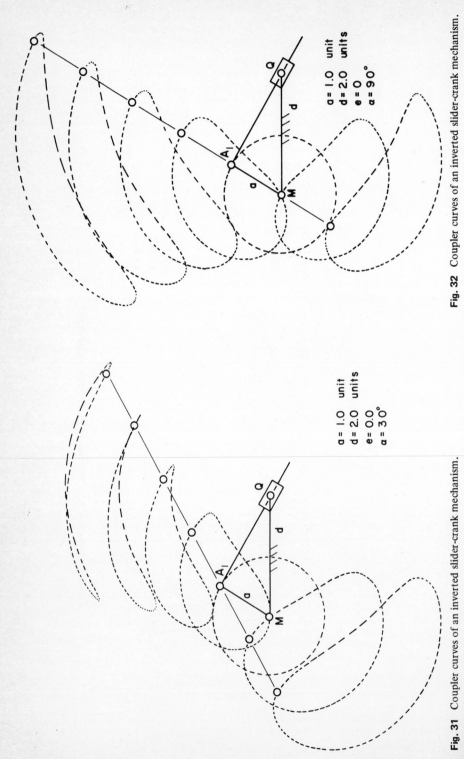

a = 1.0 unit
d = 2.0 units
e = 0
α = 90°

Fig. 32 Coupler curves of an inverted slider-crank mechanism.

a = 1.0 unit
d = 2.0 units
e = 0.0
α = 30°

Fig. 31 Coupler curves of an inverted slider-crank mechanism.

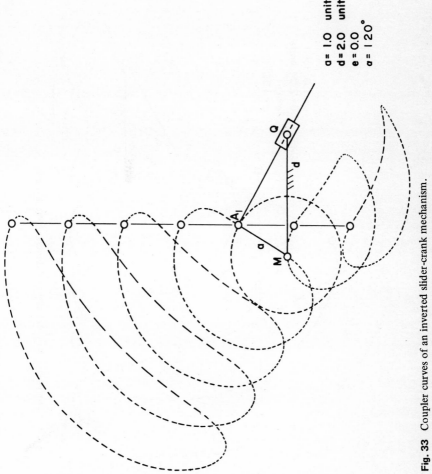

a = 1.0 unit
d = 2.0 units
e = 0.0
α = 120°

Fig. 33 Coupler curves of an inverted slider-crank mechanism.

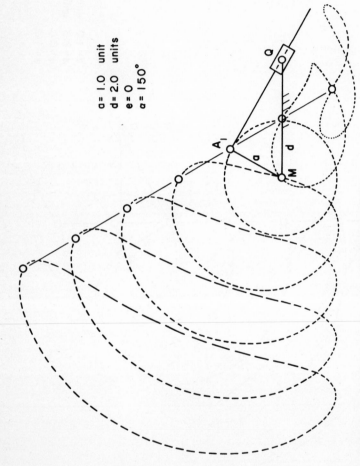

a = 1.0　unit
d = 2.0　units
e = 0
α = 150°

Fig. 34 Coupler curves of an inverted slider crank mechanism.

We know that the inverted slider-crank mechanism functions as a drag-link mechanism when $a > d + e$. For this reason, it is interesting to observe that the coupler curves of the inverted slider-crank mechanism, with the condition $a > d + e$, also appear to be similar to those of the four-link drag-link mechanism. See Fig. 35.

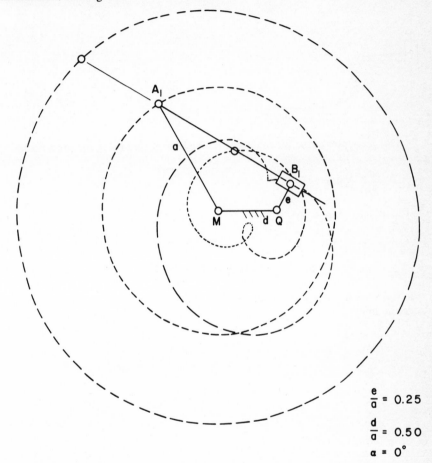

$$\frac{e}{a} = 0.25$$

$$\frac{d}{a} = 0.50$$

$$\alpha = 0°$$

Fig. 35 Coupler curves of an inverted slider-crank mechanism.

Competency Items
- List several coupler curves which have two circular arc segments of equal radius of curvature and a duration of at least 30° of input crank rotation.
- List the common properties of the coupler curves of crank-rocker and inverted slider-crank mechanisms.
- List the common properties of the coupler curves of drag-link and inverted slider-crank mechanisms.
- List the common properties of the coupler curves of slider-crank and inverted slider-crank mechanisms.

Performance Test #1

1. Derive the relationship to calculate the coupler coordinates of the double slider mechanism shown below.

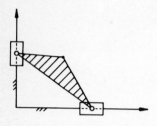

Plot a coupler curve for the point C which is located using the proportion

$$\frac{AC}{AB} = 0.5$$

and angle $CAB = 30°$.

Performance Test #2

Find the proportions of a slider-crank mechanism whose coupler curve is a straight line.

Unit XV. Application of Coupler Curves in Design of Six-link Mechanisms

Objectives

1. To name different types of coupler driven six-link mechanisms
2. To name different types of motion programs
3. To design six-link mechanisms with a single dwell
4. To design six-link mechanisms with double dwells
5. To design six-link mechanisms with double strokes
6. To design a six-link mechanism with an output link moving with a constant velocity for a finite interval of time
7. To design six-link mechanisms with output links having $720°$ rotation corresponding to $360°$ rotation of the input link

Objective 1

To name different types of coupler driven six-link mechanisms

Activities

- Read the material provided below, and as you read:
  a. Enumerate and sketch 7 six-link mechanisms derived from a four-link mechanism.
  b. Enumerate and sketch 7 six-link mechanisms derived from a slider-crank mechanism.
  c. Enumerate and sketch 7 six-link mechanisms derived from an inverted slider-crank mechanism.
- Check your answers to the competency items with your instructor.

Reading Material

The coupler curves examined in the previous unit can be used to synthesize six-link mechanisms named *coupler driven six-link mechanisms*. In this section, we will examine all possible six-link mechanisms that can be designed using the coupler curves of four-link mechanisms.

Figure 1 shows the 7 six-link mechanisms obtained using the output from a coupler point of a four-link mechanism with four revolute pairs. There is a total of seven kinematic pairs of one degree of freedom in a six-link mechanism. Figure 1a describes a six-link mechanism with seven revolute pairs. Figure 1b, c, and d describes six-link mechanisms with six revolute pairs and one prism pair. Figure 1e, f, and g describes six-link mechanisms with five revolute and two prism pairs.

Figure 2 shows the 7 six-link mechanisms obtained using the output from a coupler point of a slider-crank mechanism. Similarly, Fig. 3 shows the 7 six-link mechanisms obtained using the output from a coupler point of an inverted slider-crank mechanism, Type I.

These 21 six-link mechanisms can be classified by the two basic types of motion they execute:

Rotary to rotary. The input and output cranks are both performing rotary motion. See a, b, and c of Figs. 1–3.

Rotary to linear. The input crank performs rotary motion while the output crank performs linear or translatory motion. See d, e, f, and g of Figs. 1–3.

Competency Items

- Write two applications for each of the six-link mechanisms shown in Figs. 1, 2, and 3.
- State how the coupler curve shape will affect the motion of these six-link mechanisms.
- List all possible six-link mechanisms that can be derived from the double-slider mechanism shown below.

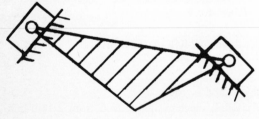

Objective 2

To name different types of motion programs

Activities

- Read the material provided below, and as you read:
  a. Name different types of motion programs that can be generated using the coupler curve of a four-link, slider-crank, and inverted slider-crank mechanism.
  b. Define R-R, D-R-R, D-R-D-R motion programs.
- Check your answers to the competency items with your instructor.

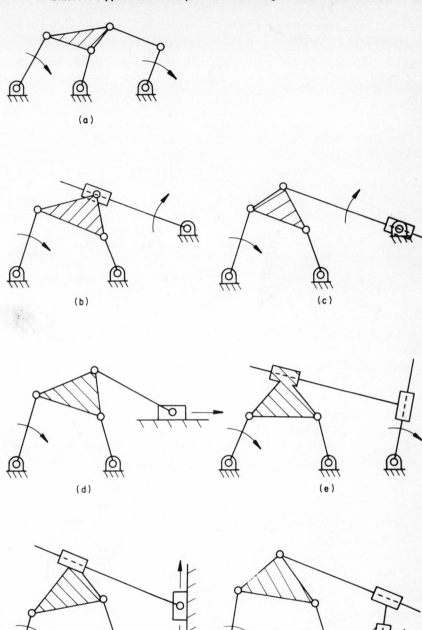

Fig. 1 Coupler driven six-link mechanisms derived from a four-revolute, four-link mechanism.

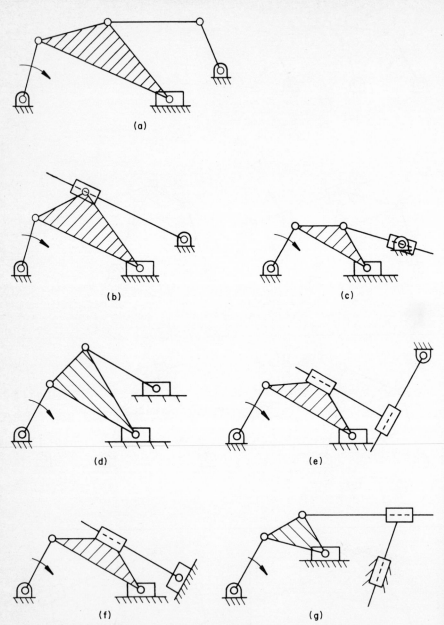

Fig. 2 Coupler driven six-link mechanisms derived from a slider-crank mechanism.

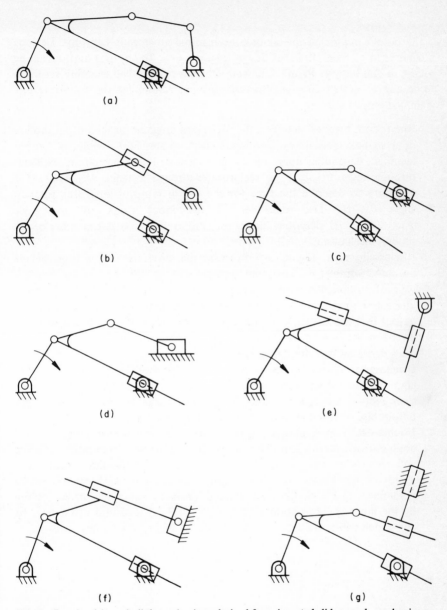

Fig. 3 Coupler driven six-link mechanisms derived from inverted slider-crank mechanism, Type I.

Reading Material

The output motion of the six-link mechanisms surveyed in Objective 1 can be programmed to meet different design requirements. The output motion can be rotary or translatory. In the following Objective 3, we will examine synthesis technique to design coupler-driven six-link mechanism for the following programmed output motion:

1. Rise-return. Figure 4 describes the rise-return program for the output motion of a six-link mechanism. The input crank rotates $360°$, while the output member, depending upon the type of the six-link mechanism, oscillates through angle $(\theta_{out})_{max}$ or reciprocates through distance $(S_{out})_{max}$. It is necessary to describe the time for the rise in terms of input link rotation angle $\theta = \beta$. The remainder of the input crank rotation angle $\theta = (360° - \beta)$ describes the return motion of the output member of the six-link mechanism.

2. Rise-dwell-return. Figure 5–7 describe the dwell-rise-return type of the motion program for the output member of a six-link mechanism. For this type of motion, the output member does not have any motion for a finite interval of time even though the input link continues to rotate. In Fig. 5, the output link dwells at the beginning of the motion. In Fig. 6, the output link dwells at the end of its maximum displacement. In Fig. 7, the output link has a dwell halfway during its return cycle.

3. Dwell-rise-dwell-return. Figure 8 describes the dwell-rise-dwell return program for the output motion of a six-link mechanism. Mechanisms executing such motion are called double-dwell mechanisms. The two dwells of the output link correspond to input link rotation angles θ_{D1} and θ_{D2}.

4. Double-stroke mechanisms. Figure 9 describes the motion program for a double-stroke mechanism. The output link, if it has a translatory motion, will reciprocate two times for every revolution of the input crank. The length of the stroke and time required to complete each stroke can be predefined. In Fig. 9, the time required for the first complete stroke is given by the input crank rotation angle $\theta = \beta_1 + \beta_2$. The second stroke must be completed within the input crank rotation of angle $[360° - (\beta_1 + \beta_2)]$.

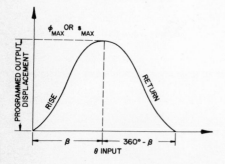

Fig. 4   Rise-return program.

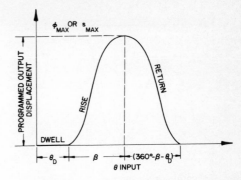

Fig. 5 Dwell-rise-return program.

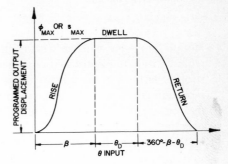

Fig. 6 Rise-dwell-return program.

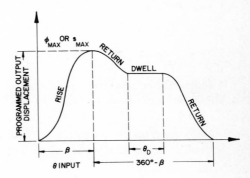

Fig. 7 Rise-return-dwell-return program.

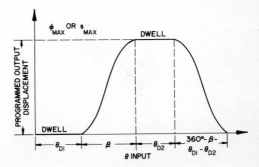

Fig. 8 Dwell-rise-dwell-return program.

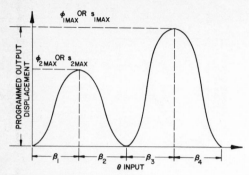

Fig. 9 Double-stroke motion.

5. Motion programs with instantaneous dwells. Quite often, a designer is required to have his mechanism an instantaneous dwell at either ends of its stroke. Such an instantaneous dwell condition forces the output link to move rather more slowly in the vicinity of the dwells. This slow motion of the output link is then gainfully utilized to perform some additional operation.

6. Motion program for larger oscillation. The output link of a crank-rocker four-link mechanism cannot oscillate through an angle greater than 180°. If it is desired that the output link mechanism oscillates through a larger angle, then synthesis of coupler-driven six-link mechanisms proves to be a practical approach.

7. Motion program for constant velocity. In many cases, the stop-start condition of the output link is not a desirable design situation. For such cases, a designer often solves his problem by forcing a condition so that the output link moves with a constant velocity for a finite interval of time. Here again, the synthesis of coupler-driven six-link mechanisms is one possible approach.

8. Motion program with hesitation. Hesitation is an instantaneous dwell without reversing the direction of motion.

Competency Items

*   The motion programs presented in this objective are very similar to those studied in the unit on cams. What differences do you notice between the cam and the linkage motion programs? When would you use one or the other?
*   Write the boundary conditions relating displacement, velocity, and acceleration for a rise-return motion program of an output link.
*   Write the boundary conditions relating displacement, velocity, and acceleration for a dwell-rise-return motion program of an output link.
*   Write the boundary conditions relating displacement, velocity, and acceleration for a dwell-rise-dwell-return motion program of an output link.

Objective 3

To design six-link mechanisms with a single dwell

Activities
- Read the material provided below, and as you read:
  a. Synthesize six-link mechanisms using the coupler curve of a four-link mechanism to generate a D-R-R motion program.
  b. Develop alternate methods.
- Check your answers to the competency items with your instructor.

Reading Material
We will examine four different ways to synthesize a six-link mechanism for an output motion having a single dwell as described in Figs. 5-7.

The six-link mechanism designed in Fig. 10, has an output link PD which dwells at its extreme position PD_2. This mechanism can be designated using the following steps.

Method I
1. Consider a coupler curve having circular arc C_2C_3. Find radius C_2D_2 of this circular arc and center D_2. Make $C_2D_2 = C_1D_1$.
2. Find extreme position D_4 by drawing tangent circles with C_1D_1 as radius. The tangent circle should not intersect the coupler curve at some other position.
3. Locate fixed center P on the bisector of D_2D_4. If the output link must oscillate through some specified angle β, then construct isosceles triangle D_2PD_4 such that angle $D_2PD_4 = \beta$.
4. It is a good practice to plot the output angular displacement of link PD versus the input displacement of input crank MA. We observe that when coupler point C_1 traces circular arc C_2C_3, point D_2 becomes a center of rotation and as a consequence, output link PD experiences a dwell.

Method II
1. Consider a coupler curve with a straight line segment. See coupler curve in Fig. 10b. C_2C_3 can be considered as a straight line for all practical purpose.
2. Extend line C_2C_3. Locate fixed center P on this line so that angle $C_3PC_4 = \beta$.
3. Place a slider pair at C_2 with its axis coincident with line C_2C_3. Note that the six-link mechanism is shown in its first position. The input output displacement plot shows the dwell at one end of its output oscillation.
4. Note that when coupler point C_1 traces the straight line segment of the coupler curve, output link PC experiences a dwell.

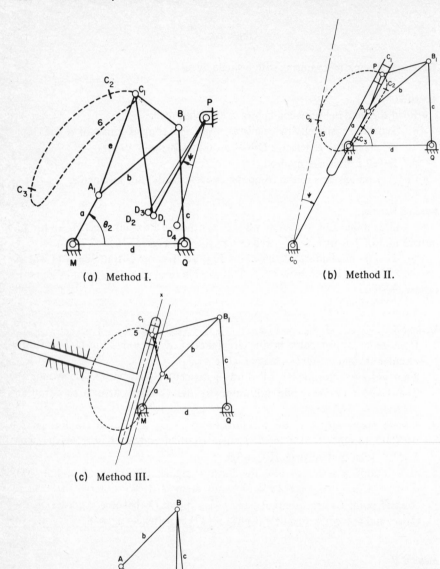

(a) Method I.

(b) Method II.

(c) Method III.

(d) Method IV.

Fig. 10 Synthesis of a six-link mechanism with a single dwell.

Method III

1. Consider a coupler curve with a straight line segment x, see curve in Fig. 10c.
2. Place a slider pair with its axis parallel to line x. Place a revolute pair at joint C_1 as shown in Fig. 10c.
3. The coupler curve is approximately symmetric. Place a prism pair with its axis coincident with the line of symmetry.
4. When coupler point C_1 traces the straight line segment, the output slider pair experiences a dwell.

Method IV

1. Consider figure-eight type of a coupler curve with either a circular or a straight line segment. See coupler curve in Fig. 10d.
2. Fixed center P is located in the straight line segment such that angle C_2PC_3 becomes equal to the desired angle of oscillation. PC_3 and PC_2 are two tangents. Also, the revolute pair at P is the output pair.
3. Place a slider pair at P with its axis coincident with straight line PC_1. Place a revolute pair at C_1.
4. We observe that as coupler point C_1 traces the straight line segment of the coupler curve, the output revolute pair experiences a dwell.

Competency Items

• Design a six-link mechanism with rotational output which has a dwell of duration $20°$ of input crank rotation (minimum). Can you design this in more than one way?
• Design a six-link mechanism with translational output which has a dwell of duration $20°$ of input crank rotation (minimum). Can you design this in more than one way?

Objective 4

To design six-link mechanisms with double dwells

Activities

• Read the material provided below, and as you read:
  a. Design six-link mechanisms for a D-R-D-R output motion program using the coupler curves of a four-link mechanism.
  b. Devise alternate methods.
• Check your answers to the competency items with your instructor.

Reading Material

There are several ways by which a six-link mechanism can be synthesized to have its output link experience double dwells. We will demonstrate these synthesis techniques.

Method I
1. Select a coupler curve having more than one circular arcs. See coupler curve in Fig. 11a.
2. Find the centers D_2 and D_3 of the circular arcs x and y. Note that the two circular arcs must have the same radius. The length of link CD is equal to this radius.
3. Locate fixed centers P on the perpendicular bisector of chord $D_2 D_3$ so that angle $D_2 P D_3 = \beta$, which is the required angle of oscillation of the output link.
4. We observe that as the coupler point traces the circular arcs x and y, output link PD experiences dwells at two positions, PD_2 and PD_3.

Method II
1. Consider a coupler curve with two straight line segments. See curve in Fig. 11b.
2. The two straight line segments are t_1 and t_2.
3. Locate fixed center P as the point of intersection of lines t_1 and t_2.
4. Place a revolute pair at P and a revolute pair and a prism pair at C_1. The prism pair translates on the link rotating about center P.
5. We observe that as coupler point C_1 traces straight line segments t_1 and t_2, output link PC experiences two dwells.

Method III
1. Consider a figure-eight type of coupler curve. See curve of Fig. 11c.
2. Locate the two circular arcs x and y. The two circular arcs are required to have the same radius. The length of link CD is equal to this radius.
3. Join centers X_0 and Y_0 of circular arcs X and Y.
4. Place a slider pair with its axis coincident with $X_0 Y_0$. Place revolute pairs at C_1 and D_1.
5. We observe that as the coupler curve traces circular arcs X and Y, the slider block experiences dwells in positions X_0 and Y_0.

Method IV
1. Consider figure-eight type of coupler curve having two intersecting line segments x and y. See curve 6 in Fig. 11d.
2. Locate fixed center P at the point of intersection of these straight line segments.
3. Place revolute pairs at C_1 and P. Place a slider pair at P so that its axis is coincident with $C_1 P$. The revolute pair at P is the output pair.
4. When coupler point C_1 traces straight line segments x and y, the output revolute pair experiences two dwells.

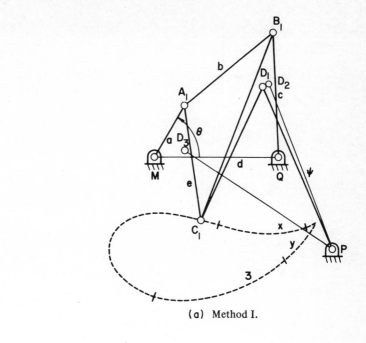

(a) Method I.

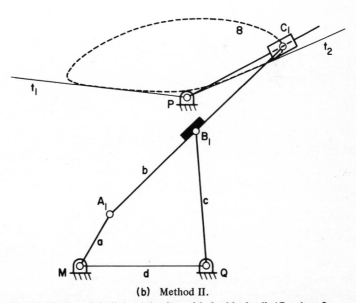

(b) Method II.

Fig. 11 Design of six-link mechanism with double dwell. (*Continued*)

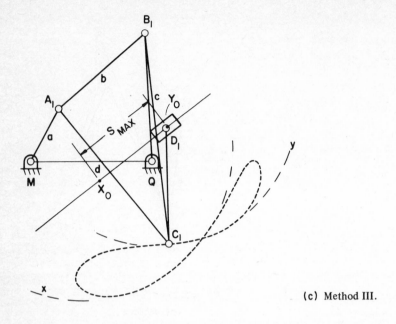

(c) Method III.

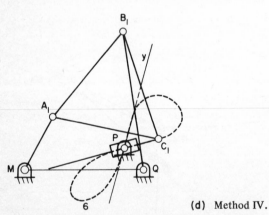

(d) Method IV.

Fig. 11 Design of six-link mechanism with double dwell.

Competency Items
- Design a mechanism with rotational output for the following motion program: 120° Rise, 60° Dwell, 180° Return.
- Design a mechanism with translational output for the motion program just designed. The length of the output stroke is 2.0 in.

Objective 5

To design six-link mechanisms with double strokes

Activities

- Read the material provided below, and as you read:
  a. Design six-link mechanisms with an output link executing a double stroke.
  b. Demonstrate the application of a figure-eight curve.
- Check your answers to the competency items with your instructor.

Reading Material

A six-link mechanism with double strokes of equal amplitudes requires that its output link either oscillate twice through the angle of oscillation β, or reciprocate twice through the distance s_{max}. A six-link mechanism satisfying such output requirements can be designed using the coupler curves of a four-link mechanism. In this section, we will demonstrate three different methods for synthesizing a six-link mechanism with double strokes of equal amplitudes.

Method I

1. Select figure-eight type of coupler curve of a four-link mechanism. See curve 6 in Fig. 12a.
2. Draw two tangent lines to the coupler curve. Note that X, Y, Z, and W are the points of tangency.
3. Locate fixed center P as the point of intersection of the two tangent lines.
4. We observe that as coupler point C passes through points X, W, and Z, the output link completes the first stroke. The second stroke is completed when the coupler point passes through points Z, Y, and X.
5. The amplitude of the stroke is determined by angle of oscillation XPY.

Method II

1. Select figure-eight type of coupler curve. See Fig. 12b.
2. Draw two tangent lines l_1 and l_2 to the coupler curve. The line l_2 is tangent to the curve at points X and Z. The line l_1 is tangent at points Y and W.
3. With X and Z as centers and any radius R_1 draw two arcs of a circle, intersecting in point D_3.
4. With Y and W as center and R_1 as radius, draw two arcs of a circle, intersecting in point D_2.
5. Locate fixed center P on the perpendicular bisector of chord D_2D_3 so that angle D_2PD_3 is the required angle of oscillation.
6. Let $CD = R_1$. Place revolute pairs at joints C, D, and P.
7. The first stroke of output link PD is completed when coupler point C passes through points X, W and Z. The second stroke is completed when the coupler point passes through point Z, Y and X.

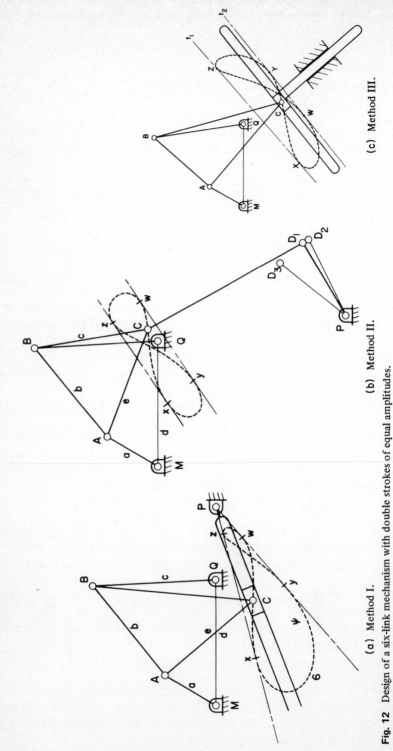

(a) Method I.

(b) Method II.

(c) Method III.

Fig. 12 Design of a six-link mechanism with double strokes of equal amplitudes.

Method III

1. Select figure-eight type of coupler curve. In order to fulfill the requirement of equal stroke, it is necessary that the coupler curve be symmetric. See Fig. 12c.
2. Draw two tangent lines t_1 and t_2 to the coupler curve. We observe that lines t_1 and t_2 are tangent at points X and Z, Y, and W. Also, if the coupler curve is symmetric, then lines t_1 and t_2 will be parallel.
3. At joint C, place a revolute pair and a prism pair with its axis parallel to tangent lines t_1 or t_2. Place a prism pair with its axis coincident with line of symmetry t_3.
4. We observe that as coupler point C passes through points X, Y, and Z, the output prism pair completes its first stroke. The second stroke is completed when coupler point C passes through the points Z, W, and X.

Competency Items

- Design a six-link mechanism which has a double stroke output for one input crank rotation. The output may be either rotational or translational.
- State the types of coupler curves you could use to obtain double stroke outputs of unequal amplitudes.
- Design a six-link mechanism having the output motion program shown below.

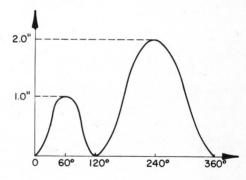

Objective 6

To design a six-link mechanism with an output link moving with a constant velocity for a finite interval of time

Activities

- Read the material provided below, and as you read:
  a. Design a six-link mechanism with an output link having a constant velocity.
  b. Demonstrate the application of coupler curves.
- Check your answers to the competency items with your instructor.

Reading Material

In some cases, it is desirable that a moving object be moved with some specified constant velocity rather than bringing it to a stop (dwell) for a finite interval of time. For such cases, a coupler driven six-link mechanism can be designed. The following construction will lead to such design.

1. Select any coupler curve. See Fig. 13. We are considering coupler curve 4.
2. Mark several points C_1, C_2, C_3, ..., C_7 on the coupler curve corresponding to the equal increment $\Delta\theta_2$ of the input crank rotation.
3. With any radius CD, and centers C_1, C_2, C_3, ..., C_7 draw arcs of a circle. These arcs are marked as Z_1, Z_2, ..., Z_7.
4. Make an overlay of output sector $X_1 P X_7$ divided into equal intervals $\Delta\phi$ of the output link rotation. These increments of $\Delta\phi$ are marked as X_1, X_2, X_3, ..., X_7.
5. Place the overlay on the top of the circular arcs until the circular arcs $X_1, X_2, ..., X_7$ of the output sector intersect with the corresponding circular arcs Z_1, Z_2, Z_3, ..., Z_7. The points of intersection are marked as $D_1, D_2, D_3, ..., D_7$.
6. Place the revolute pairs at C_1, D_1, and P. Fix P to the ground. Then output link PD of six-link mechanism MAB-CDP moves with a constant velocity. We observe that accuracy of the result depends upon the accuracy of the graphical construction.

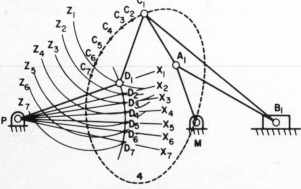

Fig. 13 Design of a six-link mechanism for constant velocity output for a finite interval of time.

Competency Items

- Design a six-link mechanism with a translating output which has a constant velocity through 45° of input crank rotation. What input crank speed would result in the constant velocity of magnitude 20 in./sec?
- Design a six-link mechanism with all revolute pairs. The output link must rotate with a constant velocity for approximately 60% of its forward stroke.

Objective 7

To design six-link mechanisms with output link having $720°$ rotation corresponding to $360°$ rotation of the input link

Activities

- Read the material provided below, and as you read:
  a. Synthesize a six-link mechanism with an output link having $720°$ rotation.
  b. Demonstrate the application of coupler curves of a drag-link mechanism.
- Check your answers to the competency items with your instructor.

Reading Material

Because of the special characteristic of the coupler curves of a drag-link mechanism, it is possible to design a coupler driven six-link mechanism in which the output link rotates $720°$ corresponding to $360°$ rotation of the input crank. The following construction will illustrate the design technique.

1. Select a coupler curve with a double loop. See curve 2 in Fig. 14.
2. Locate fixed center P at the center of the inner loop.
3. Connect P and C_1. At C_1, place a revolute pair and a prism pair with its axis coincident with PC_1.
4. We observe that as coupler point C_1 traces each loop, the output link rotates $360°$. Since there are two loops, the output link rotates $720°$ corresponding to $360°$ rotation of the input crank.

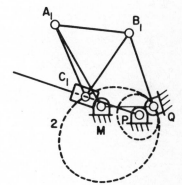

Fig. 14 Design of a six-link mechanism having the output link rotating $720°$ corresponding to $360°$ rotation of the input link.

Competency Items

- Find other types of coupler curves which could be used to synthesize a six-link mechanism in which the output link rotates $720°$ corresponding to $360°$ rotation of its input link.
  Design an eight-link mechanism in which the output link oscillates $360°$ for $360°$ rotation of the input link.

Performance Test #1

Design and build models of six-link mechanisms of the type shown below in Fig. A to obtain the motion programs shown in Figs. B–D. The types of coupler curves that you need are shown in the figures that follow Fig. D.

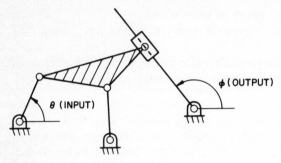

Fig. A

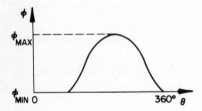

Fig. B

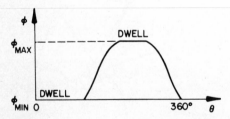

Fig. C

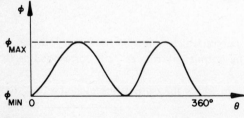

Fig. D

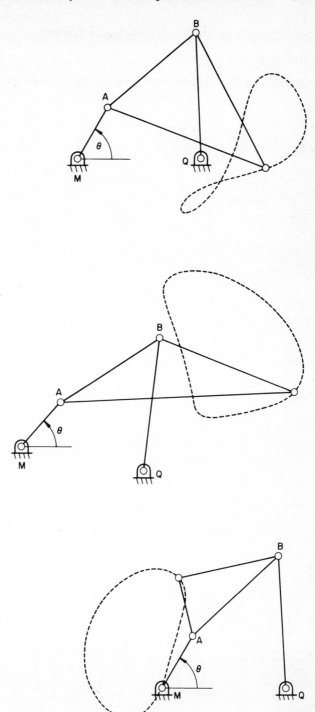

Performance Test #2

1. Using coupler curve of a four-link mechanism, design a six-bar linkage of the type of Fig. A. The required motion program is shown in Fig. B.

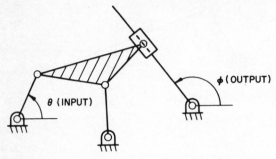

Fig. A

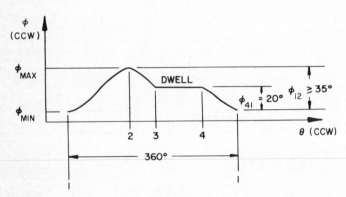

Fig. B

2. Using coupler curve of a four-link mechanism, design a six-bar linkage of the type of Fig. C. The required motion program is shown in Fig. D.

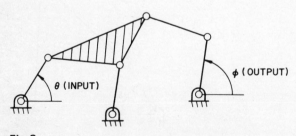

Fig. C

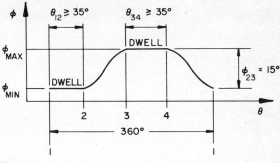

Fig. D

Performance Test #3

Design a six-link mechanism of the type shown in Fig. A to obtain a double stroke oscillating output according to the motion program given below.

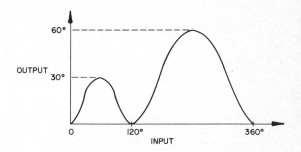

Supplementary Reference

10.

Unit XVI.   Coupler Cognate

Mechanisms

1. To design coupler cognate mechanisms of four-link and slider-crank mechanisms
2. To design six-link mechanisms for generation of parallel motion

Objective 1

To design coupler cognate mechanisms of four-link and slider-crank mechanisms

Activities
- Read the material provided below, and as you read:
  - a. Define a coupler cognate mechanism.
  - b. Construct a coupler cognate mechanism.
  - c. Construct Cayley's diagram.
  - d. State the properties of a coupler cognate mechanism.
- Check your answers to the competency items with your instructor.

Reading Material

Part I: Build all possible four-link mechanisms which draw identical coupler curves

1. Given: A source four-link mechanism $MABQ$ with C as coupler point on AB. (See Fig. 1.).
2. Required: To construct (as shown in Fig. 1), 2 four-link mechanisms $MGFO$ with coupler point C on GF and $QDEO$ with coupler point C on DE such that these mechanisms draw the same coupler curve as the source four-link mechanism $MABQ$ with C as coupler point on AB.
3. Construction:
   - a. Draw parallelograms $MACG$ and $QBCD$.
   - b. Construct triangles FGC and ECD similar to triangle CAB, that is,

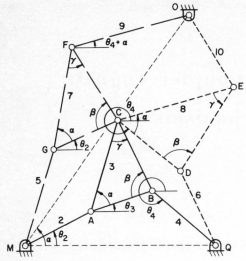

Fig. 1

$$\frac{GF}{GC} = \frac{AC}{AB} = \frac{CE}{CD}$$

and angle FGC = angle ECD = angle $CAB = \alpha$.
Check your construction by observing

Angle GFC = angle CED = angle $ACB = \gamma$
Angle FCG = angle CBA = angle $EDC = \beta$

c. Construct parallelogram $FCEO$. $MGFO$ and $QDEO$ are the left and right cognate mechanisms to the source mechanism $MABQ$.

4. Important Results:
    a. Check your construction to examine if angle OMQ = angle CAB and

$$\frac{MO}{MQ} = \frac{AC}{AB}$$

    b. Links of the three cognate mechanisms are numbered as shown in Fig. 1. Let ω_i describe the angular velocity of the ith link. Because of the parallelograms $MACG$, $QBCD$, and $FCEO$, $\omega_2 = \omega_7 = \omega_{10}$, $\omega_5 = \omega_3 = \omega_6$, $\omega_4 = \omega_8 = \omega_9$.
    c. Cayley's diagram—Let the points M and Q in Fig. 1 be stretched so that A and B lie on MQ. Then G and F will lie on MO, and D and E will lie on QO to give Cayley's diagram shown in Fig. 2. Note that cognate construction is easy to understand from Cayley's diagram.
    d. The cognate construction for the case in which point C lies on BA extended can be studied from Cayley's diagram shown in Fig. 3. Note that as C lies on BA extended for the source mechanism, C lies on FG

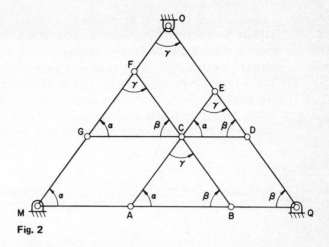

Fig. 2

extended for the cognate mechanism $MGFO$. But, point D of the cognate mechanism $QDEO$ lies on EC extended.

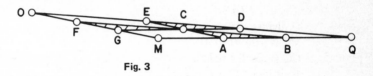

Fig. 3

Figure 4 shows the cognate construction for the case in which C lies on BA extended. Note that for the cognate mechanism $QDEO$, parallelogram $QBCD$ is drawn to locate point D. Then, points E and O are

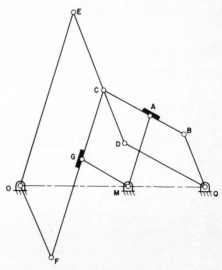

Fig. 4

located so that D-C-E is proportioned as B-A-C and Q-M-O. For the cognate mechanism $MGFO$, parallelogram $MACG$ is drawn to locate G. Point F is located so that C-G-F is proportioned as B-A-C. If point F is correctly located, then $OFCE$ must be a parallelogram.

e. The cognate construction for the case in which point C lies within AB can also be studied from Cayley's diagram. See Fig. 5. Note that as C lies within AB for the source mechanism, C lies on GF extended for cognate mechanism $MGFO$, and on DE extended for cognate mechanism $QDEO$.

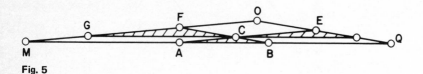

Fig. 5

Figure 6 shows the cognate construction for the case in which C lies within AB.

NOTE:

For cognate MGFO
    O divides MQ as C divides AB.
    MG is parallel to AC.
    GC is parallel to MA.
    F divides GC as C divides AB.

For cognate QDEO
    QD is parallel to CB.
    DC is parallel to QB.
    E divides CD as C divides AB.

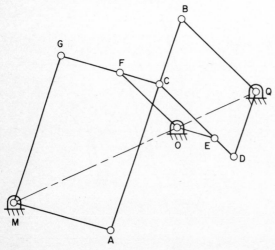

Fig. 6

Part II: Build all possible slider-crank mechanisms which draw
identical coupler curves

1. Given: A source slider-crank mechanism MAB with a slider at B. The slider is sliding along axis x-x. Coupler point C is selected on AB. Angle CAB = α. See Fig. 7.

2. Required: To construct, as shown in Fig. 7, an alternate slider-crank mechanism MGF with a coupler point C such that the constructed mechanism draws the same coupler curve as the source mechanism.

3. Construction:

    a. Construct parallelogram $MACG$.

    b. Construct triangle FGC similar to triangle CAB. That is, angle FGC = angle CAB = α and

    $$\frac{GF}{GC} = \frac{AC}{AB}$$

    c. Draw axis y-y passing through F and making angle α with x-x axis. Line y-y is the axis of the slider at F.

    MGF is the cognate mechanism to the source mechanism MAB.

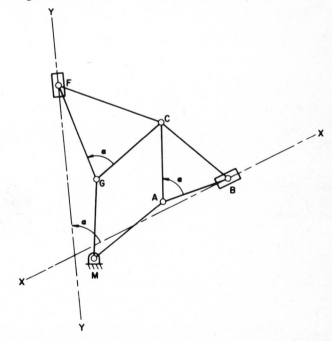

Fig. 7

Part III. Proof for the existence of the cognate four-link
mechanism

Using the cognate construction we have located pivot point O. If we prove that this pivot point O is stationary with reference to fixed pivots M and Q while the three mechanisms draw identical coupler curves with the common coupler point C, then we have established the existence of the cognate mechanisms.

To prove that point O is stationary let us examine Fig. 1. Let

$$k = \frac{AC}{AB} = \frac{GF}{GC} = \frac{CE}{CD}$$

From Fig. 1c we observe that:

$$\mathbf{MO} = \mathbf{MG} + \mathbf{GF} + \mathbf{FO}$$

But

$$\mathbf{MG} = \mathbf{AC} = kABe^{j(\theta_3 + \alpha)}$$

$$\mathbf{GF} = kGCe^{j(\theta_2 + \alpha)} = kMAe^{j(\theta_2 + \alpha)}$$

and

$$\mathbf{FO} = \mathbf{CE} = kCD\, e^{j(\theta_4 + \alpha)} = kBQe^{j(\theta_4 + \alpha)}$$

Hence,

$$MO = k\, e^{j\alpha}(ABe^{j\theta_3} + MAe^{j\theta_2} + BQe^{j\theta_3}) = k\, e^{j\theta}MQ$$

Since k, α, and MQ are of fixed quantity for a source four-link mechanism $MABQ$, point O is in fixed relationship with MQ. We observe that,

$$\frac{MO}{MQ} = k$$

and

$$\text{angle } OMQ = \alpha$$

Competency Items
- Construct the coupler cognate mechanism of the source four-link mechanism shown below:

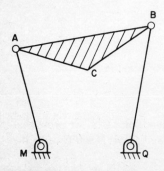

- Identify the type of cognate mechanism.
- Construct the coupler cognate mechanism of the source slider-crank mechanism shown to the right.

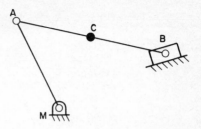

Objective 2

To design six-link mechanisms for generation of parallel motion

Activities

- Read the material provided below, and as you read:
  a. Synthesize six-link mechanisms for parallel motion generation.
  b. Demonstrate the practical application of the cognate mechanisms.
- Check your answers to the competency items with your instructor.

Reading Material

Figure 8 shows a source four-link mechanism $MABQ$ and its two cognate mechanisms $MGFO$ and $QDEO$. Let us now examine the source mechanism $MABQ$ and the cognate mechanism $QDEO$. Point C in each mechanism will trace the same coupler curve. Because of the cognate construction, angular velocities of links MA and OE are equal.

Let us now draw line x-x passing through Q and parallel to line MO. Locate point Q' on line x-x, so that $QQ' = OM$. At M draw ME' parallel to OE so that $ME' = OE$. At Q' draw line $Q'D'$ parallel to QD so that $Q'D' = QD$. At E' draw line $E'C'$ parallel to EC so that $E'C' = EC$.

The above construction has permitted us to reconstruct the cognate mechanism $QDEO$ in its new location $Q'D'E'O'$. Its coupler point C now occupies location C'. (O' and M' are coincident points.)

Since the new location of $O'E'D'Q'$ is obtained by moving $OEDQ$ parallel to itself and parallel to lines OM and QQ', CC' must be parallel to OM.

Points C of the source mechanism $MABQ$ and C' of the shifted cognate mechanism $O'E'D'Q'$ will continue to draw identical coupler curves at the same rate, provided links MA and $O'E'$ are rotating with the same angular velocity. Since the cognate construction does satisfy the equality of angular velocities, links MA and $O'E'$ can be combined to form ternary link MAE'.

Since C and C' are now tracing the coupler curve at the same rate, CC' is of fixed length. Hence we can connect CC' and remove the redundant links QD' to obtain Watt's six-link mechanism $MABQ - CC'E'$.

The designed six-link mechanism is shown in Fig. 9. Because of the construction, link CC' will always move parallel to itself. Note that any point

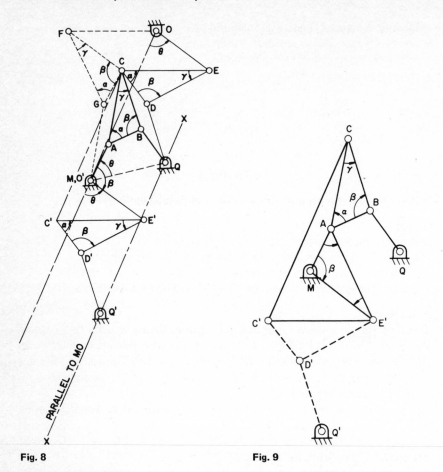

Fig. 8   Fig. 9

selected as a coupler point on link CC' will also draw the same curve as the points C and C' draw.

Competency Items

• Design a six-link mechanism for generation of parallel motion from the mechanism shown below.

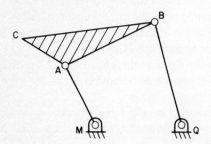

• Build a cardboard model to demonstrate the generation of parallel motion for the mechanism just designed.

Performance Test #1

The mechanism shown in the figure below is used to urge sludge along a sluice. The total assembly is an eight-link mechanism. Redesign this mechanism so that the total assembly is a Watt's six-link mechanism with two fixed pivots.

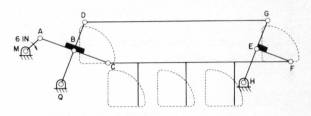

Performance Test #2

A box of diapers is to be displaced from position 1 to position 2. At these positions, the box is expected to have instantaneous dwell. Design a six-link mechanism to guide the box on a parallel path.

Performance Test #3

Design a linkage suspension for a trailer so that it is able to maintain its horizontal level even when two wheels are on an uneven surface.

Hint: Select a four-link mechanism with a coupler point that draws a straight line and design a six-link mechanism for parallel motion.

Performance Test #4

1. Obtain all possible coupler cognate mechanisms of geared five-bar mechanism with gear ratio +1.
2. Show how the results of coupler cognate mechanisms of a four-link mechanism can be applied in synthesizing a four-link mechanism for coordinating displacements of its input and output links.

Performance Test #5

Obtain a coupler cognate mechanism of a geared five-bar where the gears are mounted in the fixed link and one of the two coupler links to obtain constrained motion.

Supplementary Reference

2(6.4, 6.5)

Answers

Unit I

Performance Test #1—Objectives 1, 2, 3

1. a. A kinematic pair consisting of the elements, a spherical ball and a slotted cylinder, has three degrees of freedom. There are three constraints placed on the motion of the spherical ball when it is constrained to move in a slotted cylinder.

    b. This pair is capable of exerting motion which has three components consisting of two rotational motions about two independent axes and one translational motion along an axis.

2. A kinematic pair with four degrees of constraints has two degrees of freedom. From Table I we observe that there are three kinematic pairs which have two degrees of freedom. These are
    - a cylinder pair
    - a cam pair
    - a slotted spherical pair

3. a. From the geometry of kinematic chains, we observe:
    - number of binary links = 4
    - number of ternary links = 4
    - total number of binary and ternary links = 8
    - number of revolute pairs = P_1 = 10
    - number of loops = 3

    b. From the geometry of kinematic chains we observe:
    - number of binary links = 6
    - number of ternary links = 5
    - total number of binary and ternary links = 11
    - number of revolute pairs = P_1 = 14
    - number of loops = 4

4. a. Let us number the links and joints as shown in the figure. Note that links 8, 9, and 10 are connected using double joints 7 and 8. Also, links 3, 4, and 5 are connected using double joints 9 and 10. From the geometry of the mechanism, we count $N = 10$, $P_1 = 13$, $P_2 = 0$. Application of Gruebler's mobility criterion yields

$$
\begin{aligned}
F &= \text{degrees of freedom of mechanism} \\
&= 3(N - 1) - 2P_1 - P_2 \\
&= 3(10 - 1) - 2 \times 13 - 0 \\
&= 27 - 26 \\
&= 1
\end{aligned}
$$

Since $F = 1$, the linkage will have constrained motion when any one of the nine moving links are driven using an external force.

b. Let us number the links and joints as shown in the figure. From the geometry of the mechanism we count $N = 9$, $P_1 = 12$, $P_2 = 0$.

Application of Gruebler's mobility criterion yields

$$F = 3(N - 1) - 2P_1 - P_2$$
$$= 3(9 - 1) - 2 \times 12 - 0$$
$$= 0$$

The linkage is a structure. Application of an external force to any one of the links 2 to 9 will not produce any relative motion between the links of the linkage.

5. The two kinematic chains have one spring and one cam-follower. The rules in Objective 3 state that a spring is equivalent to two binary links connected by a revolute pair. Also, a cam-follower is equivalent to one binary link with revolute pairs at each end. The corresponding equivalent chains are shown in Fig. A1a and b.

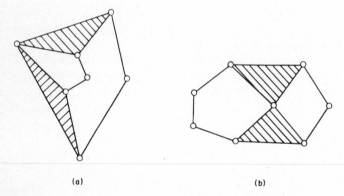

(a)   (b)

Fig. A1

6. Let us number the links as shown in Fig. A2a. We observe that the kinematic chain has six links and seven revolute pairs. The unique mechanisms are obtained by grounding one of the six links each time and by accepting those which are unique in their connectivities. Note that if we ground links 3 or 1, we only obtain one unique mechanism. Similarly, grounding links 2 or 4, we obtain only one unique mechanism. Finally, grounding links 5 and 6, we obtain only one unique mechanism. Thus, there are only three unique mechanisms even though the chain has six links. This is due to the fact that the geometry of the chain has a symmetry. The three inversions are shown in Fig. A2b–d.

Performance Test #2

1. A helical pair executes rotational and translation motion. However, its translational motion is dependent upon its rotational motion. The governing

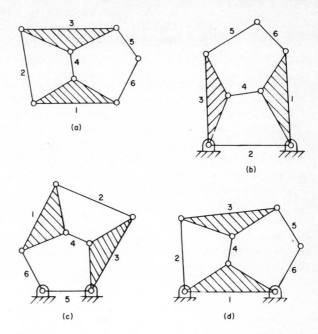

Fig. A2

relationship between rotational and translational motions defines the pitch value of the helical pair. Because of this relationship, a helical pair is attributed to have one degree of freedom.

2. a. A plane pair has two translational motions and one rotational motion.
    b. A slotted spherical pair has two rotational motions about two intersecting axes.
    c. A sphere groove pair has three rotational motions and one translational motion.

3. a. Since a plane pair has three degrees of freedom, it has three degrees of constraints.
    b. Since a slotted spherical pair has two degrees of freedom, it has four degrees of constraints.
    c. Since a sphere groove pair has four degrees of freedom, it has two degrees of constraints.

4. A higher kinematic pair has either a line or a point contact between its two contacting elements. From Table 1 we observe that all the kinematic pairs of classes IV and V are higher pairs.

5. Number the links of the linkage to count
    a. number of loops = 4
    b. number of binary links = 9
    c. number of ternary links = 2
    d. number of other links = 1
    e. number of joints = 15 (note that links 9, 10, and 12 are connected using double joints)
    f. number of degrees of freedom

Unit I

$$F = 3(N - 1) - 2P_1$$
$$= 3(12 - 1) - 2 \times 15$$
$$= 3$$

Since the linkage has three degrees of freedom, it will require three external inputs to produce a constrained motion.

6. A linkage or a mechanism is obtained by grounding any one of the links of a kinematic chain. A linkage is expected to have any number of degrees of freedom. A mechanism, however, has one degree of freedom.

7. Here we need to calculate degrees of freedom of the given linkage. For this we determine: number of links = 8, number of revolute pairs = 11, and apply Gruebler's mobility criterion to determine

$$F = 3(N - 1) - 2P_1$$
$$= 3(8 - 1) - 2 \times 11$$
$$= -1$$

Since the linkage has negative one degree of freedom, the linkage is a *super*structure.

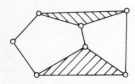

Fig. A3

8. The kinematic chain shown in Question 8 has one prism pair and one cam pair. Using the rules of Objective 3, we note that a prism pair can be replaced by one link with one revolute pair, and a cam pair can be replaced by a binary link with revolute pairs at each end. The resultant chain with all revolute pairs is shown in Fig. A3.

9. A spring is equivalent to two binary links connected by a revolute pair. A belt-pulley is equivalent to two binary links connected to a ternary link using revolute pairs. From the figure A4a, we observe that links 7 and 8 can be replaced by a spring, and links 3, 4, and 5 can be replaced by a belt-pulley. The resultant chain with a spring and a belt-pulley is shown in Fig. A4b.

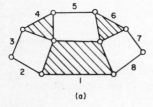 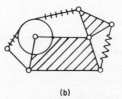

(a)   (b)

Fig. A4

Performance Test #3

1. A cylinder plane pair has two rotational and two translational motions.

2. a. Elbow appears to function as a spherical pair which has three rotational motions about three independent axes.
    b. The piston cylinder as it acts in an internal combustion engine is a prism pair which has only a translational motion.
3. Both pairs have two degrees of freedom and execute one rotational and one translational motion. However, in a cylinder pair, the axis about which it executes rotational motion is the same along which it executes translational motion. In a cam pair, the two axes are perpendicular to one another.
4. A lower kinematic pair has either area or surface contact with its two elements. Class III kinematic pairs have a spherical pair and a plane pair which are lower kinematic pairs.
5. There are no unique answers to this question. One of the several solutions is presented in Fig. A5. It has five binary links, four ternary links, and one quaternary link connected by using 14 revolute pairs.

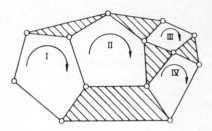

Fig. A5

6. Number the links of the linkage to count:
    a. number of binary links = 7
    b. number of ternary links = 2
    c. number of quaternary links = 2
    d. number of other links = 0
    e. number of joints = 15
    f. number of degrees of freedom

$$F = 3(N - 1) - 2P_1$$
$$= 3(11 - 1) - 2 \times 15$$
$$= 0$$

The linkage in Question 6 is a structure.
7. Let us number the links of the linkage to count

$$N = 5$$
$$P_1 = 5 \quad \text{(including revolute and prism pairs)}$$
$$P_2 = 1 \quad \text{(including a cam pair)}$$

Hence, the degrees of freedom are determined from

$$F = 3(N - 1) - 2P_1 - P_2$$
$$= 3(5 - 1) - 2 \times 5 - 1$$
$$= 1$$

The linkage in question will have a constrained motion when any one of the links is driven by an external force.

8. The linkage shown in Question 8 has two springs and a belt-pulley. Using the rules of the objective, we note that a belt-pulley can be replaced by a combination of a ternary link and two binary links. A spring can be replaced by a pair of binary links connected by a revolute pair. The resultant chain is shown in Fig. A6.

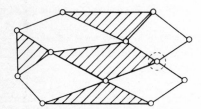

Fig. A6

9. The kinematic chain in Fig. A7a is symmetric about link 6. Hence, we will obtain identical inversions if we were to ground links 2, 1, 7, or 3, 4, 5. In addition, we will obtain a unique inversion when we ground link 6. Hence, a total of four unique mechanisms are possible. These are shown in Fig. A7b–e.

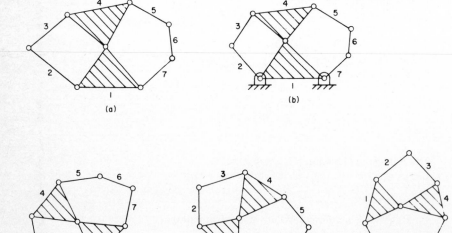

Fig. A7

Unit II

Performance Test #1–Objectives 1, 2, 3, 4

1. Velocity of the particle is obtained by taking time derivative of the path. Thus $V = dy/dt = (d/dt)\{1/2[(a^2 t^2) + at]\} = a^2 t + a$. Substituting $a = 1$ and $t = 0.5$, we get $V = 1.5$ units/sec.

2. Here again the velocity of the particle is obtained by taking time derivative of the path. Thus, $V = dy/dt = (d/dt)(a \sin 2\pi t) = 2\pi a \cos 2\pi t$. Substituting $a = 1$, $t = 0.5$, we get $V = -2\pi$ units/sec.

3. a. $\mathbf{R}_1 + \mathbf{R}_2 = 3e^{j(30°)} + 4e^{j(60°)}$
   $= 3[\cos 30° + j \sin 30°] + 4[\cos 60° + j \sin 60°]$
   $= [3 \cos 30° + 4 \cos 60°] + j[3 \sin 30° + 4 \sin 60°]$
   $= 4.60 + j3.5$

   Let $\mathbf{R}_1 + \mathbf{R}_2 = \mathbf{R}_3 = r_3 e^{j\theta_3}$ where $r_3 = (4.60^2 + 3.5^2)^{1/2} = 5.78$ and $\theta_3 = \tan^{-1}(3.5/4.6) = 33.41°$

   b. $\mathbf{R}_1 \times \mathbf{R}_2 = 3e^{j(30°)} \times 4e^{j(60°)}$
   $= 12e^{j(30° + 60°)} = 12e^{j(90°)}$
   $= 12[\cos 90° + j \sin 90°]$
   $= j12$

   c. $\mathbf{R}_1/\mathbf{R}_2 = [3e^{j(30°)}]/[4e^{j(60°)}]$
   $= 0.75e^{j(30 - 60°)}$
   $= 0.75e^{-j(30°)}$

4. Velocity of the particle $= l \cdot \omega = 5 \times 5 = 25$ in./sec.

5. From Fig. A8, we observe that $\mathbf{V} = \mathbf{V}_y$ and $\mathbf{V}_x = 0$. $\mathbf{V}_r$ and $\mathbf{V}_t$ components are obtained by projecting $\mathbf{V}$ along $\mathbf{R}$ and perpendicular to $\mathbf{R}$. By measuring the length of $\mathbf{V}_r$ and $\mathbf{V}_t$, we obtain their magnitudes thus $V_t = 8.77$ units/sec and $V_r = 4.8$ units/sec. Note that $V = (V_t^2 + V_r^2)^{1/2} = 10$ units/sec.

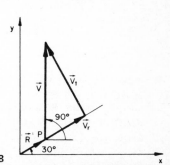

Fig. A8

6. The particle is located using the position vector $\mathbf{R} = 4\mathbf{U}_x + 8\mathbf{U}_y$. Hence the coordinates of the particle are $x = 4$ and $y = 8$. The y component of the

velocity of the particle is $dy/dt = x(dx/dt)$. Since $dx/dt = 2$ in./sec, the y component of the velocity of the particle at $x = 4$ is $y = 4 \times 2 = 8$ in./sec.

The total velocity V is the vectorial sum of x and y components. That is $V = (\dot{x}^2 + \dot{y}^2)^{1/2} = (4 + 64)^{1/2} = 8.25$ in./sec.

Performance Test #2—Objectives 1, 2, 3, 4

1. The velocity of the particle is obtained by taking time derivative. Thus, $dy/dt = 4a^3 t - 2\pi a^2 \cos 2\pi t$. For $a = 2$, and $t = 1$, $dy/dt = 4(2)^3(1) - (2)^2(2\pi)(1) = 6.87$ units/sec.

2. a. $R_1 - R_2 = 2e^{j(45°)} - 5e^{j(30°)}$
   $$= 2[\cos 45° + j \sin 45°] - 5[\cos 30° + j \sin 30°]$$
   $$= 2\left[\frac{\sqrt{2}}{2} + j\frac{\sqrt{2}}{2}\right] - 5\left[\frac{\sqrt{3}}{2} + j\frac{1}{2}\right]$$
   $$= -2.92 - j1.09$$

   Let $R_1 - R_2 = R_3 = r_3 e^{j\theta 3}$. Then $r_3 = (2.92^2 + 1.09^2)^{1/2} = 3.12$ units, $\theta = \tan^{-1}(1.09/2.92) = 28°$. Hence $R_3 = -3.12e^{j28°}$ or $R_3 = 3.12e^{j208°}$ since the negative sign in front of r_3 is equivalent $180°$ rotation of R_3.

   b. $R_2/R_1 = 5e^{j(30°)}/2e^{j(45°)} = 2.5e^{j(-15°)}$

3. a. From Fig. A9, we observe that $V_t = (V^2 - V_r^2)^{1/2} = (6^2 - 3^2)^{1/2} = 5.19$ in./sec.

   b. Also, from Fig. A9, $\theta = 180° + 60° + \tan^{-1}(V_t/V_r)$. But, $\tan^{-1}(V_t/V_r) = \tan^{-1}(5.19/3) = 60°$. Hence, $\theta = 300°$ and $V = 6e^{j300°}$.

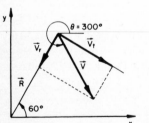

Fig. A9

4. Since $y = 4x^2 - x/2$, $dy/dx = 8x - (1/2)$. For $dy/dx = 31/2$, from the above relationship we get $x = 2$. Hence, when $x = 2$, $y = 4(2)^2 - (2/2) = 15$. The position vector R is obtained by the x and y coordinates of the particle. Hence, $R = 2U_x + 15U_y$. Velocity of the particle is obtained by taking time derivative. Thus, $dy/dt = 8x(dx/dt) - 1/2(dx/dt)$. Substituting $\dot{x} = 4$ in./sec, we get $\dot{y} = 8(2)(4) - 1/2(4) = 62$ in./sec. Hence, the velocity vector can be described as

$$V = \dot{R}_x U_x + \dot{R}_y U_y$$
$$= 4U_x + 62U_y.$$

Performance Test #3—Objectives 1, 2, 3, 4

1. Let $R_1 = -6e^{j25°}$
    $R_2 = 4e^{-j45°}$
    and $R_3 = R_1 \times R_2$
    $\qquad = (-6e^{j25°}) \times (4e^{-j45°})$
    $\qquad = -24e^{-j20°}$

    Let $R_3 = R_x U_x + R_y U_y$ where R_x and R_y are the x and y components. Hence

    $R_x = -24 \cos(-20°) = -22.6$ units
    $R_y = -24 \sin(-20°) = 8.23$ units

2. From the schematic diagram (Fig. A10), we note that $\phi + \theta = 330°$.

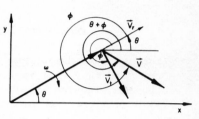

Fig. A10

$V_r = V \cos\phi$ or $\cos\phi = V_r/V = 5/10 = 1/2$. Hence $\phi = 60°$ or $300°$.

Since ω is in a clockwise direction, V_t must be in the direction shown, and $\phi = 300°$.

Since $\phi + \theta = 330°$ and $\phi = 300°$ then $\theta = 30°$.

Also $V_t = V \sin\phi = 10 \sin 300° = -8.66$ in./sec.

But $V_t = r \cdot \omega$ or $r = V_t/\omega = -8.66/-3.0 = 2.89$ in.

3. Since $y = 2x^3 - 54$

    $$\frac{dy}{dx} = \frac{dy/dt}{dx/dt} = \frac{\dot{R}_y}{\dot{R}_x} = 6x^2$$

    for $x = 3$, $(R_x = 3)$, $dy/dx = 54.0$

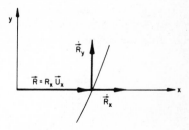

Fig. A11

    Hence $\dot{R}_x = \dfrac{\dot{R}_y}{dy/dx} = \dfrac{27}{54} = \dfrac{1}{2} = 0.5$

    When $x = 3$, $y = 2x^3 - 54 = 2(3)^3 - 54 = 0$.

    Since $R = R_x$, $R_x = r$, $\dot{R}_x = \dot{r}$ and $\dot{R}_y = V_t = r\omega$ then $\omega = \dot{R}_y/R_x = 27/3 = 9$ rad/sec.

    Figure A11 shows R, $\dot{R}_x$, $\dot{R}_y$.

4. Since the points A and B are moving relative to one another, $V_B = V_A + V_{B/A}$.
   Using this relationship, we draw the vector polygon shown below (Fig. A12). From the vector polygon we measure $V_A = 16.8$ in./sec.

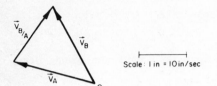

Scale: 1 in = 10 in/sec

Fig. A12

Performance Test #1—Objectives 5, 6, 7

1. We will work out the polygon technique. Using vector-velocity relationship, we note that $V_B = V_A + V_{B/A}$.
   The steps to construct vector polygon lead to the determination of angular velocity of link AB and QB. Since AB is perpendicular to MA and QB, the magnitude of $V_{B/A}$ is zero. From Fig. A13, we note that

$$\omega_{AB} = 0/2 \text{ in.} = 0 \text{ rad/sec}$$
$$\omega_{BQ} = V_B/BQ = 20 \text{ rad/sec (counterclockwise)}$$

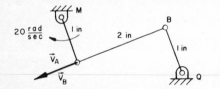

Fig. A13

2. The vector-velocity relationship to determine V_A is given by $V_A = V_B + V_{A/B}$. The magnitude and direction V_B are known. The directions of $V_{A/B}$ and V_A are perpendicular to BA and MA. By using this information, we draw the vector polygon shown in Fig. A14.

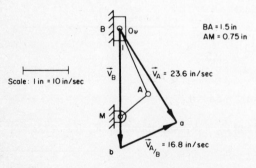

BA = 1.5 in
AM = 0.75 in

$V_A = 23.6$ in/sec

Scale: 1 in = 10 in/sec

$V_{A/B} = 16.8$ in/sec

Fig. A14

From the vector polygon, we calculate

$$\omega_{AB} = \frac{V_{A/B}}{AB} = \frac{16.8}{1.5} = 11.2 \text{ rad/sec (clockwise)}$$

$$\omega_{AM} = \frac{V_A}{AM} = \frac{23.6}{0.75} = 31.5 \text{ rad/sec (counterclockwise)}$$

3. Label the links MA and QA as links 2 and 3. We note that point A lies on links 2 and 3. The vector-velocity relationship to determine V_{A_3} is

$$V_{A_3} = V_{A_2} + V_{A_3/A_2}$$

The direction of V_{A_3} is perpendicular to QA. The direction of V_{A_3/A_2} is along QA. The direction of V_{A_2} is perpendicular to MA

$$V_{A_2} = MA \cdot \omega_2 = 1 \times 50 = 50 \text{ in./sec}$$

Since MA is perpendicular to QA

$$V_{A_2/A_3} = V_{A_2} \quad \text{and} \quad V_{A_3} = 0$$

Hence,

$$\omega_{BQ} = \frac{V_{A_3}}{AQ} = 0 \text{ rad/sec}$$

and

$$V_{A_2/A_3} = 50 \text{ in./sec along } AQ.$$

4. For the slider-crank mechanism, we determine velocity V_B using the relationship $V_B = V_A + V_{B/A}$. Let the scale be 1 in. = 10 in./sec. The velocity polygon using the vector relationship $V_A = MA \cdot \omega_2 = 1 \times 20 = 20$ in./sec is shown in Fig. A15b. The velocity V_C of point C is obtained using the vector relationship $V_C = V_B + V_{C/B}$. The velocity V_D of point D is obtained using the vector relationship $V_D = V_C + V_{D/C}$.

From the vector polygon, we calculate

$$\omega_{ABC} = \frac{V_{B/A}}{AB} = 6.5 \text{ rad/sec (clockwise)}$$

$$\omega_{CD} = \frac{V_{D/C}}{CD} = \frac{4.5}{1.36} = 3.31 \text{ rad/sec (clockwise)}$$

$$\omega_{DQ} = \frac{V_D}{DQ} = \frac{15}{0.7} = 21.4 \text{ rad/sec (counterclockwise)}$$

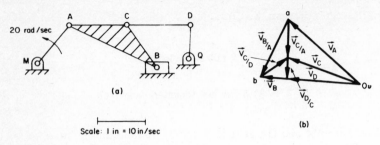

(a)

Scale: 1 in = 10 in/sec

(b)

Fig. A15

Performance Test #2—Objectives 5, 6, 7

1. The vector-velocity relationship to determine V_A is $V_A = V_B + V_{A/B}$, $V_B = QB \cdot \omega_4 = 2 \times 40 = 80$ in./sec.

   To draw the vector polygon, let the scale be 1 in. = 40 in./sec. The velocity V_B acts perpendicular to QB. V_A is perpendicular to MA. $V_{A/B}$ acts perpendicular to BA. The velocity polygon is shown in Fig. A16b.

   From the vector polygon, we measure $V_A = 86$ in./sec and $V_{A/B} = 115$ in./sec. Hence $\omega_3 = V_{A/B}/AB = 115/3 = 38.4$ rad/sec (counterclockwise). Since the point C divides AB, V_C will divide also. From the vector polygon we measure $V_C = 60$ in./sec.

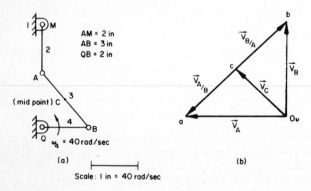

AM = 2 in
AB = 3 in
QB = 2 in

(mid point) C

ω_4 = 40 rad/sec

(a)

(b)

Scale: 1 in = 40 rad/sec

Fig. A16

2. Let us number the links as shown. The point B lies on link 3 and also on link 4 which is perpendicular to link 3. We can write the following vector-velocity relationship

$$V_{B_3} = V_A + V_{B_3/A}$$

$$V_A = QA \cdot \omega_2 = 20 \times 3 = 60 \text{ in./sec}$$

V_A acts perpendicular to QA. V_{B_3} acts along AB and $V_{B_3/A}$ acts perpendicular to AB. The velocity polygon (Fig. A17b) is drawn using the scale 1 in. = 40 in./sec.

From the vector polygon, we measure

V_{B_3} = 52 in./sec
V_C = 56 in./sec
$V_{B_3/A}$ = 29 in./sec

Hence

$$\omega_3 = \frac{V_{B_3/A}}{AB} = \frac{29}{6} = 4.8 \text{ rad/sec (clockwise)}$$

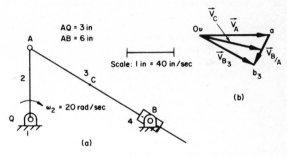

AQ = 3 in
AB = 6 in

Scale: 1 in = 40 in/sec

(b)

ω_2 = 20 rad/sec

(a)

Fig. A17

3. We know V_M. We are required to determine V_A, V_B, V_C, and angular velocities ω_3, ω_4, and ω_5. The vector-velocity relationships are

$V_A = V_M + V_{A/M}$
$V_B = V_A + V_{B/A}$
$V_C = V_B + V_{C/B}$

The vector polygon (Fig. A18b) is drawn using the scale 1 in. = 30 in./sec.

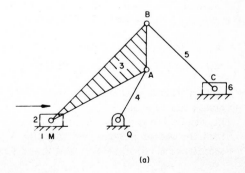

(a)

Fig. A18

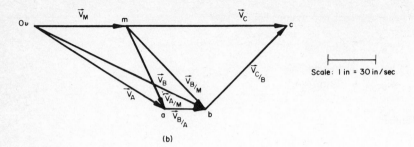

(b)

Fig. A18

From the vector polygon, we measure

V_A = 99 in./sec
V_B = 121.5 in./sec
V_C = 162 in./sec
$V_{B/A}$ = 25.2 in./sec
$V_{C/B}$ = 74.5 in./sec

From these results we calculate

$$\omega_3 = \frac{V_{B/A}}{AB} = \frac{25.2}{1.0} = 25.2 \text{ rad/sec (clockwise)}$$

$$\omega_4 = \frac{V_A}{AQ} = \frac{99.0}{1.2} = 82.5 \text{ rad/sec (clockwise)}$$

$$\omega_5 = \frac{V_{C/B}}{BC} = \frac{74.5}{2.0} = 37.3 \text{ rad/sec (counterclockwise)}$$

Performance Test #3—Objectives 5, 6, 7

1. To calculate the angular velocity ω_2 of link 2, we must first determine V_A. For this purpose, we write the vector-velocity relationship

$$V_A = V_B + V_{A/B}$$
$$V_B = QB \cdot \omega_4 = 2 \times 12 = 24 \text{ in./sec}$$

V_B acts perpendicular to QB. $V_{A/B}$ acts perpendicular to BA. V_A acts perpendicular to MA.

The vector polygon (Fig. A19b) is drawn using the scale 1 in. = 10 in./sec.
From the vector polygon, we measure V_A = 16.8 in./sec and V_B = 16.8 in./sec. Hence $\omega_2 = V_A/MA = 16.8/2 = 8.4$ rad/sec (clockwise).

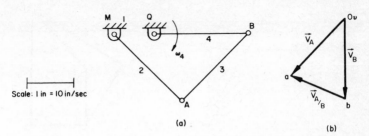

(a)

(b)

Fig. A19

Scale: 1 in = 10 in/sec

2. Velocity V_B is determined using the vector-velocity relationship $V_B = V_A + V_{B/A}$. The vector polygon (Fig. A20b) is drawn using the scale 1 in. = 2 in./sec. The magnitudes of V_B and $V_{B/A}$ are determined by directly measuring the lengths of V_B and $V_{B/A}$ from the vector polygon. Thus V_B = 4.23 in./sec and $V_{B/A}$ = 4.00 in./sec.

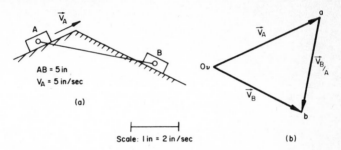

AB = 5 in
V_A = 5 in/sec

(a)

(b)

Scale: 1 in = 2 in/sec

Fig. A20

3. In order to find ω_5, we must determine V_C. The point C lies on links 3 and 5. Hence,

$$V_{C_3} = V_B + V_{C_3/B}$$

or $V_{C_3} = V_A + V_{C_3/A}$

$$V_{C_5} = V_{C_3} + V_{C_5/C_3}$$

and

The direction of V_{C_5} is perpendicular to DE. The direction of V_{C_5/C_3} is along DE. The complete velocity polygon is shown in Fig. A21b. It is drawn using the scale 1 in. = 10 in./sec.

From the vector polygon we calculate $\omega_5 = V_{C_5}/DC = 22.5/2.28 = 9.87$ rad/sec (clockwise) and V_{C_5/C_3} = 15.0 in./sec.

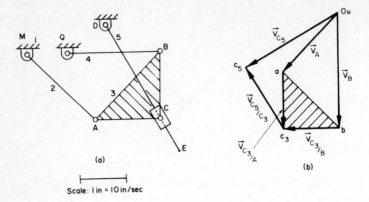

(a)

Scale: 1 in = 10 in/sec

Fig. A21

Unit III

Performance Test #1–Objectives 1, 2, 3

1. Since

$$\mathbf{R} = (x + 2)e^{j\theta}$$
$$\mathbf{V} = d\mathbf{R}/dt = (d/dt)[(x + 2)e^{j\theta}]$$
$$= \dot{x}e^{j\theta} + j(x + 2)\omega e^{j\theta}$$

and

$$\mathbf{A} = d\mathbf{V}/dt = (d/dt)[\dot{x}e^{j\theta} + j(x + 2)\omega e^{j\theta}]$$
$$= \ddot{x}e^{j\theta} + j\dot{x}\omega e^{j\theta} + j\dot{x}\omega e^{j\theta} + j(x + 2)\alpha e^{j\theta} - (x + 2)\omega^2 e^{j\theta}$$
$$= [\ddot{x} - (x + 2)\omega^2]e^{j\theta} + j[2\dot{x}\omega + (x + 2)\alpha]e^{j\theta}$$

Hence, the radial and transverse components are

$$\mathbf{A}' = [\ddot{x} - (x + 2)\omega^2]e^{j\theta}$$
$$\mathbf{A}^t = j[2\dot{x}\omega + (x + 2)\alpha]e^{j\theta} = [2\dot{x}\omega + (x + 2)\alpha]e^{j(90° + \theta)}$$

The horizontal and the vertical components of $\mathbf{A}$ are determined by substituting for $e^{j\theta} = \cos\theta + j\sin\theta$ and separating the real and the complex parts. Thus,

$$\mathbf{A} = [\ddot{x} - (x + 2)\omega^2]\cos\theta - [2\dot{x}\omega + (x + 2)\alpha]\sin\theta$$
$$+ j[\ddot{x} - (x + 2)\omega^2]\sin\theta + \{2\dot{x}\omega + (x + 2)\alpha\}\cos\theta]$$

Hence,

$$\mathbf{A}_x = \{[\ddot{x} - (x + 2)\omega^2]\cos\theta - [2\dot{x}\omega + (x + 2)\alpha]\sin\theta\}e^{j(0°)}$$
$$\mathbf{A}_y = [\ddot{x} - (x + 2)\omega^2]\sin\theta + \{2\dot{x}\omega + (x + 2)\alpha\}\cos\theta]e^{j(90°)}$$

2. Since $y = (25 - x^2)^{1/2}$, we get $x^2 + y^2 = 25$. Hence, the particle is moving on a circle with radius equal to 5 units. At $x = 2.5$, $y = [25 - (2.5)^2]^{1/2} = 4.33$. Hence, $\tan\theta = 4.33/2.5 = 1.732$ or $\theta = \tan^{-1}(1.732) = 60°$.

The radial and the transverse components are calculated using

$$\mathbf{A}' = (\ddot{r} - r\omega^2)e^{j\theta} = -5 \times (2)^2 e^{j60°} = -20e^{j60°} = -20e^{j60°} \text{ units/sec}^2$$
$$\mathbf{A}^t = (2\dot{r}\omega + r\alpha)e^{j(90° + \theta)} = (5) \times (-2) e^{j(150°)} = -10e^{j150} \text{ units/sec}^2$$

(Note that for the particle moving on a circle, $\dot{r} = \ddot{r} = 0$.)

$$\mathbf{A}'_x = -20\cos 60° = -10 \text{ units/sec}^2$$
$$\mathbf{A}'_y = -20\sin 60° = -17.32 \text{ units/sec}^2$$
$$\mathbf{A}^t_x = -10\cos 150° = 5 \text{ units/sec}^2$$
$$\mathbf{A}^t_y = -10\sin 150° = -8.66 \text{ units/sec}^2$$

Hence,

$$A_x = A'_x + A'_x = -10 + 5 = -5 \text{ units/sec}^2$$
$$A_y = A'_y + A'_y = -17.32 - 8.66 = -25.98 \text{ units/sec}^2$$

3. The acceleration of a particle moving on a path is given by $A = \ddot{s}\mathbf{U}_t + (\dot{s}^2/\rho)\mathbf{U}_n$ where ρ is the radius of curvature of the path at the position of the particle. Since $x = \pi/4$, $y = 2\sin(\pi/2) = 2$, $dy/dx = 4\cos 2x = 0$, $d^2y/dx^2 = -8\sin 2x = -8$. Hence,

$$\rho = \frac{[1 + (dy/dx)^2]^{3/2}}{d^2y/dx^2} = \frac{[1 + 0]^{3/2}}{-8} = -0.125 \text{ units}$$

and

$$\mathbf{A} = (0)\mathbf{U}_t - \left(\frac{100}{0.125}\right)\mathbf{U}_n$$

From the above we note $\mathbf{A}^t = (0)\mathbf{U}_t$ and $\mathbf{A}^n = (-800)\mathbf{U}_n$ units/sec^2. Similarly we calculate at $x = \pi/2$ and $\mathbf{A}^t = (0)\mathbf{U}_t$ and $A^n = (0)\mathbf{U}_n$.

4. If we assume that A is moving relative to B, then $\mathbf{A}_A = \mathbf{A}_B + \mathbf{A}_{B/A}$. If $A_B = 0$, then $\mathbf{A}_A = \mathbf{A}_{B/A} = A'_{B/A} + A'_{B/A}$. Where $A'_{B/A} = A_A \cos 45° = 282 \times 0.707 = 197.4$ units/sec^2 and $A'_{B/A} = A_A \sin 45° = 282 \times 0.707 = 197.4$ units/sec^2; but $r\omega^2 = 197.4 = A'_{B/A}$. Hence $\omega = (197.4/2)^{1/2} = 9.93$ rad/sec. Also, $r\alpha = A'_{B/A} = 197.4$; hence $\alpha = 197.4/2 = 98.7$ rad/sec^2 (clockwise). Since $\mathbf{A}_B$ is assumed to be zero, $\mathbf{V}_B$ must remain constant.

Performance Test #2—Objectives 1, 2, 3

1. Since

$$\dot{y} = Kt^2 - 3$$
$$\ddot{y} = 2Kt$$

at $t = 5$ sec, $\ddot{y} = A = 60$ in./sec$^2 = 2K(5)$. Hence $K = 60$ in./sec^2/(2)(5 sec) = 6 in./sec^3. Displacement is obtained by integrating $\dot{y}$. Thus $y = (6t^3/3) - 3t + C$ where C is the integrating constant. Its value is determined by using the conditions that at $t = 0$, $y = 4$ in. $= (6/3)(0)^2 - 3(0) + C$. Hence $C = 4$ in. and $y = 2t^3 - 3t + 4$. At $t = 5$,

$$y = 2(5)^3 - 3(5) + 4$$
$$= 250 - 15 + 4$$
$$= 239 \text{ in.}$$

2. The acceleration components of the particle are obtained by finding $\dot{\mathbf{R}}$ and $\ddot{\mathbf{R}}$

$$\mathbf{R} = (x - 1)e^{j\theta}$$
$$\dot{\mathbf{R}} = \dot{x}e^{j\theta} + j(x - 1)\omega e^{j\theta}$$
$$\ddot{\mathbf{R}} = \ddot{x}e^{j\theta} + j\dot{x}\omega e^{j\theta} + j\dot{x}\omega e^{j\theta} + j(x - 1)(\alpha e^{j\theta} - j\omega^2 e^{j\theta})$$

From above, the radial and the transverse components of $\ddot{R}$ are

$$\ddot{R}^r = [\ddot{x} + (x - 1)\omega^2]e^{j\theta}$$
$$\ddot{R}^t = [(x - 1)\alpha + 2\dot{x}\omega]e^{j(90° + \theta)}$$

where

$$\ddot{R} = \ddot{R}^r + \ddot{R}^t$$

The horizontal and the vertical components of $\ddot{R}$ are

$$\ddot{R}_x = [(\ddot{x} + (x - 1)\omega^2)\cos\theta - \{(x - 1)\alpha + 2\dot{x}\omega\}\sin\theta]e^{j(0°)}$$
$$\ddot{R}_y = [\{\ddot{x} + (x - 1)\omega^2\}\sin\theta + \{(x - 1)\alpha + 2\dot{x}\omega\}\cos\theta]e^{j(90°)}$$

3. The normal and tangential ocmponents of **A** are given by

$$A = \ddot{s}U_t + \frac{\dot{s}^2}{\rho}U_n = A_n + A_t.$$

For

$$y = \frac{1}{2}x^2 + 2x$$

$$\frac{dy}{dx} = x + 2,\ \text{at}\ x = 1,\ \frac{dy}{dx} = 1 + 2 = 3$$

$$\frac{d^2y}{dx^2} = 1$$

Since

$$\rho = \frac{[1 + (dy/dx)]^{3/2}}{d^2y/dx^2}$$

At $x = 1$,

$$\rho = \frac{[1 + (3)^2]^{3/2}}{1} = 10^{3/2}$$

Hence

$$A_n = \frac{\dot{s}^2}{\rho}U_n = \frac{(10)^2}{(10)^{3/2}} = 3.16\,U_n\ \text{units/sec}^2$$

$$A_t = \ddot{s}U_t = 20\,U_t\ \text{units/sec}^2$$

4. We observe that $A_A = A_r + A_n$ where

$$A_t = A\sin30° = 100 \times 0.5 = 50\ \text{ft/sec}^2$$
$$A_n = A\cos30° = 100 \times 0.866 = 86.6\ \text{ft/sec}^2$$

Since $A_n = \omega^2 \cdot QA$

$$\omega = (A_n/QA)^{1/2}$$
$$= (86.6/10)^{1/2}$$
$$= 2.94 \text{ rad/sec (clockwise)}$$

and

$$V_A = \omega \cdot QA$$
$$= 2.94 \times 10$$
$$= 29.4 \text{ ft/sec}$$

Hence

$$\mathbf{V}_A = 29.4 \, e^{j(90°)} \text{ ft/sec}$$

Performance Test #3—Objectives 1, 2, 3

1. Since

$$y = \frac{1}{3}ct^4 + bt$$
$$V = dy/dt = \frac{4}{3}ct^3 + b$$

and

$$A = d^2y/dt^2 = 4ct^2$$

at $t = 3$, and for $c = 2$,

$$A = 4ct^2 = 4 \times 2 \times (3)^2 = 72 \text{ in./sec}^2$$

2. a. Since $y = (x^2/2) - \frac{3}{2}$, at $x = \sqrt{3}$, $y = \frac{1}{2}(\sqrt{3})^2 - \frac{3}{2} = 0$. Hence

$$r = R_x = \sqrt{3} \quad \text{and} \quad \theta = 0.$$

   b. Tangential component of acceleration is $\ddot{s} = 50\sqrt{3}$. Normal component is $\dot{s}^2/\rho$. But

$$\rho = \frac{[1 + (dy/dx)^2]^{3/2}}{d^2y/dx^2} = \frac{[1 + (\sqrt{3})^2]^{3/2}}{1} = 8 \text{ in.}$$

   Hence

$$\dot{s}^2/\rho = (20)^2/8 = 400/8 = 50 \text{ in./sec}^2$$

   c. The path $y = \frac{x^2}{2} - \frac{3}{2}$ is plotted in Fig. A22 shown below. We note that $R = \sqrt{3}\,U_x + (0)U_y$. Let U_t and U_n be the unit vectors tangent to and normal to the path at point P. Let A be the acceleration of point P. Let γ, ψ, and β be the angles as shown. Then

$$\beta = \tan^{-1}(dy/dx) = \tan^{-1}(\sqrt{3}) = 60°$$
$$\gamma = \tan^{-1}[(\dot{s}^2/\rho)/\ddot{s}] = \tan^{-1}[50/(50\sqrt{3})] = 30°$$

and

$$\psi = \gamma + \beta$$
$$= 30° + 60°$$
$$= 90°$$

Hence $A_y = A$ and $A_x = 0$ where $\mathbf{A} = A_x\mathbf{U}_x + A_y\mathbf{U}_y$. We can now calculate the magnitude of A using

$$A = \frac{\dot{s}^2/\rho}{\sin\gamma} = \frac{50}{0.5} = 100 \text{ in./sec}$$

Hence $\mathbf{A} = 100\ \mathbf{U}_y$.

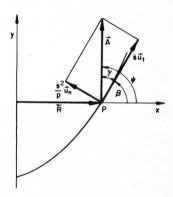

Fig. A22

d. Figure A23 shows the path of the particle and radial and transverse components of its velocity. We observe that $\phi = 180° + \beta = 240°$. We note that

$$V_r = \dot{r} = V\cos\phi$$
$$= \dot{s}\cos\phi$$
$$= 20\cos 240°$$
$$= -10 \text{ in./sec}$$

$$V_t = \dot{s}\sin\phi$$
$$= 20\sin 240°$$
$$= 20(-\sqrt{3}/2)$$
$$= -17.32 \text{ in./sec}$$

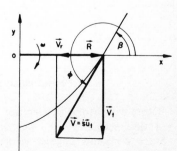

Fig. A23

Hence $\omega = V_t/r = -(17.32/1.73) = -10$ rad/sec (clockwise).

e. The radial and transverse components of acceleration $\mathbf{A}$ of particle P are shown in Fig. A24. From the figure, we observe $A_t = A = 100$ in./sec^2 and $A_r = 0$.

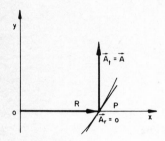

Fig. A24

f. Since

$$A_t = r\alpha + 2\dot{r}\omega$$

$$\alpha = \frac{A_t - 2\dot{r}\omega}{r} = \frac{100 - (2)(-10)(10)}{1.73}$$

$$= 57.80 \text{ rad/sec}^2 \text{ (counterclockwise)}.$$

g. Since

$$A_r = \ddot{r} - \omega^2 r = 0$$

$$\ddot{r} = \omega^2 r = 100 \times 1.73 = 173 \text{ in./sec}^2.$$

Figure A25 shows all the four components of $\mathbf{A}$ of particle P.

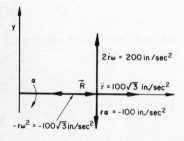

Fig. A25

Performance Test #1—Objectives 4, 5

1. Before we perform acceleration analysis we need to perform velocity analysis of the mechanism and determine angular velocities of links AB and QB. Since we know angular velocity of link MA, we begin with writing the

velocity vector relationship.

$$V_B = V_A + V_{B/A}$$
$$V_A = MA \cdot \omega_{MA} = 1.25 \times 20 = 25 \text{ in./sec}$$

Figure A26b shows the velocity polygon which is drawn using the scale 1 in. = 20 in./sec.

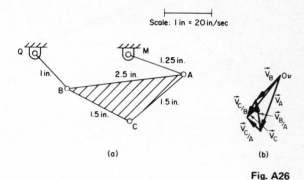

Scale: 1 in = 20 in./sec

(a)

(b)

Fig. A26

From the vector polygon, we determine

$$V_{B/A} = 13 \text{ in./sec}$$
$$V_B = 14.5 \text{ in./sec}$$
$$V_{C/A} = 8 \text{ in./sec}$$
$$V_{C/B} = 8 \text{ in./sec}$$
$$V_C = 22.5 \text{ in./sec}$$

$$\omega_{AB} = V_{B/A}/AB = 13/2.5 = 5.2 \text{ rad/sec}$$
$$\omega_{BQ} = V_B/QB = 14.5/1 = 14.5 \text{ rad/sec}$$

The acceleration analysis is performed by drawing the acceleration polygon using the following vector relationship

$$A_B^r + A_B^t = A_A^r + A_A^t + A_{B/A}^r + A_{B/A}^t$$

We can calculate the following quantities.

$$A_A^r = (MA)(\omega_{MA})^2 = 1.25 \times (20)^2 = 500 \text{ in./sec}^2$$
$$A_{B/A}^r = (AB)(\omega_{AB})^2 = 2.5 \times (5.2)^2 = 67.6 \text{ in./sec}^2$$
$$A_B^r = (QB)(\omega_{QB})^2 = 1 \times (14.5)^2 = 210 \text{ in./sec}^2$$
$$A_A^t = (MA)(\alpha) = 1.25 \times 15 = 18.75 \text{ in./sec}^2$$

The acceleration polygon is shown in Fig. A27. The acceleration polygon is drawn using the scale 1 in. = 100 in./sec^2. From the polygon, we determine $A_{B/A}^t = 270$ in./sec^2 and $A_B^t = 310$ in./sec^2. Hence

Scale: 1 in.= 100 in./sec²

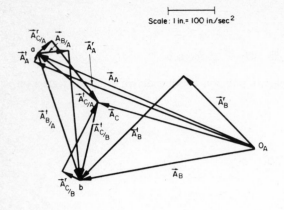

Fig. A27

$$\alpha_{BA} = \frac{A^t_{B/A}}{AB} = \frac{270}{2.5} = 108 \text{ rad/sec}^2 \text{ (counterclockwise)}$$

$$\alpha_{BQ} = \frac{A^t_B}{BQ} = \frac{310}{1} = 310 \text{ rad/sec}^2 \text{ (clockwise)}$$

The acceleration of point C is determined by drawing the vector polygon using the vector equations $\mathbf{A}_C = \mathbf{A}_A + \mathbf{A}_{C/A}$ and $\mathbf{A}_C = \mathbf{A}_B + \mathbf{A}_{C/B}$.
From Fig. A41, we determine

$$A^r_{C/A} = 1.5 \times (5.2)^2 = 40.5 \text{ in./sec}^2$$
$$A^r_{C/B} = 1.5 \times (5.2)^2 = 40.5 \text{ in./sec}^2$$
$$A_C = 343 \text{ in./sec}^2$$

2. In order to determine acceleration of link MA, we must first determine angular velocities of links AB and MA. These angular velocities are determined by drawing a velocity polygon using the vector relationship $\mathbf{V}_A = \mathbf{V}_B + \mathbf{V}_{A/B}$. Figure A28b shows the velocity polygon which is drawn using the scale 1 in. = 5 in./sec. From the velocity polygon, we measure $V_A = 4.2$ in./sec and $V_{A/B} = 2.5$ in./sec. Hence,

$$\omega_{AB} = \frac{V_{A/B}}{AB} = \frac{2.5}{10} = 0.25 \text{ rad/sec (clockwise)}$$

$$\omega_{MA} = \frac{4.2}{6.0} = 0.70 \text{ rad/sec (counterclockwise)}$$

The acceleration polygon shown in Fig. A28c is drawn using the vector relationship

$$\mathbf{A}^r_A + \mathbf{A}^t_A = \mathbf{A}_B + \mathbf{A}^r_{A/B} + \mathbf{A}^t_{A/B}$$

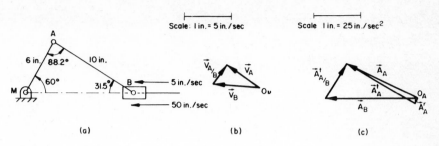

Fig. A28

Hence

$$A'_A = 6 \times (0.7)^2 = 2.44 \text{ in./sec}^2$$

$$A'_{A/B} = 10 \times (0.25)^2 = 0.625 \text{ in./sec}^2 (\cong 0)$$

From the polygon, we measure $A'_A = 41$ in./sec^2, $A'_{A/B} = 22.4$ in./sec^2, and $A_A = 41.5$ in./sec^2. Hence

$$\alpha_{AB} = \frac{A'_{A/B}}{AB} = \frac{22.4}{10} = 2.24 \text{ rad/sec}^2 \text{ (clockwise)}$$

$$\alpha_{MA} = \frac{41}{6} = 6.83 \text{ rad/sec}^2 \text{ (counterclockwise)}$$

Performance Test #2—Objectives 4, 5

1. The velocity polygon is constructed using the vector relationship $V_{A_3} = V_{A_2} + V_{A_3/A_2}$ where A_2 and A_3 are two coincident points (Fig. A29a).

$$V_{A_2} = \omega_2 \cdot MA = (0.57) \times \frac{100 \times 2 \times 3.14}{60} \times (12)$$
$$= 71.64 \text{ in./sec}$$

The velocity polygon shown in Fig. A29b is drawn using the scale 1 in.=36 in./sec. From the velocity polygon we determine $V_{A_3} = 51.6$ in./sec and $V_{A_3/A_2} = 50.16$ in./sec. Hence

$$\omega_3 = \frac{V_{A_3}}{QA_3} = \frac{51.6}{24} = 2.15 \text{ rad/sec (clockwise)}$$

The acceleration polygon shown in Fig. A29c is drawn using the vector relationship

$$A'_{A_2} + A^t_{A_2} = A'_{A_3} + A^t_{A_3} + A'_{A_2/A_3} + A^t_{A_2/A_3} + 2\omega_3 \times V_{A_2/A_3}$$

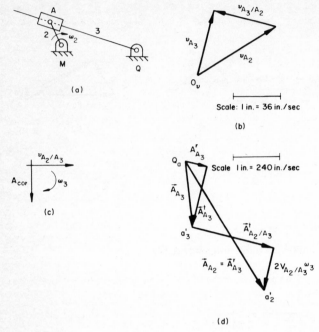

Fig. A29

We calculate

$$A'_{A_2} = \omega_2^2 \cdot MA = 753.6 \text{ in./sec}^2$$

$$A^t_{A_2} = 0$$

$$A'_{A_3} = \frac{V^2_{A_3}}{QB} = \frac{(51.6)^2}{2 \times 12} = 111.0 \text{ in./sec}^2$$

$$2V_{A_2/A_3}\omega_3 = 2 \times 50.16 \times 2.15$$
$$= 215.76 \text{ in./sec}^2$$

$$A'_{A_2/A_3} = 0$$

Direction of coriolis acceleration is determined using Fig. 29c. From the acceleration polygon shown in Fig. 29c

$$\alpha_3 = A^t_{A_3}/QB = 309.6/24 = 12.9 \text{ rad/sec}^2 \text{ (counterclockwise)}$$

2. The velocity analysis is performed by drawing the velocity polygon using the velocity-vector relationships

$$\mathbf{V}_B = \mathbf{V}_A + \mathbf{V}_{B/A}$$
$$\mathbf{V}_C = \mathbf{V}_B + \mathbf{V}_{C/B}$$

and

$$V_C = V_A + V_{C/A}$$
$$V_A = \omega_2 \cdot AM = 50 \times 0.8 = 40 \text{ in./sec}$$

The velocity polygon shown in Fig. A30b is drawn using the scale 1 in. = 20 in./sec.

From the velocity polygon, we measure $V_B = 30$ in./sec and $V_{B/A} = 19.5$ in./sec. The acceleration analysis is performed by drawing the acceleration polygon, shown in Fig. A30c, using the vector relationships

$$A_B^r + A_B^t = A_A^r + A_A^t + A_{B/A}^r + A_{B/A}^t$$
$$A_C^r + A_C^t = A_B^r + A_B^t + A_{C/B}^r + A_{C/A}^t$$

We calculate

$$A_A^r = AM \cdot \omega_2^2 = (0.8) \times (50)^2 = 2000 \text{ in./sec}^2$$

$$A_A^t = AM\alpha_2 = 0.8 \times 100 = 80 \text{ in./sec}^2$$

$$A_{B/A}^r = BA \cdot \omega_3^2 = (V_{B/A})^2/BA = (19.5)^2/2 = 190 \text{ in./sec}^2$$

$$A_B^r = BQ \cdot \omega_4^2 = (V_B)^2/BQ = (30.5)^2/1.5 = 620 \text{ in./sec}^2$$

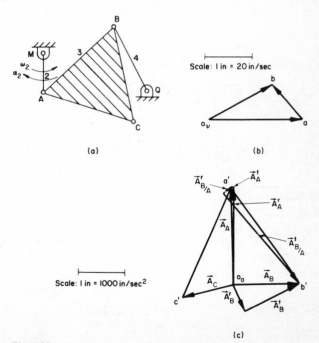

(a)

Scale: 1 in = 20 in/sec

(b)

Scale: 1 in = 1000 in/sec²

(c)

Fig. A30

The point C' in the acceleration polygon is located by constructing triangle $a'b'c'$ similar to triangle ABC. From the polygon $A_C = 1070$ in./sec² and A'_B = 1290 in./sec². Hence

$$\alpha_4 = \frac{A^t_B}{QB} = \frac{1290}{1.5} = 859 \text{ rad/sec}^2$$

$$\alpha_3 = \frac{A^t_{B/A}}{AB} = \frac{2470}{2} = 1240 \text{ rad/sec}^2 \text{ (clockwise)}$$

Performance Test #3–Objectives 4, 5

1. Since the velocity of A is known, we can write the following velocity-vector relationship: $\mathbf{V}_B = \mathbf{V}_A + \mathbf{V}_{B/A}$. The velocity polygon shown in Fig. A31b is drawn using the scale 1 in. = 2 in./sec. The acceleration polygon shown in Fig. A31c is drawn using the vector relationship

$$\mathbf{A}^r_B + \mathbf{A}^t_B = \mathbf{A}_A + \mathbf{A}^r_{B/A} + \mathbf{A}^t_{B/A}$$

We calculate

$$A^r_B = QB \cdot \omega^2_4 = \frac{V^2_B}{QB} = \frac{(8.15)^2}{1} = 66.4 \text{ in./sec}^2$$

$$A^r_{B/A} = AB \cdot \omega^2_3 = \frac{V^2_{B/A}}{AB} = \frac{(3.75)^2}{2.5} = 5.62 \text{ in./sec}^2$$

From the polygon, we measure

$$A^t_B = 27.5 \text{ in./sec}^2$$

$$A_B = 71.5 \text{ in./sec}^2$$

Hence

$$\alpha_4 = \frac{A^t_B}{QB} = \frac{27.5}{1} = 27.5 \text{ rad/sec}^2 \text{ (clockwise)}$$

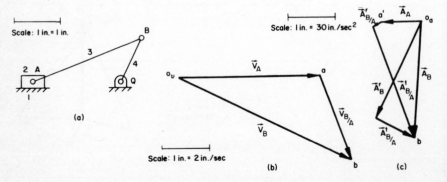

Scale: 1 in.= 1 in.

2 A

3

4

B

Q

(a)

Scale: 1 in. = 30 in./sec²

$\vec{V}_A$

$\vec{V}_B$

$\vec{V}_{B/A}$

Scale: 1 in. = 2 in./sec

(b)

$\overline{A}^r_{B/A}$ a' $\overline{A}_A$ o_a

$\overline{A}^r_B$ $\overline{A}^t_{B/A}$ $\overline{A}_B$

$\overline{A}^t_{B/A}$ b

(c)

Fig. A31

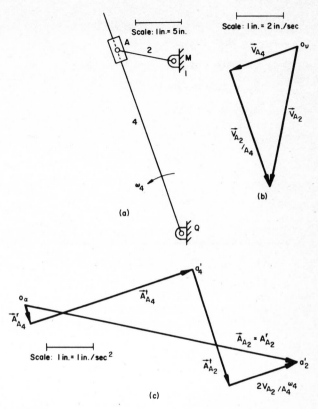

Fig. A32

Figure A32b and c shows the velocity and acceleration vector polygons. We note from Fig. A32a that

$$\mathbf{V}_{A_2} = \mathbf{V}_{A_4} + \mathbf{V}_{A_2/A_4}$$

$$\mathbf{A}^r_{A_2} + \mathbf{A}^t_{A_2} = \mathbf{A}^r_{A_4} + \mathbf{A}^t_{A_4} + \mathbf{A}^r_{A_2/A_4} + \mathbf{A}^t_{A_2/A_4} + 2\boldsymbol{\omega}_4 \times \mathbf{V}_{A_2/A_4}$$

$$\mathbf{A}^t_{A_2} = 0$$

$$A^r_{A_4} = \frac{V^2_{A_4}}{AQ} = \frac{(3)^2}{20.15} = 0.447 \text{ in./sec}^2$$

$$2V_{A_2}/A_4 \cdot \omega_4 = 2V_{A_2}/A_4 \cdot \frac{V_{A_4}}{QA} = 2 \times 5.1 \times (3/20) = 1.52 \text{ in./sec}^2$$

and

$$A^r_{A_2/A_4} = 0$$

Hence

$$\alpha_4 = \frac{A_B^t}{AQ} = \frac{3.6}{4.0} = 0.9 \, \text{rad/sec}^2$$

$$A_{A_2} = 5.95 \, \text{in./sec}^2$$

$$A_{A_4} = 3.6 \, \text{in./sec}^2$$

Unit IV

Performance Test #1

1. The displacement, velocity and acceleration relationships are similar to those given by Eqs. (61), (62), (68), (69), (74), and (75) of Objective 2.

Performance Test #2

1. The displacements and velocity relationships are obtained in a manner described in Objective 1 of Unit IV.

Performance Test #3

1. The displacement and velocity relationships are obtained from Eqs. (86), (92), and (93) by letting $e = 0$.

Unit V

Performance Test #1

1. Refer to Objective 3 of Unit V. We note $SR = \omega_2/\omega_1 = r_1/r_2$ where $C = r_1 + r_2$. Let cylinder 1 be the driver. Then $1 + (r_2/r_1) = (1/SR) + 1$ or $(r_1 + r_2)/r_1 = (1/SR) + 1$. Hence

$$r_1 = \frac{C}{1 + (1/SR)} = \frac{C(SR)}{1 + SR}$$

2. The speed ratio is given by $SR = MF/QF = 1.2/0.5 = 2.4$.

Performance Test #2

1. Draw a common normal at the point of contact between the cam and its follower. The common normal intersects center line MQ in F. Hence, from Fig. A33.

$$\omega_3 = \omega_2 \cdot \frac{MF}{FQ} = (10) \times \frac{0.42}{2.51} = 1.67 \text{ rad/sec (clockwise)}$$

The motion is due to sliding since the point of contact does not lie on the center line.

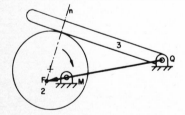

Fig. A33

2. From Objective 3, Unit V, we note

$$\tan \alpha_B = \frac{\sin \gamma}{\dfrac{\omega_B}{\omega_A} - \cos \gamma}$$

Since $\gamma = 60°$, $\dfrac{\omega_B}{\omega_A} = 0.5$ $\tan \alpha_B = \dfrac{0.5}{0.5 - 0.5}$

or $\alpha_B = 90°$ and $\alpha_A = \alpha_B - \gamma = 90° - 60° = 30°$.

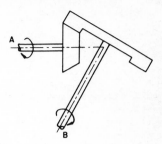

Fig. A34

Performance Test #3

1. Note that the point of contact lies on the center line MQ. Hence, the two bodies are transmitting motion by a rolling contact. From Fig. A35 $\omega_2 / \omega_3 = QP/MP = N_2 / N_3$. Therefore, $N_2 = N_3 \cdot (QP/MP) = 60 \times (1.30/2.12) = 36.8$ revolutions per minute.

   In positions marked as $2'$, $3'$ of the bodies 2 and 3, the motion will be transmitted from body 2 to body 3 via sliding contact. Note that the point of contact does not lie on center line MQ for this particular situation.

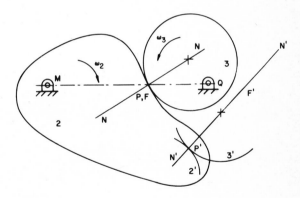

Fig. A35

2. From Fig. A36 we note $SR = \omega_2/\omega_1 = r_1/r_2$ or $(r_2/r_1) - 1 = 1/SR - 1$, or $c/r_1 = 1/SR - 1$, and $r_1 = c \cdot SR/(1 - SR)$.

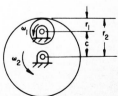

Fig. A36

3. From Fig. A37, we note

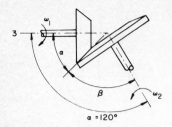

Fig. A37

$$\tan \beta = \frac{\sin \gamma}{\dfrac{\omega_2}{\omega_1} + \cos \gamma} = \frac{\sin 120°}{1.75 + \cos 120°}$$

$$= \frac{\sin 60°}{1.75 - \cos 60°} = \frac{\sin 60°}{1.25} = 0.6925$$

Hence, $\beta = 34.7°$ and $\alpha = \gamma - \beta = 120° - 35.7° = 85.3°$.

Unit VI

Performance Test #1

1. Since $P_D = 3.0$, $N_1 = 27$, $N_2 = 18$

$$R_1 = \frac{N_1}{2P_D} = \frac{27}{2 \times 3} = 4.5 \text{ in.}$$

$$R_2 = \frac{N_2}{2P_D} = \frac{18}{2 \times 3} = 3.0 \text{ in.}$$

$$R_{1ADD} = R_1 + \frac{1}{P_D} = 4.83$$

$$R_{2ADD} = R_2 + \frac{1}{P_D} = 3.33$$

Figure A38 shows that the pressure angle $14.5°$ will cause interference since B_2 is inside D_2. The pressure angle $20°$ will satisfy the necessary conditions.

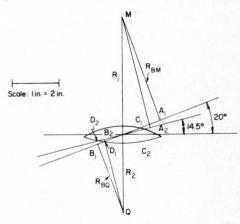

Scale: 1 in. = 2 in.

Fig. A38

2. Since $R_1 = 3.0$ in., $P_D = 2.0$

$$R_{1\,add} = R_1 + 1/P_D = 3.5 \text{ in.}$$

From Fig. A39, we measure $R_{2\,min} = 3.125$ in. Hence, $N_2 = 2R_2 P_D = 12.5$. Since N_2 must be an integer value, R_2 must be greater than or equal to 3.25 in. Hence,

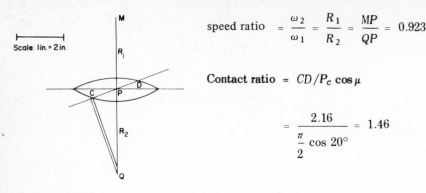

speed ratio $= \dfrac{\omega_2}{\omega_1} = \dfrac{R_1}{R_2} = \dfrac{MP}{QP} = 0.923$

Scale: 1 in. = 2 in.

Contact ratio $= CD/P_c \cos\mu$

$$= \dfrac{2.16}{\dfrac{\pi}{2} \cos 20^\circ} = 1.46$$

Fig. A39

Performance Test #2

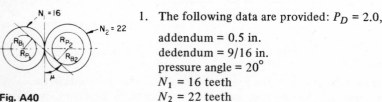

Fig. A40

1. The following data are provided: $P_D = 2.0$,
   addendum = 0.5 in.
   dedendum = 9/16 in.
   pressure angle = 20°
   $N_1 = 16$ teeth
   $N_2 = 22$ teeth

a. R_{P_1} and R_{P_2} are determined from the definition of dimetral pitch $P_D = N/2R_P$. Hence,

$$R_P = \dfrac{N}{2P_D}$$

$$R_{P_1} = \dfrac{N_1}{2P_D} = \dfrac{16}{2\,(2)} = 4 \text{ in.}$$

$$R_{P_2} = \dfrac{N_2}{2P_D} = \dfrac{22}{4} = 5.5 \text{ in.}$$

b. R_{B_1} and R_{B_2} are determined using the relationship $R_B = R_P \cos\mu$.

$$R_{B_1} = R_{P_1} \cos\mu = 4 \cos 20^\circ = 3.76 \text{ in.}$$
$$R_{B_2} = R_{P_2} \cos\mu = 5.5 \cos 20^\circ = 5.16 \text{ in.}$$

c. Circular pitch is determined from the definition

$$P_C = \dfrac{\pi D}{N} = \dfrac{3.14 \times 1}{2} = 1.57 \text{ in.}$$

$$P_B = P_C \cos\mu = 1.57 \cos 20^\circ = 1.47 \text{ in.}$$

e. Length of the path of contact CD is determined in the following manner.

$R_{\text{add}_1} = R_{A_1} = R_{P_1} + \text{add.} = 4.0 + 0.5 = 4.5 \text{ in.}$

$R_{\text{add}_2} = R_{A_2} = R_{P_2} + \text{add.} = 5.5 + 0.5 = 6 \text{ in.}$

$$CD = (R_{A_1}^2 - R_{B_1}^2)^{1/2} + (R_{A_2}^2 - R_{B_2}^2)^{1/2} - (R_{P_1} + R_{P_2})\sin\mu$$

$$= (20.2 - 14.1)^{1/2} + (36 - 26.6)^{1/2} - 3.25$$

$$= 2.47 + 3.07 - 3.25$$

$$= 2.29 \text{ in.}$$

f. Contact ratio is determined using the relationship

$$G_c = \frac{CD}{P_C \cos\mu} = \frac{CD}{P_B} = \frac{2.29}{1.47} = 1.56$$

2. Since $N_1 = 30$, $P_D = 3.0$, addendum $= 3/8$ in., and $\mu = 20°$

$$R_{P_1} = \frac{N_1}{2P_D} = \frac{30}{(2)(3)} = 5 \text{ in.} = MP$$

$$P_C = \frac{\pi D_1}{N_1} = 1.05$$

From Fig. A41, we determine the smallest pinion $R_{P_2} = 2.52$ in. $= QP$. Since N_2 must be integral,

$$N_2 = 2(R_{P_2})(P_D)$$

$$= 2(2.52)(3) = 15.1$$

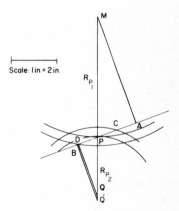

Scale: 1 in. = 2 in.

Fig. A41

So we must use $N_2 = 16$. Hence $Q'P = R'_{P_2} = 16/(2 \times 3) = 2.67$ in. Contact ratio is calculated using

$$G_c = CD/(P_C \cos\mu) = 1.68/(1.05 \cos 20°) = 1.70$$

3. When the point of contact of two teeth is at the pitch point, the teeth are instantaneously in rolling contact, but at all other times they act in sliding contact; thus, a combination.

Performance Test #3

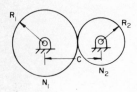

Fig. A42

1. We have the following: $C = 6$ in., $P_D = 3$ in., $SR = 1.4/1$. Since

$$SR = \frac{\omega_2}{\omega_1} = \frac{R_1}{R_2} = 1.4$$

$$R_1 = 1.4\, R_2 \text{ but } R_1 + R_2 = C = 6$$

Hence $1.4\, R_2 + R_2 = 6$ or $R_2 = \dfrac{6}{2.4} = 2.5$ in. and $R_1 = 3.5$ in.

Since $P_D = \dfrac{N}{D}$ $N = P_D \cdot D = 2RP_D$

$$N_1 = 2R_1 \cdot P_D = 2 \times 3.5 \times 3 = 21$$

and

$$N_2 = 2R_2 P_D = 2 \times 2.5 \times 3 = 15$$

2. The following data are given:

$D_1 = 5$ in.
$D_2 = 6$ in.
$P_D = 4.0$
$\mu = 20°$

Hence

$$R_1 = \frac{D_1}{2} = \frac{5}{2} = 2.5 \text{ in.}$$

$$R_2 = \frac{D_2}{2} = \frac{6}{2} = 3.0 \text{ in.}$$

Addendum $= 1/P_D = 1/4 = 0.25$ in. From Fig. A43, we measure $CD = 1.21$ in.

We now calculate the contact ratio using

$$G_c = \frac{CD}{P_C \cos\mu}$$

$$P_C = \frac{\pi}{P_D}, \; G_c = \frac{(CD)(P_D)}{\pi \cos\mu} = \frac{(1.21)(4)}{\pi \cos 20°} = 1.45$$

This contact ratio is greater than 1.4:1; it will prove to be a satisfactory value.

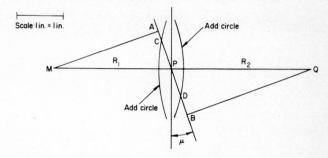

Fig. A43

3. We know the following data:

D_1 = 4 in.
D_2 = 6 in.
$\mu = 20°$

Hence

$$R_1 = \frac{D_1}{2} = \frac{4}{2} = 2 \text{ in.}$$

and

$$R_2 = \frac{D_2}{2} = \frac{6}{2} = 3 \text{ in.}$$

From Fig. A44, we measure

$$R_{add_1} = MD = 2.55 \text{ in.}$$
$$R_{add_2} = QD = 3.31 \text{ in.}$$

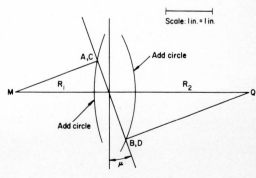

Fig. A44

The maximum addendums are

$$\text{add}_1 = R_{\text{add}_1} - R_1 = 2.55 - 2.0 = 0.55 \text{ in.}$$
$$\text{add}_2 = R_{\text{add}_2} - R_2 = 3.31 - 3.0 = 0.31 \text{ in.}$$

Unit VII

Performance Test #1

1. Read Unit VI, Objective 3.
2. The speed ratio for a compound gear train is given as

$$SR = \frac{\omega_6}{\omega_1} = \frac{\text{product of radii of driver gears}}{\text{product of radii of driven gears}} = \frac{R_5 R_3 R_1}{R_6 R_4 R_2}$$

3. From Fig. A45, we note

$$R_1 + R_{3A} = R_2 = R_{3B} + R_4$$
$$V_A = \omega_1 R_1$$
$$V_B = \omega_2 R_2$$
$$V_A = V_B + V_{A/B}$$
$$V_{A/B} = -\omega_3 R_{3A} = V_A - V_B = \omega_1 R_1 - \omega_2 R_2$$

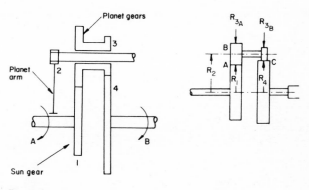

Fig. A45

Note that the direction of ω_3 will be opposite to that of ω_1.

$$V_C = \omega_4 R_4$$
$$V_C = V_B + V_{C/B}$$
$$V_{C/B} = -\omega_3 R_3 B = V_C - V_B = \omega_4 R_4 - \omega_2 R_2$$

$$\omega_3 = \frac{V_{A/B}}{R_{3A}} = \frac{V_{C/B}}{R_{3B}} = \frac{-\omega_1 R_1 - \omega_2 R_2}{R_{3A}} = \frac{\omega_4 R_4 - \omega_2 R_2}{R_{3B}}$$

By simplifying and rearranging the above equalities, we get

$$\omega_1 R_1 R_{3B} = \omega_4 R_4 R_{3A} + \omega_2 R_2 (R_{3B} - R_{3A})$$

If the gear 4 is fixed, then the speed ratio is

$$SR = \frac{\omega_2}{\omega_1} = \frac{R_1 R_{3B}}{R_2(R_{3B} - R_{3A})} = \frac{R_1(R_2 - R_4)}{R_2(R_1 - R_4)}$$

Performance Test #2

1. A reverted compound gear train is shown in Fig. A46. We note that

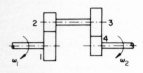

$$SR = \frac{\omega_4}{\omega_1} = \frac{N_1 N_3}{N_2 N_4} = \frac{1}{30}$$

$$= \left(\frac{1}{5}\right) \times \left(\frac{1}{6}\right)$$

Fig. A46

Let $N_1/N_2 = 1/5$ and $N_3/N_4 = 1/6$. Therefore, $N_2 = 5N_1$, $N_4 = 6N_3$. Since

$$P_D = \frac{N}{2R}$$

$$R_1 = \frac{N_1}{2P_D} = \frac{14}{2 \times 7} = 1 \text{ in.}$$

$$R_2 = \frac{N_2}{2P_D} = \frac{70}{2 \times 7} = 5 \text{ in.}$$

Hence, $R_1 + R_2 = 1 + 5 = 6$ in. Since $R_1 + R_2 = R_3 + R_4$, $N_1 + N_2 = N_3 + N_4$ or $N_1 + 5N_1 = N_3 + 6N_3$. Hence $N_1/N_3 = 7/6 = 14/12$. Let $N_1 = 14$, then $N_3 = 12$. And $N_2 = 5N_1 = 5 \times 14 = 70$, $N_4 = 6N_3 = 6 \times 12 = 72$.

2. Figure A47 shows all the notations that we will use in solving this problem.

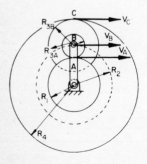

$$V_A = \omega_1 R_1$$
$$V_C = \omega_4 R_4$$
$$V_B = \omega_2 R_2$$

We note from the motion of the gears

$$V_A = V_B + V_{A/B}$$
$$V_C = V_B + V_{C/B}$$
$$V_{C/B} = V_C - V_B$$

Fig. A47

But

$$V_{A/B} = -V_{C/B}\left(\frac{R_{3A}}{R_{3B}}\right)$$

Hence,

$$V_A = V_B - (V_C - V_B)\frac{R_{3A}}{R_{3B}}$$

$$= V_B\left(\frac{R_{3B} + R_{3A}}{R_{3A}}\right) - V_C\left(\frac{R_{3A}}{R_{3B}}\right)$$

Substituting for V_A, V_B, and V_C their scalar quantities to obtain the equation of motion

$$\omega_1 R_1 = \omega_2 R_2\left(\frac{R_{3B} + R_{3A}}{R_{3B}}\right) - \omega_4 R_4\frac{R_{3A}}{R_{3B}}$$

Let us assume counterclockwise rotation as a positive sense. Then $\omega_1 = -10$ rpm; $\omega_2 = 20$ rpm. Then

$$\omega_4 = \frac{20R_2(R_{3B} + R_{3A})}{R_4 R_{3A}} + \frac{10R_1 R_{3B}}{R_4 R_{3A}}$$

$$= \frac{10}{R_4 R_{3A}}[2R_2(R_{3B} + R_{3A}) + R_1 R_{3B}] \text{ (counterclockwise)}$$

Performance Test #3

1. Since

$$P_D = \frac{N}{D} = \frac{N}{2R}$$

$$R_3 = \frac{N_3}{2P_D} = \frac{18}{2 \times 4} = 2.25 \text{ in.}$$

$$R_4 = \frac{N_4}{2P_D} = \frac{80}{2 \times 4} = 10.0 \text{ in.}$$

$$R_6 = \frac{N_6}{2P_D} = \frac{48}{2 \times 4} = 6.0 \text{ in.}$$

$$R_7 = \frac{N_7}{2P_D} = \frac{54}{2 \times 4} = 6.75 \text{ in.}$$

From the geometry of the gear arrangement, we write $R_3 + R_4 + R_5 + 2R_6 + R_7 = 33.5$. Hence,

$$R_5 = 33.5 - [2.25 + 10.0 + 2(6.0 + 6.75)] = 2.5 \text{ in.}$$
$$D_5 = 2R_5 = 5 \text{ in.}$$
$$N_5 = D_5 P_D = (5)(4) = 20$$

The speed ratio can be calculated from

$$SR = \frac{N_1 N_3 N_5}{N_2 N_4 N_7} = \frac{\omega_7}{\omega_1} = \frac{25}{900}$$

Hence,

$$N_1 = \frac{\omega_7}{\omega_1} \cdot \frac{N_2 N_4 N_7}{N_3 N_5}$$

$$= \left(\frac{25}{900}\right) \times \left(\frac{72 \times 80 \times 54}{18 \times 20}\right)$$

$$= 24$$

2. The speed ratio for the gear train is

$$SR = \frac{\omega_D}{\omega_A} = \frac{R_A R_C}{R_B R_D} = \frac{N_A N_C}{N_B N_D} = 12$$

Let $R_A/R_B = N_A/N_B = 3$ and $R_C/R_D = N_C/N_D = 4$. Then

$$R_A = 3R_B \quad \text{and} \quad N_A = 3N_B$$
$$R_C = 4R_D \quad \text{and} \quad N_C = 4N_D$$

Since $R_A + R_B = 8$, $3R_B + R_B = 8$. Hence, $R_B = 2$ in. Since $R_C + R_D = 8$ in., $4R_D + R_D = 8$ in. Hence $R_D = 1.6$ in. Since

$$N = 2R \cdot P_D$$
$$N_B = 2R_B \cdot P_D = 2 \times 2 \times 8 = 32$$
$$N_A = 3N_B = 3 \times 32 = 96$$
$$N_D = 2R_D \cdot P_D$$
$$= 2 \times 1.6 \times 10$$
$$= 32$$

and

$$N_C = 4 N_D$$
$$= 4 \times 32$$
$$= 128$$

3. The speed ratio for the given condition is

$$SR = \frac{\omega_3}{\omega_4} = \frac{2(N_1 + N_2)}{N_3}$$

$$= 1.5 = \frac{2(60 + N_2)}{N_3}$$

or

$$1.5N_3 - 2N_2 = 120$$

From the geometry of gear arrangement, we write $R_1 + 2R_2 = R_3$. Since

$$P_{D_1} = P_{D_2} = P_{D_3}$$
$$N_1 + N_2 = N_3$$
$$60 + 2N_2 = N_3$$

From the two simultaneous equations involving N_2 and N_3 as unknowns, we get $N_3 = 120$ and $N_2 = 30$.

Unit VIII

Performance Test #1

1. a. Since we are not given any requirement on the size of the cam (or the base circle), length of the follower, and eccentricity, we will arbitrarily choose these parameters. Figure A48a shows the layout of the cam. The following steps which demonstrate the principle of inversion are used to obtain the cam profile.

    i. Let us assume
       base radius = 1-13/16 in.
       length of follower arm QA = 2.0 in.
       and the cam is required to rotate in a counterclockwise direction.

    ii. Locate the pivot point Q_1, center point A_1 of the follower corresponding to the first position of cam given a $\theta = 0°$. See Fig. A48a.

    iii. Draw locus circle Q with center as M and MQ as radius, and locate on its positions Q_1, Q_2, ..., Q_{12}, $30°$ apart.

    iv. With M as a center and MA as a radius, draw the locus circle A.

    v. Locate centers A'_1, A'_2, A'_3, ..., A'_{12} on the locus circle A as shown in Fig. A48a. Note that points A_1, A_2, ..., A_{12} are selected on locus circle in the direction opposite to that of cam rotation.

    vi. Locate centers A_1, A_2, ..., A_{12} on the arc of circles with Q_1, Q_2, ..., Q_{12} as centers and QA'_1, QA'_2, ..., QA'_{12} as radii. The angles $A'_1Q_1A_1$, $A'_2Q_2A_2$, ..., $A'_{12}Q_{12}A_{12}$ measure the angular displacement ϕ of the follower corresponding to cam rotation angle θ.

    vii. With A_1, A_2, A_3, ..., A_{12} as centers and radius equal to the radius of the roller follower, draw circles. Draw a continuous curve that is tangent to the roller circles to obtain the cam profile as shown in Fig. A48a.

   b. The cam profile for a flat face follower is shown in Fig. A48b. The following steps are used.

    i. Select a base circle radius equal to 2 in. and draw the base circle.

    ii. Divide the base circle in 12 equal parts and mark them as A'_1, A'_2, ..., A'_{12}.

    iii. Draw radial lines MA'_1, MA'_2, ..., MA'_{12}.

    iv. Measure distances A'_1A_1, A'_2A_2, ..., $A'_{12}A_{12}$ along radial lines MA'_1, MA'_2, ..., MA'_{12} so that these distances measure follower travel according to the displacement program.

    v. Draw lines at A_1, A_2, ..., A_{12} perpendicular to radial lines MA_1, MA_2, ..., MA_{12}.

    vi. Draw a curve that is tangent to the lines drawn in item μ to obtain the cam profile shown in Fig. 48b.

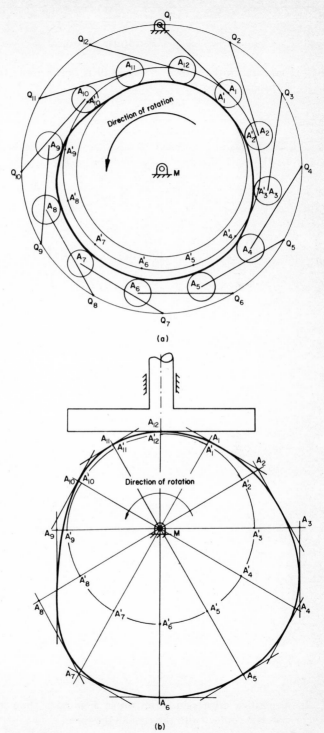

Fig. A48

Performance Test #2

1. The cam profile shown in Fig. A49 is drawn using the steps described in Performance Test #1

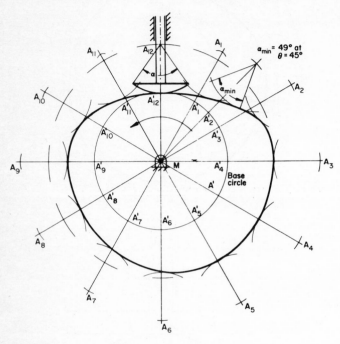

Fig. A49

Performance Test #3

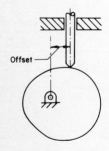

1. An example of a knife edge follower is shown in Fig. A50.

Fig. A50

2. A positive drive cam is shown in Fig. A51. Note that the follower will maintain contact with the cam all the time.

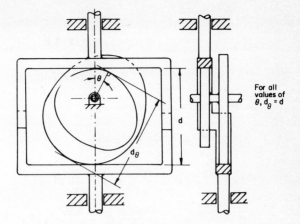

Fig. A51   A disk cam with positive drive.

3. When $\beta = 90°$, we will obtain a disk cam with translating follower as shown in Fig. A52.

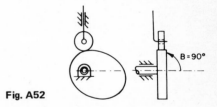

Fig. A52

4. Figure A53 shows the pressure angle μ. It is the angle between the radial line and a common normal at the point of contact.

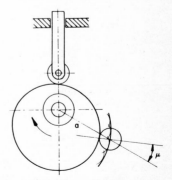

Fig. A53

5. The maximum pressure angle becomes smaller as the prime circle radius becomes larger.
6. The pressure angle is zero when cam experiences dwell. This is due to the fact that at dwelling condition there is no displacement of the follower.

Unit IX

Performance Test #1

1. The 3-4-5 polynomial yields

$$s = \frac{10H}{\beta^3}\theta^3 - \frac{15H}{\beta^4}\theta^4 + \frac{6H}{\beta^5}\theta^5$$

$$v = \frac{30H\omega}{\beta^3}\theta^2 - \frac{60H\omega}{\beta^4}\theta^3 + \frac{30H\omega}{\beta^5}\theta^4$$

$$a = \frac{60H\omega^2}{\beta^3}\theta - \frac{180H\omega^2}{\beta^4}\theta^2 + \frac{120H\omega^2}{\beta^5}\theta^3$$

$$J = \frac{60H\omega^2}{\beta^3} - \frac{360H\omega^2}{\beta^4}\theta + \frac{360H\omega^2}{\beta^5}\theta^2$$

To find A_{max}, let

$$J = 0 = \frac{60H\omega^2}{\beta^3}\left(1 - 6\frac{\theta}{\beta} + 6\frac{\theta^2}{\beta^2}\right)$$

$$0 = 6\left(\frac{\theta}{\beta}\right)^2 - 6\left(\frac{\theta}{\beta}\right) + 1$$

$$\frac{\theta}{\beta} = \frac{1}{2} \pm \frac{\sqrt{3}}{6}$$

$$\theta = 0.212\beta,\ 0.788\beta$$

$$A_{max} = \frac{60H\omega^2}{\beta^3}\left[0.212\beta - \frac{3}{\beta}(0.212\beta)^2 + \frac{2}{\beta^2}(0.212\beta)^3\right]$$

$$= \frac{60H\omega^2}{\beta^2}[0.212 - 0.135 + 0.019]$$

$$= \frac{60H\omega^2}{\beta^2}[0.096]$$

$$A_{max} = 5.76\frac{h\omega^2}{\beta^2} \quad \text{at} \quad \theta = 0.212\beta$$

To find V_{max}, let

$$a = 0 = \frac{60h\omega^2}{\beta^3}\theta\left[2\left(\frac{\theta}{\beta}\right)^2 - 3\left(\frac{\theta}{\beta}\right) + 1\right]$$

$$0 = 2\left(\frac{\theta}{\beta}\right)^2 - 3\left(\frac{\theta}{\beta}\right) + 1$$

$$\frac{\theta}{\beta} = \frac{1}{2}, 1 \quad \text{(also } a = 0 \text{ at } \theta = 0\text{)}$$

Let us use $\theta = \beta/2$

$$V_{max} = \frac{30H\omega}{\beta^3}\left[(\beta/2)^2 - \frac{2}{\beta}(\beta/2)^3 + \frac{1}{\beta^2}(\beta/2)^4\right]$$

$$= \frac{30H\omega}{\beta^3}\left[\beta^2\left(\frac{1}{4} - \frac{2}{8} + \frac{1}{16}\right)\right]$$

$$= \frac{30H\omega}{16\beta}$$

$$V_{max} = \frac{15H\omega}{8\beta} \quad \text{at} \quad \theta = \beta/2$$

2. The equations for the motion program will be derived for three intervals. Let T correspond to the rotation angle β. Let t denote some intermediate time. Then, the three intervals are

a. $0 \le t \le T/4$

$T/4 \le t \le 3T/4$

$3T/4 \le t \le T$

For each of these intervals, we will first write the equation describing the acceleration relationship which we then integrate with the respect to t to obtain velocity relationship. The value of the integrating constant is determined by using an appropriate boundary condition. The velocity relationship is further integrated with respect to t to obtain displacement of the follower. Here again, the value of the integrating constant is determined by using an appropriate boundary condition.

a. $0 \le t \le T/4$

$a = kt$

$A = A_{max}$ when $t = T/4$

Therefore $k = 4A_{max}/T$. Hence

$$a = \frac{4A_{max}}{T} t \tag{1}$$

Integrating Eq. (1), we get

$$v = \frac{4A_{max}}{2T} t^2 + C_1$$

C_1 is determined using the boundary condition that when $t = 0$, $V = 0$. Hence, $C_1 = 0$. Thus

$$v = \frac{2A_{max}}{T} t^2 \tag{2}$$

Integrating Eq. (2) and determining the integrating constant by using the boundary condition there are $t = 0$, $S = 0$, we get

$$s = \frac{2A_{max}}{3T} t^3 \tag{3}$$

b. $T/4 \leq t \leq 3T/4$. Equation relating time and acceleration is

$$(a - A_{max}) = \frac{A_{max} - A_{min}}{(T/4) - (3T/4)}\left(t - \frac{T}{4}\right)$$

$$a = \frac{A_{max} - A_{min}}{-T/2}\left(t - \frac{T}{4}\right) + A_{max}$$

From the motion program we note $A_{max} = -A_{min}$. Hence

$$a = -\frac{4A_{max}}{T}\frac{T}{4} + A_{max}$$

or

$$a = -\frac{4A_{max}}{T} t + 2A_{max} \tag{4}$$

Integrating Eq. (4) we get

$$v = -\frac{2A_{max}}{T} t^2 + 2A_{max}t + C_3$$

Value of C_3 is obtained by using the boundary condition that

at $t = \frac{T}{4}$, $v = \frac{2A_{max}}{T}\frac{T}{4}^2 = \frac{A_{max}T}{8}$

Thus $C_3 = -A_{max}/4$ and

$$v = -\frac{2A_{max}}{T} t^2 + 2A_{max}t - \frac{A_{max}T}{4} \tag{5}$$

Integrating Eq. (5), we get

$$s = -\frac{2A_{max}}{3T} t^3 + A_{max}t^2 - \frac{A_{max}T}{4} t + C_4$$

Value of C_4 is obtained by using the boundary condition that at

at $t = \dfrac{T}{4}$ $s = \dfrac{2A_{max}}{3T} \left(\dfrac{T}{4}\right)^3 = \dfrac{A_{max}T^2}{96}$

Thus, $C_4 = (A_{max}T^2/48)$ and

$$ s = -\dfrac{2A_{max}}{3T} t^3 + A_{max}t^2 - \dfrac{A_{max}T}{4} t + \dfrac{A_{max}T^2}{48} \tag{6} $$

c. $3T/4 \le t \le T$

$$ a = \dfrac{A_{min}}{(3T/4) - \beta} (t - T) $$

Since

$$ A_{min} = - A_{max} $$
$$ a = \dfrac{4A_{max}}{T} (t - T) \tag{7} $$

Integrating Eq. (7), we get

$$ v = \dfrac{2A_{max}}{T} t^2 - 4A_{max}\, t + C_5 $$

Value of C_5 is obtained by using the boundary condition that at $t = 3T/4$. $v = A_{max}T/8$ Thus, $C_5 = 2A_{max}T$ and

$$ v = \dfrac{2A_{max}}{T} t^2 - 4A_{max} t + 2A_{max} T \tag{8} $$

Integrating Eq. (8), we get

$$ s = \dfrac{2A_{max}}{3T} t^3 - 2A_{max}t^2 + 2A_{max}Tt + C_6 $$

Value of C_6 is obtained by using the boundary condition that at

$$ t = 3T/4, \; S = \dfrac{11A_{max}T^2}{96} $$

Thus,

$$ C_6 = -\dfrac{52A_{max}T^2}{96} $$

and

$$s = \frac{2}{3} \frac{A_{max}}{T} t^3 - 2A_{max}t^2 + 2A_{max}Tt - \frac{52A_{max}T^2}{96} \tag{9}$$

Since at $t = T$, $s = A_{max}T^2/8 = H$, and $A_{max} = 8H/T^2$.

Performance Test #2

1. a. Acceleration and velocity in simple harmonic motion program.

$$a = \frac{H}{2}\left(\frac{\pi\omega}{\beta}\right)^2 \cos\frac{\pi\theta}{\beta}$$

For $v = v_{max}$, $a = 0$. Hence, $\cos(\pi\theta/\beta) = 0$. That is, $\pi\theta/\beta = \pi/2$ or $\theta = \beta/2$. Since

$$v = \frac{H\pi\omega}{2\beta} \sin\frac{\pi\theta}{\beta}$$

$$v_{max} = \frac{H\pi\omega}{2\beta} \sin\frac{\pi\beta/2}{\beta}$$

$$= \frac{H\pi\omega}{2\beta}$$

Acceleration can be maximum when $da/dt = 0$. Hence,

$$\frac{da}{dt} = -\frac{H}{2}\left(\frac{\pi\omega}{\beta}\right)^3 \sin\frac{\pi\theta}{\beta} = 0$$

That is, $\pi\theta/\beta = 0$, π or $\theta = 0$, β. Substituting $\theta = 0$, β we get

$$a_{max} = \frac{H}{2}\left(\frac{\pi\omega}{\beta}\right)^2$$

or

$$a_{max} = -\frac{H}{2}\left(\frac{\pi\omega}{\beta}\right)^2$$

b. Acceleration and velocity in cycloidal motion program.

$$a = \frac{2H\pi\omega^2}{\beta^2} \sin\frac{2\pi\theta}{\beta}$$

When $a = 0$, $v = v_{max}$ or $v = v_{min}$. But $a = 0$ when $\theta = 0$, $\beta/2$. Since

$$v = \frac{H\omega}{\beta}\left(1 - \cos\frac{2\pi\theta}{\beta}\right)$$

$$v_{min} = \frac{H\omega}{\beta}(1 - \cos 0) = 0$$

$$v_{max} = \frac{H\omega}{\beta}\left(1 - \cos\frac{2\pi\beta/2}{\beta}\right)$$

$$= \frac{2H\omega}{\beta}$$

The maximum value of acceleration is obtained by setting da/dt equal to zero. Thus,

$$\frac{da}{dt} = \frac{4H\pi^2\omega^3}{\beta^3}\cos\frac{2\pi\theta}{\beta} = 0$$

or $2\pi\theta/\beta = \pi/2, 3\pi/2$. Hence $\theta = \beta/4, 3\beta/4$, and

$$a_{max} = \frac{2\pi H\omega^2}{\beta^2}\sin\frac{2\pi\beta/4}{\beta} = \frac{2\pi H\omega^2}{\beta^2}$$

$$a_{min} = \frac{2\pi H\omega^2}{\beta^2}\sin\frac{2\pi(3\beta/4)}{\beta} = -\frac{2H\pi\omega^2}{\beta^2}$$

2. The relationship between A_1 and A_2 is determined by equating the two areas that describe acceleration and deceleration. Thus, $\frac{1}{2}A_1 T = \frac{2}{3}A_2 T$ or $A_2 = \frac{3}{4}A_1$. The relationship between acceleration and time for the interval $T/2 < t < T$,

$$a + A_2 = k\left(t - \frac{3T}{4}\right)^2$$

or

$$a = k\left(t - \frac{3T}{4}\right)^2 - A_2 = k\left(t - \frac{3T}{4}\right)^2 - \frac{3}{4}A_1$$

At $t = T$, $a = 0$. Therefore,

$$0 = k\left(T - \frac{3T}{4}\right)^2 - \frac{3}{4}A_1 = \frac{kT^2}{16} - \frac{3}{4}A_1$$

That is,

$$k = \frac{12A_1}{T^2}$$

and

$$a = \frac{12A_1}{T^2}(4t - 3T)^2 - \frac{3}{4}A_1$$

or

$$a = A_1\left(12\frac{t^2}{T^2} - 18\frac{t}{T} + 6\right)$$

Integrating above relationship with respect to time, we get

$$v = A_1\left(4\frac{t^3}{T^2} - 9\frac{t^2}{T} + 6\right) + C_1$$

At $t = T$, $v = 0$. Hence, $C_1 = -A_1 T$. Thus

$$v = A_1\left(4\frac{t^3}{T^2} - 9\frac{t^2}{T} + 6t - T\right)$$

Integrating the above relationship with respect to time, we get

$$s = A_1\left(\frac{t^4}{T^2} - 3\frac{t^3}{T} + 3t^2 - Tt\right) + C_2$$

At $t = T/2$, $s = A_1 T^2/16$. Therefore,

$$A_1\frac{T^2}{16} = A_1 T^2\left[\frac{1}{16} - \frac{3}{8} + \frac{3}{8} - \frac{1}{2}\right] + C_2$$

$$= \frac{A_1 T^2}{16}(1 - 6 + 12 - 8) + C_2$$

Hence $C_2 = \frac{1}{8}A_1 T^2$ and

$$s = A_1\left(\frac{t^4}{T^2} - \frac{3t^3}{T} + 3t^2 - Tt + \frac{T^2}{8}\right)$$

We can now express A_1 and A_2 in terms of H and T. At $t = T$, $s = H$. Hence, $H = A_1 T^2(1 - 3 + 3 - 1 + \frac{1}{8})$ or $A_1 = 8H/T^2$ and $A_2 = 12H/T^2$.

Performance Test #3

1. The velocity and acceleration of the given motion program are obtained by taking time derivatives of the displacement program. Thus,

$$v = \frac{ds}{dt} = \left(\frac{H}{2}\right)\left(\frac{\pi\omega}{\beta}\right)\cos\left(\frac{\pi}{4} + \frac{\pi\theta}{\beta}\right)$$

$$a = \frac{d^2 s}{dt^2} = -\left(\frac{H}{2}\right)\left(\frac{\pi\omega}{\beta}\right)^2 \sin\left(\frac{\pi}{4} + \frac{\pi\theta}{\beta}\right)$$

For $v = v_{max}$, $a = 0 = \sin\left(\frac{\pi}{4} + \frac{\pi\theta}{\beta}\right)$

That is,

$$\frac{\pi}{4} + \frac{\pi\theta}{\beta} = 0, \pi$$

or

$$\theta = -\frac{\beta}{4}, \frac{3\beta}{4}$$

$$v_{max} = \left(\frac{H}{2}\right)\left(\frac{\pi\omega}{\beta}\right)\cos\left(\frac{\pi}{4} + \frac{3\pi}{4}\right)$$

$$= -\left(\frac{H}{2}\right)\left(\frac{\pi\omega}{\beta}\right)$$

The value $\theta = -\beta/4$ is not acceptable since it makes the velocity become zero.

The maximum value of acceleration is obtained by equating da/dt to zero. Thus

$$\frac{da}{dt} = \cos\left(\frac{\pi}{4} + \frac{\pi\theta}{\beta}\right) = 0$$

That is,

$$\frac{\pi}{4} + \frac{\pi\theta}{\beta} = \frac{\pi}{2}, \frac{3\pi}{2}$$

or

$$\theta = \frac{\beta}{4} \quad \text{or} \quad \theta = \frac{5}{4}\beta \text{ (not acceptable)}$$

$$A_{max} = -\left(\frac{H}{2}\right)\left(\frac{\pi\omega}{\beta}\right)^2 \sin\left(\frac{\pi}{4} + \frac{\pi}{4}\right)$$

$$= -\left(\frac{H}{2}\right)\left(\frac{\pi\omega}{\beta}\right)^2$$

The absolute values of maximum acceleration and velocity in a simple harmonic program are the same.

2. There are two intervals that we need to consider. These are:

a. $0 \leqslant t \leqslant T/2$
b. $T/2 \leqslant t \leqslant T$

For each of these intervals, we will derive the relationship to describe acceleration, velocity, and displacement programs.

a. $0 \leqslant t \leqslant T/2$

$$a = kt^2$$

At $t = 0$, $a = 0$ and at $t = T/2$, $a = A_1$. Hence $A_1 = k(T/2)^2$ or $k = 4A_1/T^2$ and

$$a = \left(\frac{4A_1}{T^2}\right)t^2 \tag{1}$$

Integrating Eq. (1) with respect to time t, we get $v = (4A_1/3T^2)t^3 + C_1$. C_1 is determined using the boundary condition that at $t = 0$, $v = 0$. Hence $C_1 = 0$, and

$$v = \left(\frac{4A_1}{3T^2}\right)t^3 \tag{2}$$

Integrating Eq. (2) with respect to time t we get

$$s = \left(\frac{A_1}{3T^2}\right)t^4 + C_2$$

C_2 is determined using the boundary condition that at $t = 0$, $s = 0$. Hence $C_2 = 0$ and

$$s = \left(\frac{A_1}{3T^2}\right)t^4 \tag{3}$$

b. $T/2 \le t \le T$. The relationship for acceleration program is

$$a = \left(\frac{-2A_2}{T}\right)t + A_2 \tag{4}$$

The relationship between A_1 and A_2 is determined by equating the area under the acceleration and deceleration. Thus, $\frac{1}{6}A_1 T = \frac{1}{2}A_2(T/2)$ or $A_2 = \frac{2}{3}A_1$. Substituting, we get

$$a = A_1\left[\frac{4}{3T}t + \frac{2}{3}\right] \tag{5}$$

Integrating Eq. (5) with respect to time, we get

$$v = A_1\left[\frac{-2}{3T}t^2 + \frac{2}{3}t\right] + C_3$$

C_3 is determined by using the condition that at $t = T$, $v = 0$. Hence, $C_3 = 0$ and

$$v = A_1\left[\frac{-2}{3T}t^2 + \frac{2}{3}t\right] \tag{6}$$

Integrating Eq. (6) with respect to time, we get

$$s = A_1\left[\frac{-2}{9T}t^3 + \frac{1}{3}t^2\right] + C_4 \tag{7}$$

C_4 is determined using the boundary condition that at $t = T/2$, $s = (A_1 T^2/48)$ as obtained from Eq. (3). Equating this quantity with that obtained from Eq. (7), we get

$$C_4 = -\frac{5}{(3)^2(4)^2}A_1 T^2$$

and

$$s = A_1\left[-\frac{2}{9T}t^3 + \frac{1}{3}t^2 - \frac{5}{144}T^2\right] \tag{8}$$

We are now in a position to determine A_1 in terms of H and T. Since at $t = T$, $s = H$, Eq. (8) yields

$$H = A_1 T^2 \left(-\frac{2}{9} + \frac{1}{3} - \frac{5}{144} \right)$$

$$= \frac{11}{144} A_1 T^2$$

or

$$A_1 = \frac{144}{11} \frac{H}{T^2}$$

Unit X

Performance Test#1

2. $v_{max} = \dfrac{2H\omega}{\beta} = \dfrac{2 \times 100 \times 60 \times 360}{60} = 20$ in./sec

$e_0 = \dfrac{v_{max}}{\omega} = \dfrac{20}{100 \times 60 \times 2 \times 3.16} = 1.9$ in.

From the design chart $\phi_{max} = 38°$.

In order to reduce this pressure angle, we must offset the follower. The relationship relating to the offset distance and pressure angle is

$$\tan\phi_{max} = \dfrac{(dy/ds) - e}{s + (r_b^2 - e^2)^{1/2}}$$

We can calculate e from the above expression since ϕ_{max}, dy/ds, r_b, and s are known. Then $e = 0.8$ in. To check minimum radius of curvature, calculate $H/R_0 = H/(r_b + r_F)$. Let us assume $r_F = 0.25$. Then $H/R_0 = 0.89$. From the design chart $\rho_{min}/R_0 \approx 0.58$ or $\rho \cong 0.58 \times 2.25 \cong 1.305$ in.

Performance Test #2

1. Given data are

$H = 0.25$ in.
$\beta = 30°$
$R_0 = 2.5$ in.

We wish to find $R_{B_{min}}$. From the design chart $H/R_0 = 0.25/2.5 = 0.1$.

$R_{F_{max}} = \rho_{min} = 1.1$ in.

From Fig. A54 we note

$R_0 = R_{F_{max}} + R_{B_{min}}$

or

$R_{B_{min}} = R_0 - R_{F_{max}} = 2.5 - 1.1 = 1.4$ in.

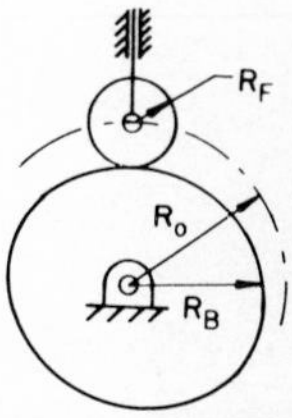

Fig. A54

Hence, minimum value for the base circle radius is 1.4 in.

2. We are given

$$H = 0.75 \text{ in.}$$
$$\beta = 45°$$
$$r_B = 2.0 \text{ in.}$$
$$e = 0.8 \text{ in.}$$

We are required to find ϕ_{max} when $e = 0.8$. We note that

$$\tan \phi_{max} = \frac{v_{max} - e\omega}{y\omega}$$

where

$$y = s + (r_B^2 - e^2)^{1/2}$$

Since ω is not known, we can simplify the above equation as

$$\tan \phi_{max} = \frac{(ds/d\theta)_{max} - e}{y}$$

For simple harmonic motion

$$s = (H/2)\left(1 - \cos\frac{\pi\theta}{\beta}\right)$$

and

$$\frac{ds}{d\theta} = \left(\frac{H\pi}{2\beta}\right)\sin\left(\frac{\pi\theta}{\beta}\right)$$

which is maximum when $\pi\theta/\beta = \pi/2$ and $(ds/d\theta)_{max} = H\pi/2\beta$. Hence

$$\tan \theta_{max} = \frac{\dfrac{\pi H}{2\beta} - e}{s + (r_B^2 - e^2)^{1/2}}$$

$$= \frac{\left(\dfrac{0.75 \times 3.14}{2.0 \times 785}\right) - 0.8}{2.175} = 0.3182$$

or

$$\phi_{max} \simeq 18°$$

Performance Test #3

1. We note that

$$e = e_{\text{critical}} = \frac{(v_f)_{\max}}{\omega}$$

For a simple harmonic motion program

$$(v_f)_{\max} = \frac{\pi H \omega}{2\beta}$$

Hence

$$e_{\text{critical}} = \frac{H\pi}{2\beta} = \frac{1 \times 3.14 \times 180}{2 \times 60 \times 3.14} = 1.5 \text{ in.}$$

2. Use the following relationships to develop the charts.

   a. For harmonic motion

$$s = \frac{H}{2}\left(1 - \cos\frac{\pi\theta}{\beta}\right)$$

   b. For cycloidal motion

$$s = \frac{H}{\pi}\left(\frac{\pi\theta}{\beta} - \frac{1}{2}\sin\frac{2\pi\theta}{\beta}\right)$$

   To calculate pressure angle use

$$\tan\phi = \frac{v_f}{(r_b + s)\omega} = \frac{ds/d\theta}{r_b + s}$$

Unit XI

Performance Test #1

1.

Link Grounded	Input Link	Type of motion
AB	AD	Drag link
AD	AB	Crank rocker
AD	CD	Rocker-rocker
CD	BC	Rocker-rocker

2. Using the steps described in Unit XI, Objective 2, the two limit positions are determined as shown in Fig. A55. MA_1B_1Q is the first limit position. MA_2B_2Q is the second limit position. Angle B_2QB_1 determines the angle of oscillation. Angle A_1MB_2 measures angle α which is needed to determine time ratio.

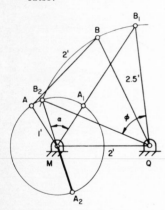

Since $\alpha = 54°$,

$$TR = \frac{180 + \alpha}{180 - \alpha} = 1.85$$

From Fig. A55, we measure $\alpha = 61°$.

Fig. A55

3. Using the steps described in Unit XI, Objective 6, a drag-link mechanism is designed as shown in Fig. A56. We can also design this mechanism using design charts. From Fig. A56, we measure $a/d = 3.6$, $b/d = 3.4$ and $c/d = 1.5$.

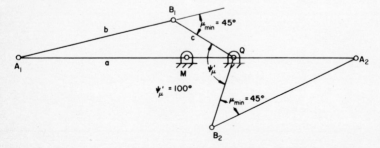

Fig. A56

Performance Test #2

1. a. We can ground either link b or d and keep link a as input.
   b. Link a should be grounded in order to obtain a drag-link mechanism.
2. The two limit positions shown in Fig. A57 are obtained when link AB becomes tangent to the circle drawn by point A. The two limit positions are MA_1B_1Q and MA_2B_2Q.

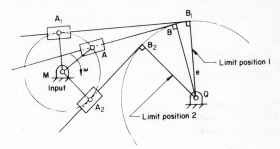

Fig. A57

3. The transmission angle in a slider crank mechanism is the angle between coupler link and a line perpendicular to the axis of the slider. The minimum and the maximum transmission angles are shown in Fig. A58.

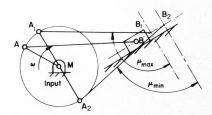

Fig. A58

4. The drag-link mechanism shown in Fig. A59 is obtained by using the design charts in Objective 6 of Unit XI. The following data are read from the charts that correspond to $\mu_{min} = 40°$, and $\psi'_\mu = 120°$.

$a/d = 2.18 \quad b/d = 1.82 \quad c/d = 1.56$

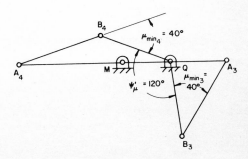

Fig. A59

Figure A59 is drawn to check the design information.
For drawing let $d = 1$ in., then $a = 2.18$ in., $b = 1.82$ in., $c = 1.56$.

Performance Test #3

1. a. According to Grashoff's criteria, a crank-rocker mechanism is obtained when the sum of the longest and the shortest link is less than or equal to the sum of the other two links and the shortest link is the driver or the crank. Since the longest link is $b = 2.0$ in. and a has to be the shortest link, the length of a should be so selected that the sum of $(a + b) \leqslant (c + d)$ or $a + 2.0 \leqslant 2.75$. Hence, the maximum length of a is 0.75 in.

   b. According to Grashoff's criteria, a drag-link mechanism is obtained when the smallest link is the ground link and when the sum of the shortest and the longest link is less than or equal to the sum of the other two links. Since $d = 1.25$ in. is expected to be the shortest link, $b + d \leqslant a + c$ or $a \geqslant b + d - c$ then $a \geqslant 1.75$ in.

2. Using the steps described in Unit XI, Objective 5, the minimum transmission angle $\mu_{min} = 63°$. See Fig. A60.

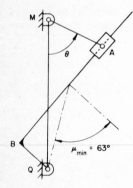

Fig. A60

3. From the design charts of crank-rocker mechanisms presented in Unit XI, Objective 7, we get $a/b = 0.4$, $b/d = 0.61$, and $c/d = 0.95$ since

   $d = 10$ in.
   $a = 4$ in.
   $b = 6.1$ in.
   $c = 9.5$ in.

   The crank-rocker mechanism is shown in Fig. A61. From the figure we note

   $\alpha = 13°$
   $\mu_{min} = 38°$
   $\mu_{max} = 127°$
   $\phi = 50°$

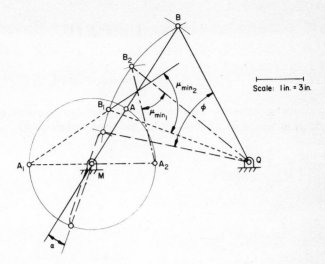

The time ratio is calculated from

$$TR = \frac{180° + \alpha}{180° - \alpha} = \frac{180° + 13}{180° - 13} = \frac{193}{187} = 1.16$$

We note that in reading the design chart we made a reading error of approximately 16 percent.

Unit XII

Performance Test #1

1. Using the steps described in Objective 2, a four-link mechanism shown in Fig. A62 is designed to satisfy the given requirements.

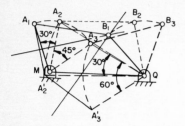

Fig. A62

2. Using the procedure described in Objective 1, a slider-crank mechanism is designed as shown in Fig. A63.

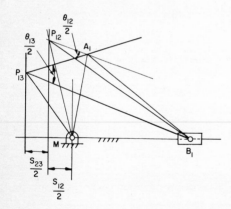

Fig. A63

Performance Test #2

1. Using the steps described in Objective 4, a slider-crank mechanism is designed as shown in Fig. A64.
2. Using the steps described in Objective 4, a four-link mechanism is designed as shown in Fig. A65.

Performance Test #3

1. Using the steps described in Objective 3, a slider-crank mechanism, shown in

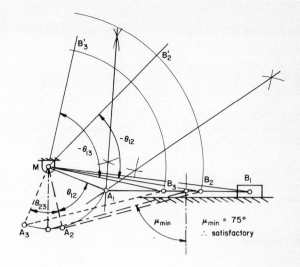

Fig. A64

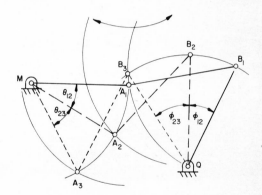

Fig. A65

Fig. A66, is designed. The minimum value of transmission angle is $69°$ which is quite satisfactory.

2. Using the steps presented in Objective 5, a four-link mechanism is designed as shown in Fig. A67.

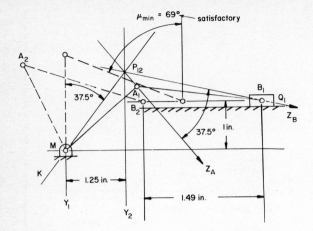

Fig. A66

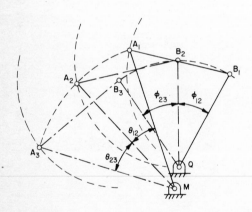

Fig. A67

Unit XIII

Performance Test #1

1. We are asked to design a four-link mechanism to coordinate displacements of its input and output links. Their displacements are governed by a functional relationship $y = x^2$, x varying between 0 and 2. The table presented below transforms these functional requirements into angular displacements.

Position	x	y	$\% \theta_2$	$\% \theta_4$	θ_2	θ_4
1	0	0	0	0	80.0°	70.0°
2	1	1	50	25	127.5°	100.0°
3	2	4	100	100	175.0°	190.0°

We note from above that the input link will rotate through 95°. With positions $\theta_{21} = 80°$, $\theta_{22} = 127.5°$, $\theta_{23} = 175°$, and $\theta_{41} = 70°$, $\theta_{42} = 100°$, $\theta_{43} = 190°$, we are in position to use the synthesis equations (2), (3), and (4) on page 300 and to solve for k_1, k_2, k_3, thus, $k_1 = 0.276$, $k_2 = 0.229$, $k_3 = 0.015$, and $a = 4.36$ in., $b = 1.03$ in., $c = 3.62$ in., and $d = 1.00$ in.

2. Using the notations of Fig. 3 on page 303, we note that three positions of point C are given by three sets of values of (r, δ). These are (4.5 in., 30°), (1.98 in., 52°), and (3.8 in., 68°). The corresponding values of θ_2 are $\theta_{21} = 80°$, $\theta_{22} = 91°$, and $\theta_{23} = 98°$. We are allowed to assume $\alpha = 40°$, $\beta = 34°$, and $\phi_1 = 95°$. We are required to design a four-link mechanism whose coupler point C will pass through the three given positions. The procedures for designing such a four-link mechanism are described on page 305. We note that we are required to obtain values of r_i for $i = 1,6$; r_1, r_2, r_3 can be obtained from Eq. (14), and r_4, r_5, r_6 can be obtained from Eq. (15).
    Simultaneous solution of the set of Eq. (18) yields:

$$l_1 = 0.6288$$
$$l_2 = 2.9314$$
$$l_3 = -9.3693$$

Similarly, simultaneous solution of the set of Eq. (19), yields:

$$m_1 = -0.1803$$
$$m_2 = 0.1526$$
$$m_3 = 1.2223$$

With the known values of l_i and m_i ($i = 1, 3$), Eq. (17) will become:

$$1.8432 \lambda^2 - 1.4326 \lambda - 0.0275 = 0$$

Hence:

$$\lambda = 1.2563 \text{ or } -53.3116$$

When $\lambda = 1.2563$, from Eq. (16), we will get:

$$k_1 = l_1 + \lambda m_1 = 0.4022$$
$$k_2 = 3.1231$$
$$k_3 = -7.8337$$

and

$$r_1 = 3.1232$$
$$r_2 = 1.4430$$
$$r_3 = 0.4023$$

From Eqs. (8) and (9) on page 303 we get,

$$\psi_1 = 96.1134°$$
$$\psi_2 = 126.0637°$$
$$\psi_3 = -39.2504°$$

Hence:

$$\phi_i = \phi_1 + (\psi_i - \psi_1)$$
$$\phi_1 = 95°$$
$$\phi_2 = -127.1772°$$
$$\phi_3 = -40.3639°$$

Equation (15) can be solved in a manner similar to the one used for Eq. (14). Equations (14) and (15) will provide two values of λ. Hence, we can obtain a maximum of four possible design solutions. These are tabulated below:

Solutions	Set 1	Set 2	Set 3	Set 4
r_1	3.1232	3.1232	-5.2058	-5.2058
r_2	1.4430	1.4430	10.2407	10.2407
r_3	0.4023	0.4023	7.5791	7.5791
r_4	-0.4796	1.9462	0.9138	28.4720
r_5	1.6795	7.2312	1.4996	31.2647
r_6	3.0453	-3.3990	3.0412	9.1687
λ Eq. (14)	1.2563	1.2563	-53.3116	-53.3116
λ Eq. (15)	-1.9893	10.9901	2.7791	261.0510

Check the above four possible solutions analytically or graphically by laying out the designed mechanism.

Performance Test #2

1. We are asked to design a four-link mechanism to coordinate displacements of input and output links according to the functional constraint $y = \log x$ for $0 < x \leqslant 10$. I suggest using the least-square technique. Let the input and output links rotate $100°$ to describe this function in ten movements. The design data can be tabulated in the following manner:

x	y	$\% x$	$\% y$	θ_2 (rad)	θ_4 (rad)
1	0.0	10.0	0.0	0.17	0.0
2	0.69	20.0	30.10	0.35	0.53
3	1.10	30.0	47.71	0.52	0.83
4	1.39	40.0	60.21	0.70	1.05
5	1.61	50.0	69.90	0.87	1.22
6	1.79	60.0	77.82	1.05	1.36
7	1.95	70.0	84.51	1.22	1.48
8	2.08	80.0	90.31	1.40	1.58
9	2.20	90.0	95.42	1.57	1.67
10	2.30	100.0	100.0	1.75	1.75

Using these data, we can calculate the coefficients of Eqs. (32), (33), and (34) given on page 310. Thus,

$2.659 k_1 - 3.149 k_2 + 3.411 k_3 = 3.286$
$3.149 k_1 + 4.030 k_2 - 5.041 k_3 = 4.847$
$3.411 k_1 - 5.041 k_2 + 10.00 k_3 = 9.700$

Using Cramer's rule, the above equations can be solved to yield:

$k_1 = 3.286$
$k_2 = -4.847$
$k_3 = 9.700$

From the above we can calculate:

$a = 5.6272$ units
$b = 1.0874$ units
$c = 6.4429$ units
$d = 1.0000$ unit

On page 309, we have defined the mean squared error. Calculate this mean squared error for this design problem. In general, this mean square is reduced significantly when Chevychev spacing is used in setting up the table of design data. Chebychev spacing is described in reference 2.

2. Here we are asked to design a four-link mechanism to guide a rigid body through its three finitely separated positions. The design data provide a description of these three positions of the rigid body by giving three sets of values for (r, ψ, δ) as (4.5 in., $80°$, $30°$), (1.98 in., $91°$, $52°$), and (3.8, in. $98°$, $68°$). We are required to obtain dimensions of r_i for $i = 1, 6$. We will assume that $\alpha = 40°$, $\beta = 34°$, $\phi = 95°$. The design of a four-link mechanism for this type of problem will require us to solve the synthesis equations (24) and (25), or Eqs. (26) and (27). Equations (26) and (27) are solved using the procedure described for Eqs. (14) and (15) on page 304. Four possible sets of solutions are obtained and are tabulated below.

Solutions	Set 1	Set 2	Set 3	Set 4
r_1	3.1232	3.1232	-5.2058	-5.2058
r_2	1.4430	1.4430	7.5791	7.5791
r_3	0.4023	0.4023	10.2407	10.2407
r_4	-0.6757	1.7031	0.9272	19.1746
r_5	1.8927	3.3218	1.4918	21.7611
r_6	2.9443	6.4529	3.0596	8.6715
λ Eq. (26)	1.2563	1.2563	-53.3100	-53.3100
λ Eq. (27)	-1.9893	10.9901	2.8368	166.2725

Performance Test #3

We are given twelve positions of a coupler point C of a four-link mechanism shown in Fig. 3, on page 303. Also given are the corresponding twelve positions of the input link MA. We are required to synthesize a four-link mechanism whose coupler point C will occupy positions so that the deviations of its positions from the prescribed position are minimum. We are asked to use the least-square technique.

Note that the data to describe positions of the coupler point are given using xy coordinates. From Fig. 3, page 303, we note that we should convert these data using polar coordinates:

$$r = (x_c^2 + y_c^2)$$

and

$$\delta = \tan^{-1} \frac{y_c}{x_c}$$

These polar coordinates are tabulated in the table that follows.

The synthesis equations are Eqs. (14) and (15). Since we are given twelve sets of values for r and δ, Eqs. (14) and (15) will permit us to calculate $r_1, r_2, r_3, r_4, r_5,$ and r_6 of the required four-link mechanism provided we assume $\alpha, \beta,$ and ϕ_1.

Position	r	δ(deg)	θ_2(deg)
1	3.86	71.57	161.00
2	4.90	65.37	131.00
3	5.87	56.53	101.00
4	6.63	46.41	71.00
5	6.93	35.85	41.00
6	6.91	27.60	11.00
7	6.26	22.14	−19.00
8	4.94	22.00	−49.00
9	3.43	25.22	−79.00
10	2.43	37.48	−109.00
11	2.27	58.13	−139.00
12	2.85	71.57	−169.00

Equations (18) and (19) are the linearized equations obtained from Eq. (14). We will apply the least-square technique to calculate l_1, l_2, l_3, m_1, m_2, and m_3. For this, however, we will define:

$$D_1 = \Sigma\{l_1[2r\cos(\alpha - \delta)] + l_2[2r\cos(\theta_2 - \delta)] + l_3 - r^2\}^2$$

and

$$D_2 = \Sigma\{m_1[2r\cos(\alpha - \delta)] + m_2[2r\cos(\theta_2 - \delta)] + m_3 - [2\cos(\theta_2 - \alpha)]\}^2$$

For minimum deviation we will set:

$$\frac{dD_1}{dl_1} = 0 \qquad \frac{dD_1}{dl_2} = 0 \qquad \frac{dD_1}{dl_3} = 0$$

and

$$\frac{dD_2}{dm_1} = 0 \qquad \frac{dD_2}{dm_2} = 0 \qquad \frac{dD_2}{dm_3} = 0$$

The conditions

$$\frac{dD_1}{dl_i} = 0$$

for $i = 1, 3$ will yield these equations:

$$l_1\Sigma[2r\cos(\alpha - \delta)]^2 + l_2\Sigma[2r\cos(\theta_2 - \delta)][2r\cos(\alpha - \delta)]$$
$$+ l_3\Sigma[2r\cos(\alpha - \delta)] = \Sigma[2r\cos(\alpha - \delta)]r^2$$

$$l_1 \Sigma [2r \cos(\alpha - \delta)][2r \cos(\theta_2 - \delta)] + l_2 \Sigma [2r \cos(\theta_2 - \delta)]^2$$
$$+ l_3 \Sigma [2r \cos(\theta_2 - \delta)] = \Sigma [2r \cos(\theta_2 - \delta)]$$

$$l_1 \Sigma [2r \cos(\alpha - \delta)] + l_2 \Sigma [2r \cos(\theta_2 - \delta)] + l_3 \Sigma(1) = \Sigma r^2$$

Since there are twelve sets of values for r and δ, and θ_2, the summation Σ is from 1 to 12. The numerical values of the coefficients can now be calculated since we know θ_2, α, δ, and r. Thus, the above equations will yield:

$$2113.07\, l_1 + 1394.00\, l_2 + 212.41\, l_3 = 6598.73$$

$$1394.00\, l_1 + 1486.11\, l_2 + 102.75\, l = 5142.72$$

$$212.41\, l_1 + 102.75\, l_2 + 24.00\, l_3 = 615.26$$

Simultaneous solution of the above linear equations will yield: $l_1 = -1.01$; $l_2 = 2.87$; and $l_3 = 22.32$.
The conditions; $dD_i/dm_i = 0$ for $i = 1, 3$ will also yield three equations.

$$m_1 \Sigma [2r \cos(\alpha - \delta)]^2 + m_2 \Sigma [2r \cos(\theta_2 - \delta)][2r \cos(\alpha - \delta)]$$
$$+ m_3 \Sigma [2r \cos(\alpha - \delta)] = \Sigma [2 \cos(\theta_2 - \alpha)][2r \cos(\alpha - \delta)]$$

$$m_1 \Sigma [2r \cos(\alpha - \delta)][2r \cos(\theta_2 - \delta)] + m_2 \Sigma [2r \cos(\theta_2 - \delta)]^2$$
$$+ m_3 \Sigma [2r \cos(\theta_2 - \delta)] = \Sigma [2 \cos(\theta_2 - \alpha)][2r \cos(\theta_2 - \delta)]$$

$$m_1 \Sigma [2r \cos(\alpha - \delta)] + m_2 \Sigma [2r \cos(\theta_2 - \delta)] + m_3 \Sigma(1)$$
$$= \Sigma [2 \cos(\theta_2 - \alpha)]$$

Since there are twelve sets of values for r, δ, and θ_2, the summation Σ is from 1 to 12. The numerical values of the coefficients can now be calculated since we know θ_2, α, δ, and r. Thus, the above equations will yield:

$$2113.07\, m_1 + 1394.00\, m_2 + 212.41\, m_3 = -1.01$$

$$1394.00\, m_1 + 1486.11\, m_2 + 102.75\, m_3 = 2.87$$

$$212.41\, m_1 + 102.75\, m_2 + 24.00\, m_3 = 22.32$$

Simultaneous solution of the above three equations will yield:

$$m_1 = 0.58$$
$$m_2 = -0.06$$
$$m_3 = -4.90$$

The parameter λ is calculated using Eq. (17).

$$\lambda = k_1 k_2 = (l_1 + \lambda m_1)(l_2 + \lambda m_2)$$

The two values of λ are:

$$\lambda_1 = 0.5364$$
$$\lambda_2 = 0.1452$$

Hence

$$k_1 = l_1 + \lambda m_1 = 2.12$$
$$k_2 = l_2 + \lambda m_2 = 2.52$$
$$k_3 = -3.98$$

or

$$r_1 = k_1 = 2.12 \text{ units}$$
$$r_2 = k_2 = 2.52 \text{ units}$$
$$r_3 = (k_3 - r_2^2 - r_1^2)^{\frac{1}{2}} = 2.63 \text{ units}$$

With the known values of $r_1, r_2, r_3, \alpha, \delta,$ and θ_2 we can calculate where $\psi_1, \psi_2, \ldots, \psi_{12}$ and $\phi_2, \phi_3, \ldots, \phi_{12}$, as described in the stepwise procedure on page 305.

Equation (15) can be solved in the same manner as Eq. (14). Since there are two sets of values for λ in Eqs. (14) and (15), we will obtain four sets of solutions which are represented below. Figure A68 shows one of the four solutions and the coupler curve traced by the designed linkage.

Parameters	Solution 1	Solution 2	Solution 3	Solution 4
r_1	2.125	2.125	7.476	7.476
r_2	2.525	2.525	1.942	1.942
r_3	2.629	2.629	3.289	3.289
r_4	−604.907	7.881	66.628	−10.701
r_5	1489.200	3.647	14.015	5.294
r_6	−2066.439	4.757	56.391	−11.370
λ Eq. (14)	5.364	5.364	14.52	14.52
λ Eq. (15)	1.250×10^6	3.815	3757.0	12.17
a	56.0°	56.0°	56.0°	56.0°
β	−6.0°	−6.0°	−6.0°	−6.0°
ϕ_1	166.0°	166.0°	166.0°	166.0°

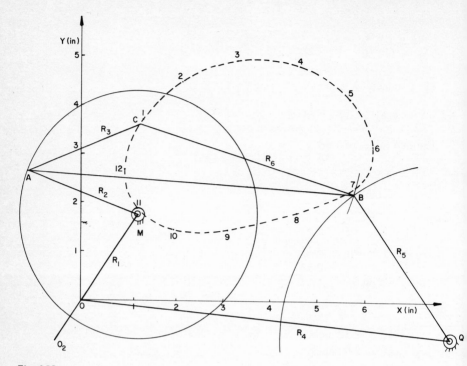

Fig. A68

Unit XIV

Performance Test #1

We are asked to obtain the coordinates of coupler point C of the double-slider mechanism, shown in Fig. A69.

Let $OA = S_1$, $OB = S_2$, $AB = b$, and $AC = a$, angle $OAB = \beta$, and angle $BAC = \alpha$, then $\tan \beta = S_2/S_1$

Since AOB is a right-angle triangle,

$$S_1^2 + S_2^2 = b^2$$

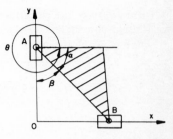

Fig. A69

Hence,

$$\tan \beta = \frac{[b^2 - S_1^2]^{\frac{1}{2}}}{S_1}$$

and

$$\theta = 270° + \beta + \alpha$$

The coordinates, X_c and Y_c of the coupler point C are:

$$X_c = a \cos \theta = a \cos(270° + \beta + \alpha)$$
$$Y_c = a \sin \theta = a \sin(270° + \beta + \alpha) + S_1$$

It is suggested that the student develop computer programs to plot a series of coupler curves of this mechanism.

Performance Test #2

Refer to Fig. 21 on page 336. We can obtain a straight line as a coupler curve when $AB = AC = MA$ and $\alpha = 180°$.

Unit XV

Performance Test #1

There are three design problems.
1. Design of a six-link mechanism to display dwell-rise-return motion as shown in Fig. A, of the problem statement. The mechanism shown below in Fig. A70 is designed using the steps described in Objective 3.

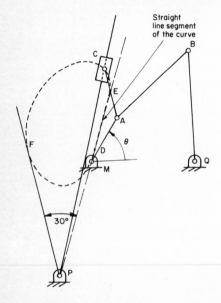

Fig. A70

We note that as the coupler point C traces the path DE on the coupler curve, the slider block at C travels along line DE and the output link PC experiences a dwell. The input crank MA rotates $30°$ corresponding to the distance DE traveled by the coupler point C. Also, the extreme position PF occupied by the output link shows that it oscillates through angle FPD which is equal to $30°$.

2. Design of a six-link mechanism to display dwell-rise-dwell-return motion programs as shown in Fig. C of the problem statement. The mechanism shown below in Fig. A71 is designed using the steps described in Objective 4.

We note from the above design that as the coupler point C traces paths DE and FG, the output link PC experiences dwells. As the coupler point moves from position C to F, the output link experiences a forward stroke corresponding to the "rise" segment of the motion program. The output link returns to its initial position as coupler point C travels from position G to D.

3. Design of a six-link mechanism to display double stroke output as shown in

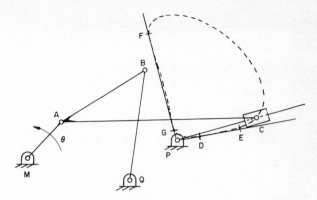

Fig. A71

Fig. D of the problem statement. The mechanism shown below in Fig. A72 is designed using the steps described in Objective 5.

We note that as the coupler moves from position C to D to F, the output link PC completes one stroke. The second stroke is executed by the output link PC when the coupler point moves from position F to E to C.

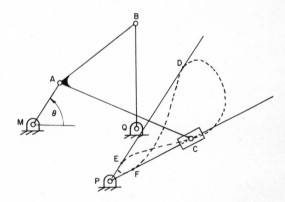

Fig. A72

Performance Test #2

1. We are asked to design a six-link mechanism that is expected to display a rise-return-dwell-return motion program. This is essentially a dwell-rise-return motion program. Objective 3 presents the steps to design such a linkage. We have selected a figure-eight type of coupler for the six-link design which is shown in Fig. A73.

We note from the above design that the cycle describing the given motion program begins at the position that corresponds to position mi of the coupler point C which traces the coupler curve as shown. As coupler point C

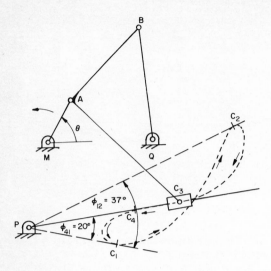

Fig. A73

passes from positions C_1 to C_2 the output link PC experiences "rise." As coupler point C passes from C_2 to C_3, C_4 and back to C_1, output link PC experiences "return." During the return segment of the cycle, output link PC has a dwelling motion. This is due to the fact that point C traces the straight-line path $C_3 C_4$.

2. We are asked to design a six-link mechanism for double dwells. Using the steps described in Objective 4, the mechanism is designed as shown in Fig. A74.

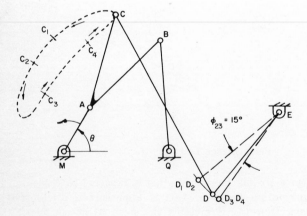

Fig. A74

Performance Test #3

We are asked to design a six-link mechanism of the type shown in Fig. A on page 372. The solution to this problem lies in the selection of a proper coupler curve. For the selection of the coupler curve, reference 10 is recommended. Use the following steps to select the proper coupler curve.

1. Draw an output sector showing oscillation of 30° and 60° as shown in Fig. A75.
2. Select a figure-eight type of coupler curve.
3. Place the output sector on the selected coupler curve so that its legs A, B, and C are tangents to the coupler curve at w, x, y, and z. The point P where the legs A, B, C intersect, becomes the fixed pivot point. Leg A can be used as the axis of the slider block which is connected with the coupler point tracing the selected coupler curve.
4. The second requirement for the proper coupler curve is that the two strokes should be completed between 120° and 240° rotation of the input crank. This can be checked by counting the dashes of the coupler curve selected from reference 10. Each dash measures 5° rotation of the input crank. Hence, we must count 24 dashes from x to y to w, and 48 dashes from w to z to y.

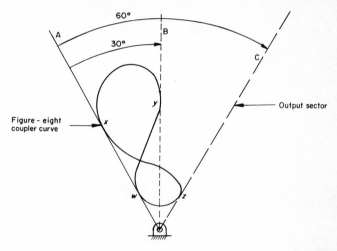

Fig. A75

Unit XVI

Performance Test #1

The mechanism in question is an eight-link mechanism which is built from a source four-link mechanism $MABQ$, with coupler point C which traces the coupler curve as shown. Our task is to design a Watt's six-link mechanism in which one of its coupler links will be used to trace a series of coupler curves as those drawn by the coupler link CF. We will use the following steps.

a. We will isolate the four-link mechanism $MABQ$ with the coupler point C, as shown in Fig. A76.

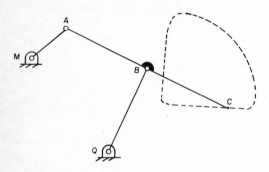

Fig. A76 Source four-link mechanism.

b. Using the steps described on page 377, we will locate the right and the left cognate mechanisms as shown in Fig. A77.

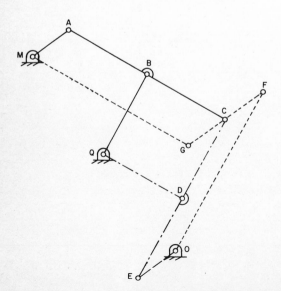

Fig. A77 Cognate construction.

c. From the cognate construction we note that the angular velocity of link *OE* will be equal to the angular velocity of link *MA*.

d. Using the steps described on page 382, we will "add" the cognate mechanism *OEDQ*, (with coupler point *C* on coupler link *QD*), to source mechanism *MABQ* (with coupler point *C* on coupler link *AB*). As a result of this "addition," we will get Watt's six-link mechanism as shown in Fig. A78. Note that points *C* and *C'* draw the same coupler curves as shown.

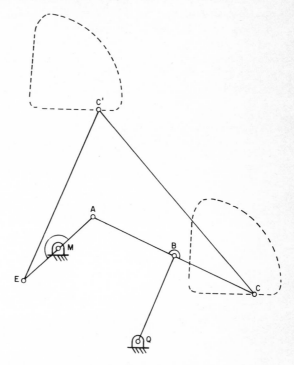

Fig. 78 Design of six-link mechanism.

e. As shown in Fig. A79, we will modify the six-link mechanism of Fig. 78 so that it will perform the same tasks as the given eight-link mechanism.

Performance Test #2

Here we are asked to design a transport mechanism that will transport the box from position A_1, B_1, C_1 and D_1, to position A_2, B_2, C_2, and D_2, keeping itself parallel to the first position. The designed mechanism should not tilt the box. There are several ways one can approach designing such a mechanism. We will, however, demonstrate the application of coupler cognate mechanisms in designing a six-link mechanism for parallel motion.

a. Since the body is required to have instantaneous dwells in its two positions, we will select a four-link mechanism having a coupler curve with two cusps at each end. The cusps at each end will provide the instantaneous dwell. The selected four-link mechanism and the coupler curve are shown in Fig. A80.

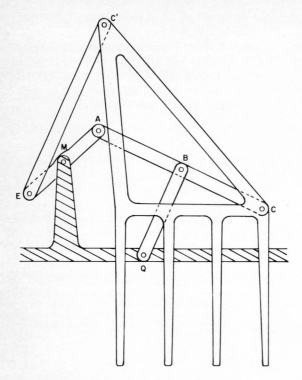

Fig. A79 Final design of a six-link mechanism.

b. Using the steps described in Objective 1 we will obtain the two cognate mechanisms for the source mechanism. Using the steps described in Objective 2 we will design the six-link mechanism to function as a transport mechanism as shown in Fig. A81.

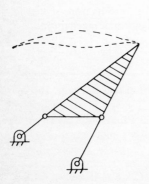

Fig. 80 Four-bar mechanism with a selected coupler curve.

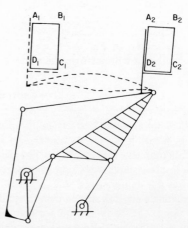

Fig. 81 Transport mechanism.

Performance Test #3

Here we are asked to design a suspension mechanism which, when properly controlled will permit us to hold a trailer body parallel to itself, even though the wheels supporting the body are placed on an uneven surface. A good design of suspension should also prevent any lateral motion of the wheels while the trailer body is maintained in the required position.

There are several ways to design such trailer suspensions. We will, however, apply the results of the coupler cognate mechanism in designing a six-link mechanism for parallel motion generation. In designing the required six-link mechanism, we will proceed in the following manner:

a. Since the trailer body is required to maintain horizontal position, or a position perpendicular to the gravity force, we will select a four-link mechanism with a coupler curve having a straight-line segment. The selected four-link mechanism with the coupler curve is shown in Fig. A82.

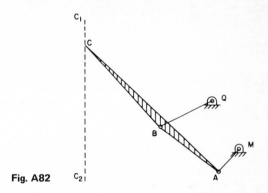

Fig. A82

b. Using the steps of Objective 1, we will obtain the cognate mechanisms. Using the steps of Objective 2, we will design the six-link mehcanism to function as the trailer suspension as shown in Fig. A83.

Performance Test #4

Figure A84 shows a geared five-link mechanism $XABCY$ with the gears at joints Y and C. The gear at joint C is attached to coupler link CB, and the gear at joint Y is attached to ground link XY. The speed ratio for this planetary gear train where the sun gear is fixed $1 + n$, and where n is the teeth ratio between the sun gear and the planet gear, assuming that they both have the same dimetral pitch.

A coupler point P is selected on the coupler link AB. As shown in Fig. A84 the coupler point P is obtained by constructing parallelograms $XAPD$, $BPQC$, $CQLY$, $QRML$, $PEFR$, and $FRNZ$ and by constructing triangles DEP, PRQ, and RNM

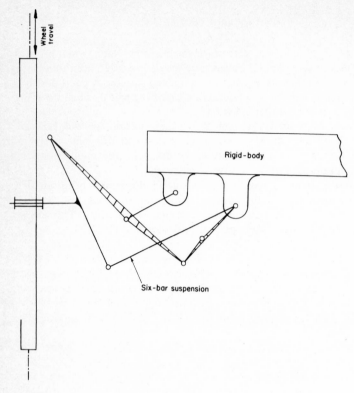

Fig. A83

similar to triangle APB. The existence of the cognate mechanism can be proved using complex number algebra.

If the fixed pivots X and Y are stretched along XY so that the points A, B, C lie on a straight line, then points D, E, F and L, M, N will lie on straight lines XZ

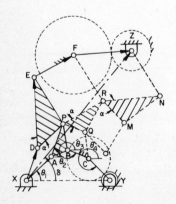

Fig. 84 Construction of the cognate geared five-link mechanism $XDEFZ$ to the source geared five-link mechanism $XABCY$, both mechanisms having the coupler point P in common.

and YZ. The resultant figure yields, as shown in Fig. A85, a generalized Cayley diagram, for this type of source mechanism.

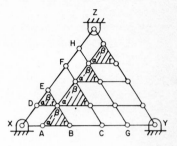

Fig. A85 Generalized Cayley diagram.

Performance Test #5

1. A geared five-bar mechanism $MABCQ$, with a gear ratio of +1 is shown in Fig. A86. The general case of coupler cognate mechanism for this case, is beyond the scope of this book. We will, however, consider the case where the knee point B becomes the coupler point. Let us draw two parallelograms $MABD$ and $QCBE$. Because of the parallelograms, construction: $\omega_{MA} = \omega_{DB}$ and $\omega_{QC} = \omega_{EB}$.

    The gears are connected to links MA and QC. The speed ratio for this simple gear train is +1. Hence $\omega_{DB} = \omega_{QC}$; that is, $\omega_{DB} = \omega_{EB}$ Since links BD and BE have the same angular velocity, the points D, B, E do not have any relative motion. Hence, the triangle DBE is a structure. We note that $MDEQ$ is a four-link coupler point B as the coincident point with the knee point of the geared five-bar mechanism. The coupler cognate mechanisms of the four-link mechanism $MDEQ$ with the coupler point B will be the coupler cognate mechanism of the gear-five-bar with gear ratio of +1 and a knee point as the coupler point.

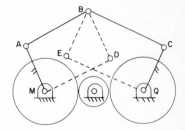

Fig. A86

2. Figure 5 on page 307 describes the synthesis problem to coordinate three positions of input link MA with the rigid body $CDEF$. In describing the synthesis problem, the design data consists of specifying coordinates of some point P within the rigid body and rotation angle ψ. Note that angle ψ and ϕ are related by a constant. That is: $\phi = \psi +$ constant.

Hence, specifying ψ is as good as specifying ϕ. Draw a parallelogram $ABQT$. Then we obtain $MATQ$ as a four-bar mechanism. Since the angular displacement of link AB is transferred to link QT, the synthesis problem coordinating displacement of input link and positions of rigid body also provides the synthesis of a four-link mechanism $MATQ$, to coordinate displacements of input link MA and output link QT. The design data are θ_2 and $180 + \psi$.

Supplementary References

1. Harrisberger, L. *Mechanization of Motion*, Wiley, 1961.
2. Hartenberg, R. S., and Denavit, J. *Kinematic Synthesis of Linkages*, McGraw-Hill, 1964.
3. Hinkle, R. T. *Kinematics of Machines*, Prentice-Hall, 1960.
4. Jensen, P. W. *Cam Design and Manufacture*, The Industrial Press, 1965.
5. Martin, G. H. *Kinematics and Dynamics of Machines*, McGraw-Hill, 1969.
6. Rothbart, H. A. *Cams*, Wiley, 1956.
7. Shigley, J. E. *Kinematic Analysis of Mechanisms*, McGraw-Hill, 1969.
8. Tao, D. C. *Fundamentals of Applied Kinematics*, Addison-Wesley, 1967.
9. Tao, D. C. *Applied Linkage Synthesis*, Addison-Wesley, 1964.
10. Hrones, J. A., and Nelson, L. *Analysis of the Four-Bar Linkages*, First Edition, MIT Press, New York, 1951.

Index

Acceleration, 63
  angular, 64
  centripetal component, 69, 76
  components in moving coordinate
    system, 70
  components in rotating coordinate
    system, 69
  coriolis, 69
  definition, 64
  direction of components of, 78
  horizontal component of, 76
  normal component of, 66
  radial component of, 66, 67
  tangential component of, 65
  transverse component of, 65
  vector relationship of, 78
  vertical component of, 76
Acceleration analysis, 77
  four-link mechanism, 82, 103
  inverted slider-crank mechanism, 87, 113
  six-link mechanism, 90
  slider-crank mechanism, 85, 109
  vector polygon technique, 82
AGMA standards, 149
Analysis of linkages:
  acceleration, 82, 86, 87, 90
  velocity, 17, 36, 39, 41, 45
Angular velocity ratio, 56
Axes of cone, 130

Bloch's synthesis technique, 312

Cam follower, 167
  jams, 224
Cam motion programs, 169, 181
  constant acceleration, 184
  constant velocity, 182
  cubic curve, 189
  cycloidal motion, 188
  polynomial, 191
  simple harmonic, 186
  trapezoidal, 194
Cam profile, 211
  flat face follower, 213

Cam profile (*Cont'd*)
  swinging follower, 211, 217
  translating follower, 215, 216
Cam terminology, 177
  base circle, 177
  pitch circle, 177
  pitch curve, 177
  pitch point, 177
  pressure angle, 178
  prime circle, 178
  trace point, 178
Cam types, 172
Cam undercutting, 218
Cayley's diagram, 376
Centro (*see* Instant center)
Collineation axis, 57
Complex numbers:
  complex part of, 28
  operator of, 27
  real part of, 28
  representation of vectors, 27
Contact ratio, 146
Coriolis acceleration, 72
Coupler cognate mechanisms, 375
  four-link mechanism, 375
  slider-crank mechanism, 379
  proof for existence, 379
Coupler curve, 315
  coordinates, 315
    for inverted slide-crank mechanism
     of, 342
    of slider-crank mechanism of, 336
  crank-rocker mechanism of, 317–325
  drag-link mechanism of, 327–335
  inverted slider-crank mechanism of,
    344–349
  slider-crank mechanism of, 338–341
Crank-rocker mechanism, 233
  design of, 257
  design charts of, 262
Cycloidal curve, 125

Dead center position, 233
  four-link mechanism of, 236

Dead center position (*Cont'd*)
   slider-crank mechanism of,
Degrees of freedom, 2
   complex number technique,
   kinematic pair of, 3
   linkages of, 5
   mechanisms of, 5
Drag-link mechanism, 233
   design of, 250
   design charts of, 254

Epicycloidal curve, 126
Epicycloidal, theory of, 208

Four-link mechanism, 5
   acceleration analysis, 82
   coupler-link, 38
   coupler cognate of, 375
   coupler curves of, 317–325, 327–335
   dead center positions of, 236
   design of, 250, 257
   follower-link (output link), 38
   Grashoff's criteria for, 231
   input link, 36
   limit positions of, 237
   synthesis of, 267, 270, 273, 299, 302
   transmission angle of, 245
   types of motion, 232
   velocity analysis, 36, 102
Freudenstein's equation, 100, 309
Freudenstein's technique, 300
Freudenstein's theorem on instant center
   and velocity ratio, 56

Gears, 153
   contact ratio, 138
   helical, 139
   hypoid, 139
   idlers, 155
   interference, 138
   mated, 138
   mating tooth, 139
   noncircular, 139
   spur, 139
   undercut, 138
   worm, 139
Gear teeth, 135
Gear tooth terminology, 135
   adendum, 136
   adendum radius, 136
   angle of action, 137
   base circle, 135

Gear tooth terminology (*Cont'd*)
   base pitch, 137
   circular pitch, 137
   clearance, 136
   dedendum, 136
   diametral pitch, 137
   line of action, 137
   pitch angle, 137
   pitch circle, 135
   pitch point, 136
   pressure angle, 137
   root circle, 135
   tooth face, 136
   tooth flank, 137
   tooth thickness, 137
   working depth, 136
Gear trains, 153
   compound, 156
   epicyclic gear trains, 161
   planetary, 161
   reverted compound, 160
   simple, 153
   speed ratio, 153
Generating circle, 126
Grashoff's criteria, 231
Gruebler, 5

Helical gears, 139
   axial pitch, 142
   contact ratio, 142
   hand of helix, 139
   helix angle, 139
   normal circular pitch, 141
   normal diametral pitch, 141
Hypocycloidal curve, 126

Instant center, 53, 54
   Freudenstein's theorem of, 56
   Kennedy's theorem of, 55
   number of, 54
Interference, 148
Inversion technique, 273
Inverted slider-crank mechanism, velocity
   analysis of, 41
Involute, 124
   curve equation, 125
   curves, 124
   profile, 124

Joint, 1
   forced-closed, 1
   form-closed, 1
   (*See also* kinematic pairs)

Kennedy's theorem, 55
Kinematic analysis, 97
  analytical method, 97
  four-link mechanism, 98
  inverted slider-crank mechanism, 110
  slider-crank mechanism, 106
Kinematic chain, 4
  loops of, 4
Kinematic link, 4
  binary link, 4
  definition, 4
  quaternary link, 4
  ternary link, 4
Kinematic pairs, 2
  cam pair, 3
  classification of, 2
  cylinder pair, 3
  definition, 1
  degrees of freedom, 1
  helical pair, 3
  higher pairs, 3
  lower pairs, 3
  number of cam joints, 3
  number of points of contact, 2
  plane pair, 3
  prism pair, 1
  revolute, 1
  spherical pair, 3
  surface contacts, 3
  (*See also* Joint)

Least square technique, 309
Limit positions, 233
  four-link mechanism of, 237
  inverted slider-crank mechanism, 243
  slider-crank mechanism of, 237
Line of contact, 130
Linkage motion-program, 356–358
  constant velocity, 358
  double strokes, 356, 358
  dwell-rise-dwell-return, 356
  dwell-rise-return, 357
  instantaneous dwell, with, 358
  large oscillation, for, 358
  rise-return, 356
  rise-return-dwell-return, 357
Linkages, 5
  constrained motion of, 5
  definition, 5
  degrees of freedom of, 5
  mobility criteria of, 5
Loops, 4

Mechanisms, 5
  definition, 5
  degree of freedom, 5
  instant centers of, 54
  inverted slider-crank, 7
  loops of, 5
  six-links, 45
  slider-crank, 7
Minimum radius of roller, 207
  oscillating follower, 207
  translating follower, 207
Mobility criteria of linkages, 5
Motion program synthesis, 198
Motion of rigid body, 32

Pitch circle, 126
Pitch point, 128
Point of contact, 130
Pole, 268
Pole technique, 270, 271
Pressure angle, 123, 207, 223, 224, 225

Racks, 143
Radius of curvature, 219
Right-hand rule for vector representation,
  20
Rigid body guidance, 305
Roberts cognate, 375
Rocker-rocker mechanism, 232
Rolling cones, 130
  external, 130
  internal, 132
  speed ratio, 131
Rolling cylinders, 124, 125
  center distance, 128
  external, 128
  internal, 129
  speed ratio of, 128
Rolling motion, 118
Rotary transmission criteria of, 118
  direct rolling in, 118
  fundamentals of, 117
  sliding contact, 118

Scalar, 18
Six-link chains:
  Stephenson, 9
  Watt, 9
Six-link mechanisms, 45
  acceleration analysis, 90
  types of, 351
Sizing of cams, 207

Slider-crank mechanism, 8
    acceleration analysis, 85
    coupler cognate of, 379
    coupler curves of, 338, 341
    dead center positions of, 241
    limit positions of, 241
    synthesis of, 271, 274
    transmission angle of, 245
    velocity analysis, 39, 108
Speed ratio, 120
Springs, 9
Structures, 6
Synthesis:
    dimensional, 267, 270, 271, 273, 274,
        299, 302
    of mechanisms: four-link mechanism,
        267
        slider-crank mechanism, 271, 274
        three positions, 299, 302
        two positions, 270, 273
    number, 1
    of six-link mechanism: constant velocity,
        with, 367
        double dwells with, 363–364
        generation of parallel motion for, 381
        single dwell with, 359, 360
        structural, 1

Tooth thickness, 145
Trace point, 126
Transmission angle in a mechanism, 245
    four-link mechanism of, 245
    inverted slider-crank mechanism of, 245
    slider-crank mechanism of, 245

Unit vector, 19

Vector polygons, 32, 34
    construction of, 33
Vectors:
    algebraic sum of, 19
    complex number representation of, 27
    cross products of, 20
    definition of, 18
    derivatives of, 21
    dot product of, 20
    magnitudes of, 18
    origin point of, 33
    polygons of, 32
    properties of, 18
    right-hand rule for, 20
    scales for, 33
    unit, definition of, 19
Velocity:
    angular, 24
    definition of, 22
    horizontal component of, 23
    radial components of, 25
    transverse component of, 25
    vector relationship for, 36
    vertical component of, 23
Velocity analysis:
    four-link mechanism of, 36, 102
    instant center method, 57
    inverted slider-crank mechanism of, 41,
        112
    slider-crank mechanism of, 39, 108
    vector polygon technique: for four-link
        mechanism, 36
        for inverted slider-crank mechanism,
        41
        for slider-crank mechanism, 38

Watt chain, 9